*Springer Monographs* in *Mathematics*

For further volumes:
http://www.springer.com/series/3733

Shaoqiang Deng

# Homogeneous Finsler Spaces

 Springer

Shaoqiang Deng
School of Mathematical Sciences
Nankai University
Tianjin
People's Republic of China

ISSN 1439-7382
ISBN 978-1-4899-9476-9      ISBN 978-1-4614-4244-8 (eBook)
DOI 10.1007/978-1-4614-4244-8
Springer New York Heidelberg Dordrecht London

Printed on acid-free paper

Springer is part of Springer Science+Business Media (www.springer.com)

*Dedicated to the memory of Professor Shiing-Shen Chern on the occasion of the centenary of his birth*

# Preface

In recent years, Finsler geometry has been developing rapidly. There are two main reasons for this development. First, the influence of S.S. Chern in the late twentieth century was enormous; his guidelines rapidly accelerated both interest and results, first in the United States, including David Bao, Zhongmin Shen, as well as other mathematicians, and then in China. Today, a large group of young and talented mathematical researchers all over the world are working on specific issues in this field. Second, new areas of application arose in the 1980s (R.S. Ingarden, P.L. Antonelli, among others) in biology, optics, quantum physics, and also in psychology, geosciences, geodesy, and other fields. The importance of the field also grew in the 1990s, when its strong relationship with complex analysis became clear. A new field, called complex Finsler geometry, was launched. In the past 20 years, much significant progress has been made in this field, and work in this subject has resulted in a number of substantial books and monographs that have greatly enhanced the study of Finsler geometry.

This monograph contains a series of results obtained by the author and collaborators in the last decade. Our main idea was to show that one can use Lie theory to study Finsler geometry, and our research shows that this method is actually applicable to many problems. Moreover, results obtained on related topics are generally simpler in this form than similar results in general Finsler geometry. In fact, in some special cases (e.g., Randers spaces), we can express curvatures using only the algebraic structures and the metric, without local coordinate systems. Such approaches are particularly welcome, since Finsler geometry has a reputation for complexity due to the extensive use of tensors and indices.

The field of Finsler geometry originated from Riemann's celebrated habilitation lecture, "On the Hypotheses, Which Lie at the Foundation of Geometry," given on June 10, 1854. A translation of this lecture can be found in M. Spivak's book *A Comprehensive Introduction to Differential Geometry*, Volume II, Chap. 4, Publish or Perish Inc., 1970. In this lecture, Riemann introduced the notion of a manifold and metric structures on a manifold. In the special case that the manifold is smooth and the metric is a quadratic differential form, Riemann successfully introduced the notions of curvature tensor and sectional curvature, and provided a complete treatment of this complicated quantity. This setting had an important impact on Einstein's general theory of relativity, and has received a great deal of attention from both mathematicians and physicists. The subject is now called Riemannian geometry in the literature.

The restriction to a quadratic form constitutes only a special case. Riemann did not regard this restriction as necessary. However, for the general case, he wrote the following: "The next simplest case would perhaps include the manifolds in which the line element can be expressed as the fourth root of a differential expression of the fourth degree. Investigation of this more general class would actually require no essentially different principles, but it would be rather time-consuming and throw proportionally little new light on the study of space." This commentary rendered the general cases dormant for a rather long period. In 1918, Paul Finsler initiated the study of variational problems in the spaces where the metric is defined by Minkowski norms. This setting was developed into a new field, which was much more complicated and difficult compared to Riemannian geometry and was eventually named Finsler geometry.

In 1926, L. Berwald introduced the notion of flag curvature, which is the natural generalization of sectional curvature in Riemannian geometry. Berwald found an important connection, called the Berwald connection, which is a very important connection in Finsler geometry. He also studied the geometric properties of a special kind of Finsler spaces—Berwald spaces. The work of Berwald has had a great impact in the study of Finsler geometry. The Berwald connection is torsion-free but not metric-compatible. In 1934, É. Cartan found another connection—the Cartan connection—which is metric-compatible but has torsion. Great progress was also made in 1943 by S.S. Chern, who found a connection that is torsion-free and almost metric-compatible. The Chern connection is the simplest in form and now appears very frequently in the literature.

As a branch of geometry, Finsler geometry has inevitably been influenced by group theory. The celebrated Erlangen program of F. Klein, posed in 1872, greatly influenced the development of geometry. Klein proposed to categorize the new geometries by their characteristic groups of transformations. To this day, this program has been extremely successful. Nowadays, every geometer expects to find an effective method for studying geometry by applying group theory. The theory of homogeneous/symmetric Riemannian spaces provides a sample for this program. Lie theory was developed in the late nineteenth century, first in the local fashion, by S. Lie, W. Killing, and É. Cartan. Global Lie groups were emphasized through the work of H. Weyl, É. Cartan, and O. Schreier during the 1920s. One of the most important applications of Lie theory to Riemannian geometry was Cartan's work on the classification of globally symmetric Riemannian spaces. The Myers–Steendrod theorem, published in 1939, extended the application of Lie theory to the scope of all homogeneous Riemannian manifolds. By now, the theory of homogeneous/symmetric Riemannian manifolds has become the basis of many branches of mathematics, including group and geometric analysis, and representation theory. The Myers–Steenrod theorem was generalized to the Finslerian case by the author and Z. Hou in 2002. This result opened a door to using

Lie theory to study Finsler geometry. In the last decade, the author and collaborators have successfully developed the theory of homogeneous/symmetric Finsler spaces. Meanwhile, there has appeared in the literature some work of other mathematicians on Finsler geometry that is closely related to Lie theory, including some classical results of H.C. Wang and Z.I. Szabó's on Berwald spaces.

The purpose of this book is to introduce the major part of the aspect of Finsler geometry that has a close relationship with Lie theory, and to bring the reader to the frontiers of the active research on related topics. The book consists of seven chapters. Chapters 1 and 2 are an introduction to Finsler geometry and Lie theory. Chapter 1 can also be used as a textbook for a course in Finsler geometry, provided the lecturer adds all the necessary details. Chapter 3 is focused on the study of isometries of Finsler space. The main result of this chapter is the generalized Myers–Steenrod theorem. In Chap. 4, the theory of general homogeneous Finsler spaces is developed. Chapter 5 deals with symmetric Finsler spaces. The main features of É. Cartan's theory on Riemannian symmetric spaces are generalized to the Finslerian case. In Chap. 6, we develop a theory of weakly symmetric Finsler spaces. Chapter 7 is devoted entirely to homogeneous Randers spaces.

Apart from the first two chapters, each chapter begins with a brief introduction to the background and the motivation of the topics under consideration. We believe that some historical elements will help readers grasp the main ideas more easily. Moreover, we discuss possible further development of the fields discussed in the chapter. We hope that this will enhance the study of homogeneous Finsler spaces to some extent.

There are some aspects of the field of Finsler geometry that have not been involved in this book. First, although we present some results on invariant complex structures on homogeneous/symmetric Finsler spaces in Sect. 5.5, a complete theory of homogeneous complex Finsler spaces has not been established, which in my opinion will be a main focus in the near future. Second, this book does not deal with any explicit applications of the theory to the real world, although we have provided a large number of carefully selected examples, which definitely have promising applications to other scientific fields.

Finally, although the author has done his best to make everything as accurate as possible, errors or mistakes, minor or major, are very likely to exist in the book. Comments and suggestions from readers, either for the improvement of the book or pointing out mistakes to the author, are very welcome.

Tianjin, China                                                                                           *Shaoqiang Deng*

# Acknowledgments

The composition of this book has benefited greatly from the help of many people. I am deeply grateful to Professors Joseph A. Wolf and Zhongmin Shen for their suggestion that this book be written. Joe not only kindly shared with me his valuable experience on how to organize the text of a scientific book, but also read the first draft of the book very carefully and gave me much advice on how to improve the presentation.

I am indebted to Professors David Bao and Zhongmin Shen for their help in my study of Finsler geometry. In the beginning of my study, I received a great deal of material on Finsler geometry from them that I could not access in China. Such help was extremely valuable, especially because they often took the initiative in providing such material. Furthermore, I was always able to obtain many useful suggestions from them on how to study Finsler geometry.

I am grateful to my supervisor, Professor Zixin Hou, for his long-term support and encouragement. I became a graduate student under his guidance when I entered Nankai University in 1988 and have been his colleague since my graduation in 1991. He has generously given me considerable support during this period, both in mathematics and in life.

I would like to thank the four anonymous reviewers and two series editors of this book, whose valuable suggestions have greatly improved the presentation. I am also grateful to the excellent work of the Springer team involved in the publication of this book, including Ann Kostant, Elizabeth Loew, and the copyeditor.

Thanks also go to some of my students for useful comments on the book, particularly to Dr. Zhiguang Hu, who helped me greatly in typing the book.

This book is dedicated to the memory of Professor S.S. Chern on the occasion of the centenary of his birth. It was his motivations and encouragement that led me to the study of Finsler geometry.

This work is partially supported by NSFC (no. 10671096, 10971104) and SRFDP of China.

# Contents

# Chapter 1
# Introduction to Finsler Geometry

In this chapter, we will give a brief introduction to Finsler geometry. Our goal here is twofold. First, although currently Finsler geometry is developing rapidly, it is not as well known as some standard parts of mathematics that are routinely taught in graduate schools. Additionally, the terminologies and techniques involved in this field are rather complicated. Therefore, a concise introduction would be helpful. On the other hand, since the notation used by Finsler geometers varies considerably, beginners in Finsler geometry sometimes may find it very difficult to adapt to so many definitions and indices. Our second purpose is therefore to set the notation and symbols that will be used in this book. The chapter will be restricted to those aspects of Finsler geometry that are related to the topics in this book. There is no attempt to make a detailed introduction here, and the content of this chapter is far from being self-contained.

## 1.1 Finsler Spaces

As pointed out by the late great geometer Chern [43], Finsler geometry is just Riemannian geometry without the quadratic restriction. In Riemannian geometry, the restriction of the metric to a tangent space is an inner product; hence tangent spaces at different points are linearly isometric to each other. However, in Finsler spaces this is no longer true. In fact, in a Finsler space, tangent spaces at different points can be very different. We first give the definitions of Minkowski norms and Minkowski spaces.

**Definition 1.1.** Let $V$ be an $n$-dimensional real vector space. A Minkowski norm on $V$ is a real-valued function $F$ on $V$ that is smooth on $V \setminus \{0\}$ and satisfies the following conditions:

1. $F(u) \geq 0, \forall u \in V$;
2. $F(\lambda u) = \lambda F(u), \forall \lambda > 0$;

S. Deng, *Homogeneous Finsler Spaces*, Springer Monographs in Mathematics, DOI 10.1007/978-1-4614-4244-8_1, © Springer Science+Business Media New York 2012

3. Fix a basis $u_1, u_2, \ldots, u_n$ of $V$, and write $F(y) = F(y^1, y^2, \ldots, y^n)$ for $y = y^1 u_1 + y^2 u_2 + \cdots + y^n u_n$. Then the Hessian matrix

$$(g_{ij}) := \left( \left[ \frac{1}{2} F^2 \right]_{y^i y^j} \right)$$

is positive definite at every point of $V \setminus \{0\}$.

The pair $(V, F)$ is called a Minkowski space.

Sometimes we simply say that $V$ is a Minkowski space if the norm is clear. We first give some examples of Minkowski norms.

*Example 1.1.* Let $\langle , \rangle$ be an inner product on $V$. Define $F(y) = \sqrt{\langle y, y \rangle}$. Then $F$ is a Minkowski norm. In this case it is called Euclidean or coming from an inner product.

*Example 1.2.* Let $\alpha$ be a Euclidean norm on an $n$-dimensional real vector space $V$ as in Example 1.1 and $\beta$ a real linear function on $V$. Define $F(y) = \alpha(y) + \beta(y)$. Then $F$ is a real function on $V$ satisfying conditions (1) and (2) of Definition 1.1. Let us determine the condition for $F$ to be a Minkowski norm. Fix an orthonormal basis of $V$ with respect to $\langle , \rangle$. Suppose $\beta(e_i) = b_i$. Then for $y = y^i e_i$ we have $F(y) = \sqrt{\sum_{i=1}^{n} (y^i)^2} + b_i y^i$. The Hessian matrix with respect to the basis $e_1, e_2, \ldots, e_n$ has determinant

$$D(y) = \left( \frac{F(y)}{\alpha(y)} \right)^{n+1}, \quad y \neq 0. \tag{1.1}$$

This fact can be found in [16, p. 284]. We leave the proof as an exercise for the reader. This formula has the interesting corollary that the Hessian matrix of such a function is positive definite if and only if it is positive, i.e., if and only if $F(y) > 0$, for $y \neq 0$. It is easily seen that this is true if and only if

$$\sum_{i=1}^{n} (b_i)^2 < 1.$$

If we define an inner product $\langle , \rangle^*$ on the dual space $V^*$ by requiring the dual basis $e_1^*, e_2^*, \ldots, e_n^*$ to be orthonormal, then $F$ is a Minkowski norm if and only if the length of $\beta$ is less than 1. Such Minkowski norms are called Randers norms. A Randers norm is Euclidean if and only if $\beta$ is zero.

*Example 1.3.* Let $(V, \langle , \rangle)$ be a Euclidean space and $V_1, V_2, \ldots, V_s$ nontrivial subspaces of $V$ such that $V$ is the orthogonal sum of the $V_i$. For a positive integer $m \geq 2$, define

$$F(y) = \sqrt{|y|^2 + \sqrt[m]{\sum_{i=1}^{s} |y_i|^{2m}}},$$

where $|\cdot|$ is the length function of $\langle,\rangle$ and $y_i$, $i = 1, 2, \ldots, s$, are uniquely determined by

$$y = y_1 + y_2 + \cdots + y_s, \quad y_i \in V_i, \quad i = 1, 2, \ldots, s.$$

It is easy to check that $F$ is a Minkowski norm on $V$. This Minkowski norm is always non-Euclidean.

We now study some fundamental properties of Minkowski norms. The following proposition shows that Minkowski norms have all the characteristics of a *norm* in the usual sense, except the absolute homogeneity.

**Proposition 1.1.** *Let $(V, F)$ be a Minkowski space. Then*

1. *For any $y \neq 0$, $F(y) > 0$.*
2. *$F(y_1 + y_2) \leq F(y_1) + F(y_2)$, with equality if and only if $y_1 = \alpha y_1$ or $y_2 = \alpha y_1$ for some $\alpha \geq 0$.*

*Proof.* In the process of the proof we will introduce the fundamental tensor of the Minkowski norm $F$. Note that the positive definiteness of the Hessian matrix amounts to saying that for any nonzero $y$, the bilinear form

$$g_y(u, v) = g_{ij}(y)u^i v^j, \quad u, v \in V,$$

is an inner product on $V$. Using Euler's theorem on homogeneous functions, one easily deduces that

$$F^2(y) = g_{ij}(y)y^i y^j = g_y(y, y), \quad y \neq 0.$$

Hence $F(y) > 0$, for $y \neq 0$. This proves (1).

By a direct computation, one easily checks that

$$g_y(u, v) = \frac{1}{2}\frac{\partial^2}{\partial s \partial t}[F^2(y + su + tv)]_{s=t=0}, \quad y \neq 0, \ u, v \in V.$$

Now for nonzero $u, v$, set $\bar{u} = \frac{u}{F(u)}$ and $\bar{v} = \frac{v}{F(v)}$. Assume first that $\bar{u} \neq \pm \bar{v}$. Consider the function

$$\varphi(t) = F^2(t\bar{u} + (1-t)\bar{v}).$$

Then $\varphi(0) = \varphi(1) = 1$ and

$$\varphi''(t) = 2g_y(\bar{u} - \bar{v}, \bar{u} - \bar{v}) > 0, \quad 0 < t < 1,$$

where $y = t\bar{u} + (1-t)\bar{v}$ (note that the assumption $\bar{u} \neq \pm \bar{v}$ implies that $y \neq 0$). This means that $\varphi$ is a concave function on the interval $[0, 1]$. Therefore for $0 < t < 1$, $\varphi(t) < 1$. Now in the above function we let

$$t = t_0 = \frac{F(u)}{F(u) + F(v)}.$$

Then we have

$$F^2(t_0\bar{u} + (1-t_0)\bar{v}) < 1.$$

From this we easily deduce that

$$F(u+v) < F(u) + F(v).$$

If $\bar{u} = -\bar{v}$, then the above inequality holds as well. If $\bar{u} = \bar{v}$, then it is obvious that

$$F(u+v) = F(u) + F(v).$$

Further, equality still holds if at least one of $u, v$ is equal to 0. This completes the proof of the proposition.                                                                    □

The tensor

$$g_y(u,v) = g_{ij}(y)u^i v^j, \quad u, v \in V,$$

is called the fundamental tensor of the Minkowski space. There is another tensor, called the Cartan tensor, defined by

$$C_y(u,v,w) = C_{ijk}(y)u^i v^j w^k, \quad u, v, w \in V,$$

where

$$C_{ijk}(y) = \frac{1}{4}[F^2]_{y^i y^j y^k}.$$

Equivalently,

$$C_y(u,v,w) = \frac{1}{4}\frac{\partial^3}{\partial t \partial s \partial r}[F^2(y + su + tv + rw)]|_{t=s=r=0}.$$

The fundamental tensor $g_y$ and the Cartan tensor $C_y$ are both symmetric. The Cartan tensor also has the following property:

$$C_y(y,u,v) = 0, \quad u, v \in V.$$

The proof of this fact is left to the reader.

Now we can define the notion of a Finsler metric.

**Definition 1.2.** Let $M$ be a (connected) smooth manifold. A Finsler metric on $M$ is a function $F : TM \to [0, \infty)$ such that

1. $F$ is $C^\infty$ on the slit tangent bundle $TM \setminus \{0\}$;
2. The restriction of $F$ to any $T_x M$, $x \in M$, is a Minkowski norm.

The pair $(M, F)$ is called a Finsler space. We first give some examples of Finsler spaces.

*Example 1.4.* Any Riemannian manifold must be a Finsler space. Hence Riemannian geometry is a special case of Finsler geometry.

*Example 1.5.* Let $(V, F)$ be a Minkowski space. As a real vector space, for any $u \in V$, the tangent space $T_u(V)$ can be canonically identified with $V$. We can then define a Finsler metric on $V$ by this identification. Usually the Finsler space is still denoted by $(V, F)$. Hence a Minkowski space is typically a Finsler space.

Let $(M, F)$ be a Finsler space. If for any $x \in M$, there is a local coordinate system $(U; x_1, x_2, \ldots, x_n)$ of $x$ such that on $U$ the expression of $F$ has no independence on $x$, then $(M, F)$ is called locally Minkowskian. Obviously, a Minkowski space is locally Minkowskian. However, the converse is not true. The author can give some counterexamples without any difficulty (e.g., some Riemannian metrics).

*Example 1.6.* Let $\alpha$ be a Riemannian metric on a connected manifold $M$ and $\beta$ a smooth 1-form on $M$. As pointed out in Example 1.2, the function $F = \alpha + \beta$ on $TM$ defines a Finsler metric on $M$ if and only if the length of $\beta$ with respect to $\alpha$ is everywhere less than 1. Finsler metrics of this type are called Randers metrics. Such metrics were introduced by Randers in 1941 [131], in his study of general relativity.

*Example 1.7.* Let us give an explicit example of a Randers metric. Let $B^n(r_\mu)$ be the open ball of $\mathbb{R}^n$ centered at the origin with radius $r_\mu = \frac{1}{\sqrt{-\mu}}$, where $\mu < 0$. Identifying any tangent space of $x \in \mathbb{R}^n$ with $\mathbb{R}^n$ in the canonical way, we define

$$F(x, y) = \frac{\sqrt{|y|^2 + \mu(|x|^2|y|^2 - \langle x, y \rangle^2)} + \sqrt{-1\mu}\langle x, y \rangle}{1 + \mu|x|^2}. \tag{1.2}$$

By definition, $F$ is a Randers metric on $B^n(r_\mu)$. When $\mu = -1$, the metric in (1.2) is called the Funk metric.

*Example 1.8.* The Randers metric in (1.2) is not reversible. We define another Finsler metric $H$ on $B^n(r_\mu)$ by

$$H(x, y) = \frac{1}{2}\left(F(x, y) + F(x, -y)\right).$$

The metric $H$ is reversible and is called the Hilbert metric. The Hilbert metric is projectively flat; see [44] for details. When $\mu = -1$, $H$ is the Klein metric on $B^n(1)$.

Now we introduce the following result of Deicke to conclude this section; see [46].

**Theorem 1.1 (Deicke).** *Let $(V, F)$ be a Minkowski space. Fix a basis of $V$ and let $g_{ij}$ and $C_{ijk}$ be the coefficients of its fundamental tensor and Cartan tensor, respectively. Then $F$ is a Euclidean norm if and only if $C_{ijk} = 0$ for all $i, j, k$, if and only if $C_k = C_{ijk}g^{jk} = 0$ for all $k$. Here $(g^{jk})$ is the inverse matrix of $(g_{ij})$.*

## 1.2 The Chern Connection and Curvature

To study geometric properties of Finsler spaces, it is important to have a method to differentiate vector fields and tensor fields. In Riemannian geometry, the Levi-Civita connection is such a tool, and it plays a fundamental role in the study of all aspects of the field. It is natural to consider the generalization of the Levi-Civita connection to Finsler geometry. Unfortunately, in Finsler geometry there exists no connection having the same fine properties as the Levi-Civita connection. In particular, there is no connection in Finsler geometry that is both torsion-free and metric compatible.

In 1943, Chern introduced a connection in Finsler geometry. This connection is now called the Chern connection in the literature. The Chern connection is torsion-free and almost metric compatible. We will introduce this important tool in the following. We stress here that there are several connections in Finsler geometry, such as the Cartan connection, the Berwald connection, and the Hashiguchi connection. In the study of Finsler geometry, it is important to use the most appropriate connection for different problems.

The Chern connection is a linear connection on a vector bundle over the slit tangent bundle $TM_o = TM \setminus \{0\}$. Let us first explain the related settings. Recall that the tangent bundle $\pi : TM \to M$ is a vector bundle over $M$. The restriction of $\pi$ to the slit tangent bundle $TM_o$ is a smooth map from $TM_o$ to $M$. Therefore we have the pullback bundle $\pi^*TM$, which is a vector bundle over the manifold $TM_o$. Intuitively, the vector bundle $\pi^*TM$ is obtained by erecting the fiber $T_x(M)$ over $(x,y) \in TM_o$.

A linear connection of $\pi^*TM$ is a smooth map from $\Gamma(\pi^*(TM))$ to

$$\Gamma(T^*(TM_o) \otimes \pi^*(TM))$$

satisfying some additional conditions. Here $\Gamma$ means the set of all smooth sections of the vector bundle. Sometimes we can also express the connection as a family of smooth maps

$$\nabla : T_{(x,y)}(TM_o) \times \Gamma(\pi^*(TM)) \to (\pi^*(TM))_{(x,y)},$$

where $(\pi^*(TM))_{(x,y)}$ means the fiber of $\pi^*(TM)$ at $(x,y)$. The image of $(\widetilde{X}, v)$ under $\nabla$ is usually denoted by $\nabla_{\widetilde{X}} v$. The condition for the connection can then be expressed as follows:

1. $\nabla_{\widetilde{X}} v$ is linear in $\widetilde{X}$;
2. $\nabla_{\widetilde{X}}(fv) = df(\widetilde{X})v + f\nabla_{\widetilde{X}} v$, for $f \in C^\infty(TM_o)$.

Now we consider the connection in a local coordinate system. Let

$$(U; x^1, x^2, \ldots, x^n)$$

be a local coordinate system of the manifold $M$. Then for any $x \in M$, the tangent vectors $\frac{\partial}{\partial x^1}, \frac{\partial}{\partial x^2}, \ldots, \frac{\partial}{\partial x^n}$ form a basis of the tangent space $T_x(M)$. Therefore each $y \in T_x(M)$ can be uniquely expressed as $y = y^i \frac{\partial}{\partial x^i}$. Thus

$$(x^1, x^2, \ldots, x^n, y^1, y^2, \ldots, y^n)$$

is a local coordinate system of the open subset $TU \setminus \{0\}$. Note that $\frac{\partial}{\partial x^1}, \frac{\partial}{\partial x^2}, \ldots, \frac{\partial}{\partial x^n}$ also form a local framing of the vector bundle $\pi^*(TM)$. Suppose

$$\nabla \frac{\partial}{\partial x^i} = \omega_j^i \otimes \frac{\partial}{\partial x^j}, \quad i = 1, 2, \ldots, n,$$

where $\omega_j^i$ are 1-forms on $TU \setminus \{0\}$. Then the connection is uniquely determined by the family of 1-forms $\omega_j^i$, together with the conditions (1) and (2) above. These forms are called the connection forms of the connection. Therefore, to characterize a linear connection on the vector bundle, it suffices to give the expression of the connection forms.

Now we are ready to introduce the Chern connection. For this we need some quantities related to the metrics. The formal Christoffel symbols of the second type are

$$\gamma_{jk}^i = g^{is} \frac{1}{2} \left( \frac{\partial g_{sj}}{\partial x^k} - \frac{\partial g_{jk}}{\partial x^s} + \frac{\partial g_{ks}}{\partial x^j} \right).$$

They are functions on $TU \setminus \{0\}$. We also define some other functions on $TU \setminus \{0\}$ by

$$N_j^i(x, y) := \gamma_{jk}^i y^k - C_{jk}^i \gamma_{rs}^k y^r y^s.$$

The quantities $N_j^i$ are called the nonlinear connection of the Finsler manifold.

**Theorem 1.2 (Chern [42]).** *The pullback bundle $\pi^*TM$ admits a unique linear connection, called the Chern connection. Its connection forms $\omega_j^i$ satisfy the following conditions:*

- *Torsion-freeness:*

$$dx^j \wedge \omega_j^i = 0;$$

- *Almost g-compatibility:*

$$dg_{ij} - g_{kj} \omega_i^k - g_{ik} \omega_j^k = 2C_{ijs} (dy^s + N_k^s dx^k).$$

*The torsion-freeness implies that $\omega_j^i = \Gamma_{jk}^i dx^k$, with $\Gamma_{jk}^i = \Gamma_{kj}^i$. The coefficients $\Gamma_{jk}^i$ are then expressed as*

$$\Gamma_{jk}^l = \gamma_{jk}^l - g^{li} (C_{ijs} N_k^s - C_{jks} N_i^s + C_{kis} N_j^s).$$

We will not pursue the proof of this theorem, which is a complicated computation using the classical Christoffel technique; see [15, 16]. Instead, we will explain in some detail the terminology appearing in the above theorem. First note that, a priori, the connection forms $\omega_j^i$, which are one-forms on $TU \setminus \{0\}$, can be expressed as

$$\omega_j^i = \Gamma_{jk}^i dx^k + Z_{jk}^i dy^k.$$

Then the torsion-freeness

$$dx^j \wedge \omega_j^i = 0$$

implies that

$$\Gamma_{jk}^i dx^j \wedge dx^k + Z_{jk}^i dx^j \wedge dy^k = 0.$$

Thus $Z_{jk}^i = 0$ and $\Gamma_{jk}^i = \Gamma_{kj}^i$. Then one can use the almost metric compatibility to get the formula for the coefficients $\Gamma_{jk}^i$.

If the Finsler metric is Riemannian, then by Deicke's theorem we have $C_{ijk} = 0$. In this case, the almost metric compatibility reads as

$$dg_{ij} - g_{kj}\omega_i^k - dg_{ik}\omega_j^k = 0.$$

This is exactly the metric compatibility of the Levi-Civita connection. Hence the Chern connection is a natural generalization of the Levi-Civita connection. Note also that in this case the Chern connection coefficients are the Riemannian metric's Christoffel symbols of the second kind.

We now introduce the notions of flag curvature and Ricci scalar in Finsler geometry. First we introduce some differential forms on the manifold $TM_o$. Let

$$\delta y^i = dy^i + N_j^i dx^j.$$

The curvature 2-forms of the Chern connection are

$$\Omega_j^i = d\omega_j^i - \omega_j^k \wedge \omega_k^i.$$

Since $\Omega_j^i$ are 2-forms on $TU_0$, they can be expanded as

$$\Omega_j^i = \frac{1}{2}R_j{}^i{}_{kl}dx^k \wedge dx^l + P_j{}^i{}_{kl}dx^k \wedge \frac{\delta y^l}{F} + \frac{1}{2}Q_j{}^i{}_{kl}\frac{\delta y^k}{F} \wedge \frac{\delta y^l}{F},$$

where $R_j{}^i{}_{kl} = -R_j{}^i{}_{lk}, Q_j{}^i{}_{kl} = -Q_j{}^i{}_{lk}$. Using the torsion-freeness we get $dx^j \wedge dw_j^i = -d(dx^j \wedge \omega_j^i) = 0$. Thus

$$dx^j \wedge \Omega_j^i = dx^j \wedge \omega_j^i - dx^j \wedge \omega_k^i \wedge \omega_j^k$$

$$= \frac{1}{2}R_j{}^i{}_{kl}dx^j \wedge dx^k \wedge dx^l + P_j{}^i{}_{kl}dx^j \wedge dx^k \wedge \frac{\delta y^l}{F}$$

$$+ \frac{1}{2}Q_j{}^i{}_{kl}dx^j \wedge \frac{\delta y^k}{F} \wedge \frac{\delta y^l}{F}$$

$$= 0.$$

From this we deduce that

$$Q_j{}^i{}_{kl} = 0,$$

that

$$P_j{}^i{}_{kl} = P_k{}^i{}_{jl},$$

and that

$$R_j{}^i{}_{kl} + R_k{}^i{}_{lj} + R_l{}^i{}_{jk} = 0. \tag{1.3}$$

Equation (1.3) is called the first Bianchi identity.

Now we are in a position to introduce the flag curvature in Finsler geometry. This is a natural generalization of the sectional curvature in Riemannian geometry. Let us first explain the notion of a flag on $M$. Given $x \in M$, a flag in the tangent space $T_x(M)$ is a pair $(P, y)$, where $P$ is a 2-dimensional subspace (a tangent plane) of $T_x(M)$ and $y$ is a nonzero vector contained in $P$. The nonzero vector $y$ is usually called the flag pole. Let

$$R_{jikl} = g_{is} R_j{}^s{}_{kl}.$$

The flag curvature of the flag $(P, y)$ is defined to be

$$K(P, y) := \frac{u^i (y^j R_{jikl} y^l) u^k}{g_y(y, y) g_y(u, u) - [g_y(y, u)]^2},$$

where $u$ is a nonzero vector in $P$ such that $y, u$ span $P$. It is easy to show that this quantity is independent of the selection of $u$.

Since the flag curvature is the most important notion in Finsler geometry, we will pause here to present some further clarification. The coefficients $R_j{}^i{}_{kl}$ can be contracted with $y$ to produce

$$R^i{}_k = y^j R_j{}^i{}_{kl} y^l.$$

The Riemann tensor is defined to be

$$\mathscr{R} = R^i{}_k \frac{\partial}{\partial x^i} \otimes dx^k.$$

In particular, if $y \in T_x(M) \setminus \{0\}$, then we can define a linear map $\mathscr{R}_y$ of $T_x(M)$ by

$$\mathscr{R}_y(u) = R^i{}_k u^k \frac{\partial}{\partial x^i}, \quad u \in T_x(M).$$

The collection $\{\mathscr{R}_y\}$ is called the Riemann tensor; see [144]. The flag curvature can then be rewritten as

$$K(P, y) = \frac{g_y(\mathscr{R}_y(u), u)}{g_y(y, y) g_y(u, u) - [g_y(y, u)]^2}.$$

The above expression can be used to define the notion of Ricci scalar. The quantity

$$\text{Ric}(y) = \frac{\text{tr}(\mathscr{R}_y)}{F^2(y)}, \quad y \in TM_o,$$

is called the Ricci scalar of the Finsler space. The Ricci scalar can be defined in another way. Let $\frac{y}{F(y)}, u_2, \ldots, u_n$ be an orthonormal basis of $T_x(M)$ with respect to the inner product $g_y$. Then it is easily seen that

$$\text{Ric}(y) = \sum_{j=2}^{n} K(P_i, y),$$

where $P_i$ is the tangent plane spanned by $y$ and $u_i$.


## 1.3  Arc-Length Variations and the Exponential Map

In this section we will give the formula for arc-length variations and define the notion of exponential map for Finsler spaces.

Let $(M, F)$ be a connected Finsler space and $\sigma : [a, b] \to M$ a smooth curve on $M$. The length of the curve $\sigma$ is defined by

$$L(\sigma) = \int_a^b F(\sigma(t), \dot{\sigma}(t)) \, dt,$$

where $\dot{\sigma}(t)$ denotes the tangent vector of the curve at $\sigma(t)$. The above definition can easily be generalized to a piecewise smooth curve in an obvious way. The distance function $d : M \times M \to \mathbb{R}^+$ can then be defined in a standard way: for $x, y \in M$, just let $d(x, y)$ be the infimum of the lengths of all the piecewise smooth curves connecting $x$ and $y$. The distance function $d$ has almost all the characteristics of a distance in a metric space. In fact, we have

1. $d(x, y) \geq 0$, with equality if and only if $x = y$.
2. $d(x, z) \leq d(x, y) + d(y, z)$, for any $x, y, z \in M$.

However, in general $d$ is not symmetric. In fact, $d(x, y) = d(y, x)$ for any $x, y \in M$ if and only if the Finsler metric is reversible. In this case, $(M, d)$ is a genius metric space. All these facts can be proved without much difficulty and can be left to the reader.

Now we consider the variation of the arc length. For this we need some notation. Let $V$ be a nowhere-zero vector field on an open subset $N$ of $M$. Then we can define an affine connection on $N$. In fact, in a local coordinate system $(U; x^1, x^2, \ldots, x^n)$ with $U \subset N$, suppose $T = T^i \frac{\partial}{\partial x^i}$, $W = W^i \frac{\partial}{\partial x^i}$. Define

$$\nabla^V_T W|_x = [T(W^i) + W^j T^k \Gamma^i_{jk}(x,V)] \frac{\partial}{\partial x^i},$$

where $x \in M$, $\Gamma^i_{jk}$ are the coefficients of the Chern connection, and $T(W^i)$ is the action of a vector field on a function. It is easy to check that $\nabla^V$ is actually an affine connection on $N$. Moreover, from the properties of the Chern connection we can easily deduce the following:

1. $\nabla^V$ is torsion-free, namely, for any vector fields $W_1, W_2$ on $N$,

$$\nabla^V_{W_1} W_2 - \nabla^V_{W_2} W_1 = [W_1, W_2].$$

2. $\nabla^V$ is almost metric compatible, namely, for any vector fields $W, W_1, W_2$ on $N$, we have

$$W g_V(W_1, W_2) = g_V(\nabla^V_W W_1, W_2) + g_V(W_1, \nabla^V_W W_2) + 2C_V(\nabla^V_W V, W_1, W_2).$$

Suppose $\sigma$ is a regular curve on $M$ with velocity vector field $T$ and $W_1$, $W_2$ are two vector fields along $\sigma$. Extending $T$ (resp. $W_1$, $W_2$) to a smooth vector field $T_1$ (resp. $W'_1$, $W'_2$) defined on a open subset of $M$ containing the whole curve $\sigma$, we can define

$$D_{W_1} W_2 = \nabla^{T_1}_{W'_1} W'_2|_\tau.$$

It is easily seen that the above definition does not depend on the extension of the vector fields.

Now we can deduce the formula of variation of arc length. Suppose $\sigma : [a,b] \to M$ is a regular curve with tangent vector field $T$. A variation of $\sigma$ is a smooth map $\tau : (-\varepsilon, \varepsilon) \times [a,b] \to M$ such that $\sigma(t) = \tau(0,t)$, $a \le t \le b$. Since $T$ is nonzero along $\sigma$ and $\tau$ is smooth, the tangent vector field of the curve $\tau(s,t)$, $a \le t \le b$, must be nonzero for $s$ small enough. Without loss of generality we can assume that the curves $\varphi_s(t) = \tau(s,t)$, $-\varepsilon < s < \varepsilon$, are all regular. Denote by $T$, $U$ the tangent vector fields of the $t$-curves and the $s$-curves, respectively (note that for $s = 0$, this $T$ coincides with the above $T$). Denote the length of the curve $\varphi_s$ by $L(s)$. Then

$$L(s) = \int_a^b F(\tau(s,t), T(s,t)) \, dt.$$

Since $F(\tau(s,t), T(s,t)) = \sqrt{g_T(T,T)}$, we have

$$L'(s) = \frac{d}{ds} \int_a^b \sqrt{g_T(T,T)} \, dt$$

$$= \int_a^b U \sqrt{g_T(T,T)} \, dt$$

$$= \int_a^b \frac{1}{2} \frac{2g_T(D_U T, T) + 2C_T(D_U T, T, T)}{\sqrt{g_T(T,T)}} \, dt$$

$$= \int_a^b \frac{g_T(D_U T, T)}{F(T)} \, dt,$$

where we have used the almost metric compatibility of the Chern connection and the fact that

$$C_T(V_1, T, V_2) = 0, \quad \forall V_1, V_2.$$

Note that $D_U T - D_T U = [T, U] = 0$ and

$$T g_T(U, T) = g_T(D_T U, T) + g_T(U, D_T T) + 2C_T(D_T T, U, T)$$
$$= g_T(D_T U, T) + g_T(U, D_T T).$$

We thus get

$$L'(s) = \int_a^b \frac{T g_T(U, T) - g_T(U, D_T T)}{F(T)} \, dt$$

$$= \int_a^b \left[ T g_T\left(U, \frac{T}{F(T)}\right) - g_T\left(U, D_T\left(\frac{T}{F(T)}\right)\right) \right] dt$$

$$= g_T\left(U, \frac{T}{F(T)}\right)\Big|_a^b - \int_a^b \left[ g_T\left(U, D_T\left(\frac{T}{F(T)}\right)\right) \right] dt,$$

where we have used again the properties of the Chern connection to handle the factor $\frac{1}{F(T)}$.

The above argument can be easily generalized to regular piecewise smooth curves. Suppose $\sigma(t)$, $a \le t \le b$, is a regular piecewise smooth curve on $M$ and Let $a = t_0 < t_1 < \cdots < t_k = b$ be a partition of the interval $[a,b]$ such that $\sigma$ is regular smooth on each subinterval $[t_{i-1}, t_i]$. Suppose $\tau(s,t)$, $-\varepsilon < s < \varepsilon$, $a \le t \le b$, is a variation of $\sigma$ such that $\tau$ is smooth on each $(-\varepsilon, \varepsilon) \times [t_{i-1}, t_i]$. Denote the length of the curve $\tau(s,t)$, $a \le t \le b$, by $L(s)$. Then a similar argument shows that

$$L'(s) = g_T\left(U, \frac{T}{F(T)}\right)\Big|_a^b - \sum_{i=1}^{k-1} g_T\left(U, \frac{T}{F(T)}\right)\Big|_{t_i^-}^{t_i^+}$$

$$- \sum_{i=1}^{k} \int_{t_{i-1}}^{t_i} g_T\left(U, D_T\left(\frac{T}{F(T)}\right)\right) dt.$$

This is the formula of the first variation of the arc length.

Now we introduce the exponential map of a Finsler space. Before this we need the notion of geodesics in Finsler geometry. The formula of the first variation of the arc length has the following corollary:

**Proposition 1.2.** *Let $\sigma(t) : a \leq t \leq b$ be a regular piecewise smooth curve on M. Then the following two conditions are equivalent:*

(a) *$L'(0) = 0$ for any piecewise smooth variation of $\sigma$ that keeps the start and end points fixed.*
(b) *$\sigma$ is smooth and $D_T(\frac{T}{F(T)}) = 0$.*

The proof of this proposition is very similar to the Riemannian case and will be omitted. A curve $\sigma$ satisfying any of the above two conditions is called a geodesic. In the Riemannian case, a geodesic can also be characterized as a curve locally minimizing the distance of the points of this curve. This is also true for Finsler spaces.

**Proposition 1.3.** *A curve $\sigma$ on M is a geodesic if and only if it locally minimizes the distance along this curve. More precisely, $\sigma$ is a geodesic if and only if for any $t_0$, there exists $\varepsilon > 0$ such that for any $t_1$ with $|t_1 - t_0| < \varepsilon$, the curve $\sigma$ is the shortest path connecting $\sigma(t_0)$ and $\sigma(t_1)$.*

For the proof of this proposition, see [16].

In the Riemannian case, a geodesic must be of constant speed, i.e., $F(T)$ must be a constant. However, this is not true in the general Finslerian case. From now on, we will consider only geodesics of constant speed in a Finsler space. It is easily seen that constant-speed geodesics can be characterized by the equation

$$D_T T = 0.$$

In a local coordinate system, this is a system of ordinary differential equations:

$$\frac{d^2\sigma^i}{dt^2} + \frac{d\sigma^j}{dt}\frac{d\sigma^k}{dt}\Gamma^i{}_{jk}(\sigma, T) = 0.$$

Since this is a system of nonlinear equations, we can get only the local existence of geodesics:

**Proposition 1.4.** *For any $p \in M$ there exist a precompact neighborhood $U$ of $p$ and a positive number $\varepsilon$ such that for any $x \in U$ and $y \in T_x(M) \setminus \{0\}$ with $0 < F(x,y) < \varepsilon$ there exists a unique constant-speed geodesic $\sigma_{(x,y)}$ that is defined on $(-2,2)$ with $\sigma_{(x,y)}(0) = x$, $\dot{\sigma}_{(x,y)}(0) = y$. Moreover, $\sigma_{(x,y)}(t)$ is smooth on $t$, $x$, and $y$.*

**Definition 1.3.** The exponential map of $(M, F)$ is defined by

$$\exp(x,y) = \begin{cases} \sigma_{(x,y)}(1), & y \neq 0, \\ x, & y = 0. \end{cases}$$

From this definition, we see that the exponential map is generally defined only in a tube neighborhood of the zero section of the tangent bundle. For an arbitrary point, the following theorem guarantees the existence of the certain special neighborhood. We first define some notation. For $x \in M$ and $r > 0$, we define

$$\mathscr{B}_x^+(r) = \{p \in M | d(x,p) < r\}, \quad \mathscr{B}_x^- = \{p \in M | d(p,x) < r\},$$

and

$$\mathscr{S}_x^+(r) = \{p \in M | d(x,p) = r\}, \quad \mathscr{S}_x^- = \{p \in M | d(p,x) = r\}.$$

We also define

$$B_x(r) = \{V \in T_x(M) | F(x,V) < r\}, \quad S_x(r) = \{V \in T_x(M) | F(x,V) = r\}.$$

**Theorem 1.3.** *Let $(M,F)$ be a connected Finsler space and $x \in M$. Then there are positive numbers $r$ and $\varepsilon$ such that* exp *is a $C^1$ diffeomorphism from the open subset $B_x(r+\varepsilon)$ to its image. Moreover, the following assertions hold:*

1. *Any radial geodesic* $\exp(tV)$, $0 \le t \le r$, $V \in T_x(M)$, $F(x,V) = 1$, *minimizes the distance among all piecewise smooth curves that share the same endpoints.*
2. *Any piecewise smooth curves on $M$ that share the same arc length and with the same start and end points as $\exp(tV)$ in (1) must be a reparametrization of the curve $\exp(tV)$.*
3. *It is true that $\exp[B_x(r)] = \mathscr{B}_x(r)^+$, $\exp_x[S_x(r)] = \mathscr{S}_x^+(r)$.*

We refer to [16] for the proof. This theorem has the following corollary.

**Corollary 1.1.** *For any $x$ in a connected Finsler space $(M,F)$, there exists a neighborhood $U$ of $x$ such that each pair of points in $U$ can be joined by a unique minimizing geodesic lying in $U$.*

Note that the uniqueness in the above corollary means that it is unique as a point set. In general, we cannot have the global definition of the exponential map. The Hopf–Rinow theorem presents some conditions for this. Before stating the theorem, we give some related definitions. A Finsler manifold $(M,F)$ is called forward geodesically complete if every geodesic $\gamma(t)$, $a \le t < b$, can be extended to a geodesic defined on $[a,+\infty)$. A sequence $\{x_i\}_{i=1}^{\infty}$ of points in $M$ is called a forward Cauchy sequence if for any $\varepsilon > 0$, there exists a natural number $N$ such that for any $N < m_1 < m_2$, we have $d(x_{m_1}, x_{m_2}) < \varepsilon$. A subset $U$ of $M$ is called forward bounded if there exist a point $x_0 \in U$ and a positive number $r$ such that $d(x_0,x) < r$, $\forall x \in U$. The manifold $(M,F)$ is called forward complete if each forward Cauchy sequence converges to a point in $M$.

**Theorem 1.4 (Hopf–Rinow).** *Let $(M,F)$ be a connected Finsler space. Then the following conditions are mutually equivalent:*

1. *$(M,F)$ is forward complete.*
2. *$(M,F)$ is forward geodesically complete.*

3. *At every point $p \in M$, the exponential map* $\exp_p$ *is defined on the whole tangent space $T_p(M)$.*
4. *At some point $p \in M$, the exponential map* $\exp_p$ *is defined on the whole tangent space $T_p(M)$.*
5. *Any closed forward bounded subset of $M$ is compact.*

*Moreover, if any one of the above conditions is satisfied, then any two points of $M$ can be joined by a minimizing geodesic.*

We will not pursue the proof of this theorem, which is similar to the Riemannian case and can be found in [16]. We remark here that one can similarly define the notions of backward completeness, backward geodesic completeness, backward bounded subsets, etc., and there is a backward version of the Hopf–Rinow theorem (see [16]). In a general Finsler space, backward completeness is not equivalent to forward completeness. However, if the Finsler metric is reversible, then there is no difference between backward completeness and forward completeness. This is the case, in particular, for a Riemannian metric.

At the final part of this section, we will introduce the formula for the second variation of the arc length. Since the general case is rather complicated, we will confine ourselves in the special case that the base curve is a (constant-speed) geodesic. Let $\sigma(t)$, $a \leq t \leq b$, be a constant-speed geodesic and let $\tau(s,t)$, $-\varepsilon < s < \varepsilon, a \leq t \leq b$, be a variation of $\sigma$. Denote the arc length of the curve $\tau(s,t)$, $a \leq t \leq b$, by $L(s)$. Then

$$L''(0) = I(U,U) + g_T\left(\nabla_U^T U, \frac{T}{F(T)}\right)\Big|_a^b - \int_a^b \frac{1}{F(T)}\left(\frac{\partial F(T)}{\partial u}\right)^2 dt,$$

where

$$I(V,W) = \int_a^b \frac{1}{F(T)}[g_T(D_T V, D_T U) - g_T(R(V,T)T,W)]\, dt$$

is the index form. Here $R(V,T))T$ can be expressed in a local coordinate system as

$$R(V,T)T = (T^j R_j{}^i{}_{kl} T^l)V^k \frac{\partial}{\partial x^i}.$$

A proof of this formula can be found in [16].

## 1.4  Jacobi Fields and the Cartan–Hadamard Theorem

In this section we will introduce the notion of Jacobi fields along a constant-speed geodesic and use it to study the relationship between the sign of the flag curvature and the behavior of the geodesic rays. Let $\sigma(t)$, $a \leq t \leq b$, be a constant-speed geodesic, with velocity vector field $T$. A vector field $V$ along $\sigma$ is called a Jacobi field if

$$D_T D_T V + R(V,T)T = 0. \tag{1.4}$$

It is obvious that if $J$ is a Jacobi field along $\sigma$ and $[c,d] \subset [a,b]$ is a subinterval, then the restriction of $J$ to $\sigma|_{[c,d]}$ is also a Jacobi field. Moreover, in a local coordinate system, (1.4.1) is a system of linear ordinary differential equations. By the theory of ordinary differential equations, we have the following propositions.

**Proposition 1.5.** *Given any $y_1, y_2 \in T_{\sigma(a)}(M)$, there is a unique Jacobi field $J$ along $\sigma$ such that $J|_{\sigma(a)} = y_1$ and $D_T J|_{\sigma(a)} = y_2$.*

**Proposition 1.6.** *The set of zero points of a nonzero Jacobi field along $\sigma$ is discrete.*

Now we will present a method to construct Jacobi fields along a geodesic.

**Proposition 1.7.** *Let $\sigma(t)$, $a \le t \le b$, be a geodesic and $\tau(s,t)$, $-\varepsilon < t < \varepsilon$, $a \le b \le b$, a smooth variation of $\sigma$ such that each $t$-curve is a geodesic (i.e., $\tau$ is a geodesic variation). Then the restriction of the transversal vector field $U$ to $\sigma$ is a Jacobi field.*

The proof of this proposition can be found in [16]. Using this proposition, we can construct Jacobi fields satisfying some prescribed initial conditions. In the following we will denote $D_T U$ by $\dot{U}$, $D_T D_T U$ by $\ddot{U}$, etc.

*Example 1.9.* Fix $x \in M$ and let $T \in T_x(M)$ be a nonzero tangent vector. Then there is $r > 0$ such that the curve $\exp_x(tT)$ is defined on $[0,r]$ and is a constant-speed geodesic. Suppose $W$ is a nonzero vector in $T_x(M)$. Then

$$\tau(s,t) = \exp_x(t(T + sW))$$

is a geodesic variation of the base curve $\exp t(T + sW)$, $0 \le t \le r$. By Proposition 1.7, the variation vector field $U$ is a Jacobi field. It is easy to check that $U(0) = 0$ and $\dot{U}(0) = W$.

*Example 1.10.* Let $\tau$ be a smooth regular curve with velocity vector field $V$. A nowhere-zero vector field $U$ along $\tau$ is called parallel if in extending $V, U$ to smooth vector fields on an open subset containing $\tau$, we have

$$\nabla_V^U U = 0.$$

Now suppose $\sigma(t)$, $a \le t \le b$, is a geodesic and $v, w \in T_x(M)$ are two nonzero tangent vectors. Let $\zeta(s)$ be a smooth curve satisfying $\zeta(0) = \sigma(0)$ and $\dot{\zeta}(0) = v$. Suppose $T_1, W$ are two parallel vector fields along the curve $\zeta$ satisfying $T(0) = \dot{\sigma}(0)$ and $W(0) = w$. Define

$$\gamma(s,t) = \exp_{\zeta(s)}(t(T_1(s) + sW(s))).$$

Then $\gamma$ is a geodesic variation of $\sigma$, and the variation vector field $U$ is a Jacobi field satisfying $U(0) = v$, $\dot{U}(0) = w$.

It should be noted that the converse of Proposition 1.7 is also true, namely, any Jacobi field along a geodesic must be the variation field of a geodesic variation. The proof of this assertion is left to the reader.

One of the important applications of the above-mentioned variations is to study the influence of the flag curvature on the behavior of geodesics. Fix $x \in M$ and consider two vectors in $T_x(M)$ with $F(T) = 1$ and $g_T(W,W) = 1$, such that $T, W$ are linearly independent. Consider the variation of the geodesic $\exp_x(tT)$:

$$\sigma(t,s) = \exp_x(t(T + sW)).$$

Let $U(t) = d(\exp_x)|_{(tT)}(tW)$. Then the length of the vector field $U(t)$ measures the rate at which the $s$th geodesic deviates from the base geodesic $\exp_x(tT)$. Suppose

$$f(t) = \|U(t)\|^2 = g_T(U(t), U(t)).$$

Then it can be obtained by a direct computation that (see [16])

$$\begin{aligned}
f^{(1)}(t) &= 2g_T(U, U^{(1)}), \\
f^{(2)}(t) &= 2g_T(U^{(1)}, U^{(1)}) + 2g_T(U, U^{(2)}), \\
f^{(3)}(t) &= 6g_T(U^{(2)}, U^{(1)}) + 2g_T(U, U^{(3)}), \\
f^{(4)}(t) &= 8g_t(U^{(3)}, U^{(1)}) + 6g_T(U^{(2)}, U^{(2)}) + 2g_T(U, U^{(4)}),
\end{aligned}$$

where

$$\begin{aligned}
U^{(1)} &= D_T U, \\
U^{(2)} &= D_T D_T U = -R(U, T)T, \\
U^{(3)} &= D_T D_T D_T U = -(D_T R)(U, T)T - R(\dot{U}, T)T, \\
U^{(4)} &= D_T D_T D_T D_T U,
\end{aligned}$$

where we have used the the fact that the variation field of a geodesic variation is a Jacobi field.

Considering the above quantities at $t = 0$, we get

$$f(t) = \|U(t)\|^2 = t^2 - \frac{1}{3}K(T,W)t^4 + O(t^5).$$

This means that if $(M,F)$ has positive flag curvature, then $f(t) < t^2$ for small $t$, and hence geodesics starting from the same point will bunch together. If $(M,F)$ has negative flag curvature, then $f(t) > t^2$ for small $t$; hence geodesics starting from the same point will be dispersive. A more accurate statement of such facts can be described by Rauch's comparison theorem; see [16] for more details.

Finally, we introduce the Cartan–Hadamard theorem. For this we need the notion of conjugate points in Finsler spaces. Suppose $(M,F)$ is a forward geodesically complete connected Finsler space and $\sigma$ is a unit-speed geodesic on $M$. Let $p \in \sigma$.

A point $q$ in $\sigma$ is called conjugate to $p$ along $\sigma$ if there is a nonzero Jacobi field $J$ along $\sigma$ such that $J$ vanishes at both $p$ and $q$. It is obvious that if $p$ is conjugate to $q$, then $q$ is conjugate to $p$.

**Theorem 1.5 (Cartan–Hadamard).** *Let* $(M,F)$ *be a forward geodesically complete connected Finsler space with every flag curvature nonpositive. Then we have the following:*

1. *Geodesics on M do not contain conjugate points.*
2. *For any* $x \in M$, *the exponential map* $\exp_x : T_x(M) \to M$ *is a* $C^1$ *covering projection.*
3. *If M is simply connected, then* $\exp_x : T_x(M) \to M$ *is a* $C^1$ *diffeomorphism.*

Using the notion of Jacobi fields, we can study the general properties of the index form of Finsler spaces. This leads to the following result.

**Theorem 1.6 (Bonnet–Myers).** *Let* $(M,F)$ *be an n-dimensional complete Finsler space. Suppose the Ricci scalar of* $(M,F)$ *satisfies the condition*

$$\mathrm{Ric} \geq (n-1)\lambda > 0.$$

*Then the following assertions hold:*

1. *Any geodesic with length at least* $\pi/\sqrt{\lambda}$ *must contain conjugate points.*
2. *M is a compact manifold.*
3. *The fundamental group of M is finite.*

We refer to [16] for the proofs of these theorems.

## 1.5  Parallel Displacements and Holonomy Groups

In this section we will introduce the notions of parallel displacements and holonomy groups in Finsler geometry. This will be useful in the next section, where we will define the notions of Berwald and Landsberg spaces.

Let $(M,F)$ be a connected Finsler space and $\sigma(t)$, $a \leq t \leq b$, a piecewise smooth curve in $M$ connecting $p$ and $q$, with velocity vector field $T$.

**Definition 1.4.** Let $U = U^i(t)\frac{\partial}{\partial x^i}|_{\sigma(t)}$ be a vector field along $\sigma$, and define the covariant derivative $\widetilde{D}_{\dot{\sigma}}U(t)$ along $\sigma$ by

$$\widetilde{D}_T U(t) = \{\dot{U}^i(t) + T^j(t)N_j^i(\sigma(t), U(t))\}\frac{\partial}{\partial x^i}\Big|_{\sigma(t)}. \qquad (1.5)$$

The vector field $U$ is said to be parallel along $\sigma$ if $\widetilde{D}_{\dot{\sigma}}U(t) = 0$.

*Remark 1.1.* Notice that the covariant derivative defined here is different from the $D_V W$ in Sect. 1.3. In [16], the derivative defined in Sect. 1.3 is called the covariant derivative with reference vector $T$, and that defined in the above definition is called the derivative with reference vector $U$. However, it is obvious that the definition of parallel vector fields here coincides with that of Example 1.10 in Sect. 1.4.

In a local coordinate system, (1.5) is a system of linear ordinary differential equations. Therefore, for any $u \in T_p(M)$, there is a unique vector field $U$ that is parallel along $\sigma$ such that $U(a) = u$. This fact enables us to define the parallel displacements along $\sigma$. In fact, one just needs to define $P_\sigma : T_p(M) \to T_q(M)$ by

$$P_\sigma(u) = U(b),$$

where $U = U(t)$ denotes the (unique) parallel vector field along $c$ with $U(a) = u$. It is easily seen that $P_\sigma$ is a $C^\infty$ diffeomorphism from $T_p M \setminus \{0\}$ onto $T_q M \setminus \{0\}$, and is positively homogeneous of degree one, i.e., $P_\sigma(\lambda u) = \lambda P_\sigma(u), \lambda > 0, u \in T_p(M)$; $P_\sigma$ is called the parallel displacement along $\sigma$. Note that by definition, the parallel displacement is in general not a linear map. However, we have the following result.

**Lemma 1.1.** *Let $U$ be a parallel vector field along a piecewise smooth curve $\sigma$. Then*

$$F(\sigma(t), U(t)) = \text{constant}.$$

For the proof, see [44].

Using the notion of parallel displacement, we can define the holonomy group of $(M, F)$ at a point $p \in M$.

**Definition 1.5.** Define

$$H_p(M) = \{P_\sigma \mid \sigma \text{ is a piecewise smooth curve starting from } p \text{ and ending at } p\}.$$

$H_p$ is called the holonomy group of $(M, F)$ at $p$.

The proof of the fact that $H_p$ is a group is easy and can be left to the reader. It is obvious that for different points $p, q \in M$, the groups $H_p$ and $H_q$ are isomorphic. Therefore, we sometimes write $H_p$ as $H$ and call it the holonomy group of $(M, F)$. From the above discussion it is easily seen that $H_p$ consists of diffeomorphisms of $T_p(M) \setminus \{0\}$. Furthermore, since for any $\sigma \in H_p$, $F(\sigma(u)) = F(u)$, $u \in T_p(M)$, $H_p$ is also a transformation group of the indicatrix

$$\mathscr{I}_p = \{y \in T_p(M) | F(y) = 1\}.$$

Let $(M_1, F_1)$ be a Finsler space and $(M, \pi)$ the universal covering manifold of $M_1$. Then we can define a Finsler metric on $M$ by $F(y) = F_1(\pi_*(y)), y \in TM$. Using a similar argument as in the Riemanian case, we can prove the following lemma.

**Lemma 1.2.** *If we endow the holonomy groups of* $(M_1, F_1)$ *and* $(M, F)$ *with the compact-open topology, then the holonomy group of* $(M, F)$ *at* $x \in M$ *can be identified with the identity component of the holonomy group of* $(M_1, F_1)$ *at* $x_1 = \pi(x)$. *In particular, the holonomy group of a simply connected Finsler space is connected.*

For the definition of compact-open topology, we refer the reader to [83].

One of the most important applications of the notion of holonomy group is to study affinely equivalent Finsler metrics. Let $M$ be a connected manifold. Two Finsler metrics $F_1$ and $F_2$ on $M$ are called affinely equivalent if their geodesics coincide as parametrized curves. Precisely, if $\sigma$ is a constant-speed geodesic of $F_1$, then it is a constant-speed geodesic of $F_2$ (but perhaps with different speeds), and vice versa. This notion is very useful in studying some special class of Finsler spaces; see the next section. Now we can use the notions of parallel displacements and holonomy groups to describe affinely equivalent Finsler metrics.

**Lemma 1.3.** *Let* $F_1$ *and* $F_2$ *be two Finsler metrics on the manifold* $M$. *Suppose for any piecewise smooth curve* $\sigma$ *and* $F_1$-*parallel vector field* $U$ *along* $\sigma$, *we have*

$$F_2(\sigma(t), U(t)) = \text{constant}.$$

*Then* $F_2$ *is affinely equivalent to* $F_1$.

The following theorem presents a method to construct Finsler metrics affinely equivalent to a prescribed metric.

**Theorem 1.7.** *Let* $(M, F)$ *be a Finsler space and* $x \in M$. *If* $\bar{F}$ *is a Minkowski norm on* $T_x(M)$ *that is invariant under the action of the holonomy group* $H_x$ *of* $(M, F)$, *then* $\bar{F}$ *can be extended to a Finsler metric* $\widetilde{F}$ *such that* $\widetilde{F}$ *is affinely equivalent to* $F$.

We refer to [44] for the proof of the above result.

## 1.6  Berwald Spaces and Landsberg Spaces

In this section we will introduce the notions of Berwald spaces and Landsberg spaces. Let $(M, F)$ be a connected Finsler space and fix a local coordinate system $(U; x^1, x^2, \dots, x^n)$. Then we have a standard coordinate system

$$(x^1, x^2, \dots, x^n, y^1, y^2, \dots, y^n)$$

on the open set $TU \setminus \{0\}$ of the slit tangent bundle $TM_o$. Recall that the fundamental tensor $\{g_y\}$ is defined by

$$g_y(u, v) = \frac{1}{2} \frac{\partial^2 F^2(y + su + tv)}{\partial s \partial t}\bigg|_{s=t=0}.$$

Define

$$g_{ij}(y) = g_y \left( \frac{\partial}{\partial x^i}, \frac{\partial}{\partial x^j} \right).$$

Then for any nonzero $y \in T_x(M)$, the matrix $(g_{ij}(y))$ is positive definite. Let $(g^{ij}(y))$ be the inverse matrix of $(g_{ij}(y))$. Similarly, for any $y(\neq 0), u, v, w \in T_x(M)$, we have the Cartan tensor $C_y$ defined by

$$C_y(u, v, w) = \frac{1}{4} \frac{\partial^3 F^2(y + su + tv + rw)}{\partial s \partial t \partial r} \Big|_{s=t=r=0}.$$

The geodesic spray $G$ is a vector field on $TM_o$ defined by

$$G = y^i \frac{\partial}{\partial x^i} - 2G^i \frac{\partial}{\partial y^i},$$

where

$$G^i(x, y) = \frac{1}{4} g^{il}(x, y)([F^2]_{x^k y^l}(x, y) y^k - [F^2]_{x^l}(x, y)), \qquad (1.6)$$

where $[F^2]_{x^k y^l}$ denotes the partial derivative of $F^2$ with respect to $x^k$ and $y^l$. The vector field $G$ is well defined and $G^i$ satisfies $G^i(x, \lambda y) = \lambda^2 G(x, y)$, $\lambda > 0$.

From the definition of a geodesic we see that a smooth curve $\tau$ is a geodesic if and only if in the local coordinate system we have

$$\ddot{\tau}^i(t) + 2G^i(\tau(t), \dot{\tau}(t)) = 0, \quad i = 1, 2, \ldots, n.$$

The Landsberg curvature $L = \{L_y \mid y \in TM \setminus \{0\}\}$ is defined by

$$L_y(u, v, w) = L_{ijk} u^i v^j w^k, \qquad (1.7)$$

where the functions $L_{ijk}(x, y)$ are defined by

$$L_{ijk}(x, y) = -\frac{1}{2} y^s g_{sm} \frac{\partial^3 G^m}{\partial y^i \partial y^j \partial y^k}. \qquad (1.8)$$

There is another expression of Landsberg curvature that is useful in practice. Let

$$N_j^i = \frac{\partial G^i}{\partial y^j}$$

and fix a geodesic $\tau$. A vector field $U$ defined by $U(t) = U^i(t) \frac{\partial}{\partial x^i}$ along $\tau$ is said to be linearly parallel along $\tau$ if it satisfies

$$\dot{U}^i(t) + U^j(t) N_j^i(\tau(t), \dot{\tau}(t)) = 0.$$

**Proposition 1.8.** *For $y(\neq 0), u, v, w \in T_x(M)$, we have*

$$L_y(u,v,w) = \frac{\mathrm{d}}{\mathrm{d}t}(C_{\dot{\sigma}(t)}(U(t),V(t),W(t))), \tag{1.9}$$

*where $\sigma(t)$ is the geodesic starting from $x$ with initial vector $y$ and $U(t), V(t), W(t)$ are the linearly parallel vectors fields along $\sigma(t)$ with initial vectors $u, v, w$, respectively.*

For the proof, see [44].

From (1.6)–(1.9) it is easily seen that $L_{\lambda y} = L_y$, for $\lambda > 0$, and that $L_y(u,v,w)$ is symmetric with respect to $u, v, w$. Further, if one of $u, v, w$ is equal to $y$, then $L_y(u,v,w) = 0$. Moreover, if the Finsler metric $F$ is reversible, i.e., $F(u) = F(-u)$ for any tangent vector $u$, then $L_{-y}(u,v,w) = L_y(u,v,w)$, $\forall y(\neq 0), u, v, w \in T_x(M)$.

**Definition 1.6.** A Finsler space $(M,F)$ is called a Berwald space if the geodesic spray coefficients $G^i$ are quadratic in $y \in TM_o$. It is called a Landsberg space if its Landsberg curvature vanishes.

From (1.7) and (1.8) it is easily seen that any Berwald space must be a Landsberg space. The question whether there exists a Landsberg space that is not a Berwald space is one of the most longstanding open problems (called by Bao the unicorn problem; see [14]) in Finsler geometry.

Next we study the fundamental properties of Berwald spaces. By the definition of parallel displacements in Definition 1.4, one easily deduces the following result.

**Proposition 1.9.** *A connected Finsler space $(M,F)$ is a Berwald space if and only if the parallel displacement along any piecewise smooth curve is a linear map, if and only if the holonomy group of $(M,F)$ at any point of $M$ consists of linear transformations.*

Using the notion of affinely equivalent Finsler metrics, we have the following theorem.

**Theorem 1.8 (Szabó [152]).** *A Finsler space $(M,F)$ is a Berwald space if and only if $F$ is affinely equivalent to a Riemannian metric on $M$.*

By Definition 1.6, the coefficients $\Gamma^i_{jk}$ of the Chern connection of a Berwald space $(M,F)$ are functions on the underlying manifold. Therefore the Chern connection is actually a linear connection on $M$. From this point of view, Szabó determined the structure of Berwald spaces.

**Theorem 1.9 (Szabó [152]).** *Let $(M,F)$ be a Berwald space. Then there exists a Riemannian metric $g$ on $M$ whose Levi-Civita connection coincides with the linear Chern connection of $F$.*

Szabó also studied the converse problem of Theorem 1.9, namely, given a Riemannian metric $g$ on $M$, whether there is a non-Riemannian Berwald space $(M,F)$

such that the Chern connection of $F$ coincides with the Levi-Civita connection of $g$. If this is the case, then the Riemannian metric $g$ is called Berwald-metrizable. He proved the following theorem.

**Theorem 1.10 (Szabó [152]).** *Let $(M,g)$ be a Riemannian metric. Then $(M,g)$ is Berwald-metrizable if and only if it is holonomy reducible or a holonomy irreducible locally symmetric space of rank $\geq 2$.*

For the proof, see Szabó's paper [152]. In Chap. 4, we will give an alternative proof of this result.

Theorem 1.10 has the following interesting corollary.

**Corollary 1.2 (Szabó).** *A 2-dimensional Berwald space is either Riemannian or locally Minkowskian.*

Finally, we mention a result of Ichijyō. Recall that the Chern connection of a Berwald space is a linear connection on the underlying manifold. Hence the two notions of parallel displacements coincide for a Berwald space. This leads to the following theorem.

**Theorem 1.11 (Ichijyō [92]).** *Let $(M,F)$ be a Berwald space. Then for any two points $p,q$ and a smooth curve $c$ connecting $p,q$, the parallel displacement along $c$ is a linear isometry between the Minkowski spaces $T_p(M)$ and $T_q(M)$.*

## 1.7  S-Curvature

In this section we will introduce the notion of S-curvature. Let us first explain briefly the merits of this definition. S-curvature was introduced by Shen in [138]. It is a quantity to measure the rate of change of the volume form of a Finsler space along geodesics. S-curvature is a non-Riemannian quantity, or in other words, any Riemannian manifold has vanishing S-curvature. It is shown in [138] that the Bishop–Gromove volume comparison theorem is true for a Finsler space with vanishing $S$-curvature.

We now give the definition of this important quantity. Let $V$ be an $n$-dimensional real vector space and $F$ a Minkowski norm on $V$. For a basis $\{b_i\}$ of $V$, let

$$\sigma_F = \frac{\mathrm{Vol}(B^n)}{\mathrm{Vol}\{(y^i) \in \mathbb{R}^n \mid F(y^i b_i) < 1\}},$$

where Vol means the volume of a subset in the standard Euclidean space $\mathbb{R}^n$ and $B^n$ is the open ball of radius 1. This quantity is generally dependent on the choice of the basis $\{b_i\}$. But it is easily seen that

$$\tau(y) = \ln \frac{\sqrt{\det(g_{ij}(y))}}{\sigma_F}, \quad y \in V \setminus \{0\},$$

is independent of the choice of basis. We call $\tau = \tau(y)$ the distortion of $(V,F)$.

Now let $(M, F)$ be a Finsler space. Let $\tau(x, y)$ be the distortion of the Minkowski norm $F_x$ on $T_x(M)$ and $\sigma$ the geodesic with $\sigma(0) = x$ and $\dot{\sigma}(0) = y$. Then the quantity

$$S(x, y) = \frac{\mathrm{d}}{\mathrm{d}t}[\tau(\sigma(t), \dot{\sigma}(t))]|_{t=0}$$

is called the S-curvature of the Finsler space $(M, F)$.

An $n$-dimensional Finsler space $(M, F)$ is said to have almost isotropic S-curvature if there exists a smooth function $c(x)$ on $M$ and a closed 1-form $\eta$ such that

$$S(x, y) = (n+1)(c(x)F(y) + \eta(y)), \quad x \in M, y \in T_x(M).$$

If in the above equation $\eta = 0$, then $(M, F)$ is said to have isotropic S-curvature. If $\eta = 0$ and $c(x)$ is a constant, then $(M, F)$ is said to have constant S-curvature.

Let us deduce a formula for S-curvature in a local coordinate system. Suppose $(U; x^1, x^2, \ldots, x^n)$ is a local coordinate system and $(x^1, \ldots, x^n, y^1, y^2, \ldots, y^n)$ is the standard coordinate system on $TU \setminus \{0\}$. It is easily seen that

$$\tau(x, y)_{y^i} = \frac{\partial}{\partial y^i}\left[\ln\sqrt{\det(g_{jk}(y))}\right] = \frac{1}{2}g^{jk}\frac{\partial g_{jk}}{\partial y^i}.$$

On the other hand, for a geodesic $\sigma$ we have

$$\ddot{\sigma} + 2G(\sigma, \dot{\sigma}) = 0,$$

where $G$ is the geodesic spray of $(M, F)$. It follows that

$$\sum_{m=1}^{n}\frac{\partial G^m}{\partial y^m} = \frac{1}{2}g^{ml}\frac{\partial g_{ml}}{\partial x^i}y^i - 2\tau_{y^i}G^i.$$

We then get

$$\begin{aligned}
S(x, y) &= y^i\frac{\partial \tau}{\partial x^i} - 2\frac{\partial \tau}{\partial y^i}G^i \\
&= \frac{1}{2}g^{ml}\frac{\partial g_{ml}}{\partial x^i}y^i - 2\tau_{y^i}G^i - y^m\frac{\partial}{\partial x^m}(\ln\sigma_F(x)) \\
&= \sum_{m=1}^{n}\frac{\partial G^m}{\partial y^m}(x, y) - y^m\frac{\partial}{\partial x^m}(\ln(\sigma_F(x))).
\end{aligned} \tag{1.10}$$

This formula was given by Shen; see [44].

Finally, we introduce the following result to conclude this section.

**Theorem 1.12 (Shen [138]).** *Let $(M,F)$ be a Finsler space of Berwald type. Then the S-curvature of $(M,F)$ is everywhere vanishing. In particular, the S-curvature of a Riemannian manifold is vanishing.*

*Proof.* Suppose $x \in M$ and $y \in T_x(M)$, $y \neq 0$. Let $\gamma$ be a geodesic with $\gamma(0) = x$ and $\dot{\gamma}(0) = y$. Select a basis $b_1, b_2, \ldots, b_n$ of the tangent space $T_x(M)$. Let $P_t$ be the parallel displacement of $(M,F)$ from $x$ to $\gamma(t)$, along the geodesic $\gamma$. Then by Theorem 1.11, $P_t$ is a linear isometry between the Minkowski spaces $(T_x(M),F)$ and $(T_{\gamma(t)}(M),F)$. Therefore, with respect to this basis, the functions $\det(g_{ij}(\gamma(t)))$ and $\sigma_F(\gamma(t))$ are both constant. Therefore, the distortion $\tau$ is constant along $\gamma(t)$. Hence $S(x,y) = 0$. $\qquad\square$

The converse of this theorem is not true. In fact, there exist many reversible non-Berwald Finsler spaces with vanishing S-curvature. We will give some examples in Chap. 6.

## 1.8 Spaces of Constant Flag Curvature and Einstein Metrics

In this section we collect some important results on Finsler spaces of constant flag curvature and Einstein–Finsler metrics. In Riemannian geometry, a connected simply connected complete Riemannian manifold of constant sectional curvature is called a space form. It is well known that for any given constant $c$, there exists exactly one space form with constant sectional curvature $c$ (up to isometries). In particular, up to homotheties, there are only three space forms. However, in Finsler geometry, this is no longer true. In fact, on the sphere there are infinitely many Randers metrics of constant positive flag curvature that are not isometric to each other; see [21] for the classification of such metrics.

We first give the definitions of Finsler spaces with special properties concerning flag curvature.

**Definition 1.7.** Let $(M,F)$ be a Finsler space. Given $x \in M$, $y \in T_x(M) \setminus \{0\}$, and a tangent plane $P$ in $T_x(M)$ containing $y$, denote the flag curvature of the flag $(P,y)$ by $K(P,y)$. If the flag curvature $K(P,y)$ has no independence on $y$, then $(M,F)$ is said to have sectional flag curvature. If $K(P,y)$ has no independence on $P$, then $(M,F)$ is said to have scalar flag curvature. If $K(P,y)$ has independence neither on $P$ nor on $y$, namely, $K$ is a scalar function on the manifold $M$, then $(M,F)$ is said to have isotropic flag curvature. In the special case that $K$ is a constant, we say that $(M,F)$ has constant flag curvature.

It is obvious that any Riemannian manifold has sectional flag curvature, and any Finsler surface has scalar flag curvature. There is much interesting research on Finsler spaces with special properties concerning flag curvature; see, for example, [13, 18, 20, 137, 141, 142].

*Example 1.11.* The metric $F$ defined in (1.2) in Sect. 1.1 is of constant flag curvature $\mu$. Since it is not reversible, it cannot be isometric to any Riemannian metric. This metric is not complete.

As in the Riemannian case, we have the following Schur's lemma.

**Theorem 1.13 ([115, 132]).** *Let $(M, F)$ be a connected Finsler space with dimension $\geq 3$. If $(M, F)$ has isotropic flag curvature, then $(M, F)$ has constant flag curvature.*

For the proof, see [16]. One can also find some characterizations of Finsler spaces of scalar flag curvature, especially for spaces of constant flag curvature. We now present Akbar-Zadeh's theorem on Finsler spaces of constant flag curvature.

In the following, we denote by $L$ the Landsberg curvature and define $A = FC$, where $C$ is the Cartan tensor. For a fixed point $x \in M$ and a positive number $r$, we set $\mathscr{B}_x(r) = \mathscr{B}_x^+(r) \cup \mathscr{B}_x^-(r)$.

We first recall the definition of the tensor norm. For simplicity, we explain this only for the tensor $A$. Let $(V, F)$ be a Minkowski space. Fix a basis of $V$ and let $A_{ijk}$ denote the coefficients of $A$ with respect to this basis. Using the Hessian matrix $(g_{ij})$ of $F$ we can raise the indices of $A$, namely,

$$A^i{}_{jk} = g^{il} A_{ljk},$$
$$A^{ij}{}_k = = g^{jl} A^i{}_{lk},$$
$$A^{ijk} = g^{kl} A^{ij}{}_l.$$

The norm of $A$ is then defined as

$$\|A\| = \max_{y \in \mathscr{I}} \sqrt{A_{ijk} A^{ijk}},$$

where $\mathscr{I}$ is the indicatrix of $F$. It is easily seen that this norm is independent of the choice of basis. Now for a Finsler space $(M, F)$ and $x \in M$, we use $\|A\|_x$ to denote the norm of $\|A\|$ of the Minkowdski space $(M, F|_{T_x(M)})$. Fix a point $O$ in $M$. We say that

$$\sup_{x \in \mathscr{B}_O(r)} \|A\|_x = o(f(r))$$

if the left side grows more slowly than $f(r)$ as $r \to \infty$. One can similarly define the above setting for $L$ or other types of tensors.

**Theorem 1.14 (Akbar-Zadeh [2]).** *Let $(M, F)$ be a connected Finsler space of constant flag curvature $\lambda$. Let $O$ be a designed origin of $M$, and $\nabla$ the Chern connection on $\pi^*(TM)$ over $TM \setminus \{0\}$.*

1. *Suppose $\lambda < 0$ and $(M,F)$ is complete. If*

$$\sup_{x \in \mathscr{B}_O(r)} \|A \text{ or } L\|_x = o\left[e^{\sqrt{-\lambda}r}\right],$$

*then $(M,F)$ must be Riemannian.*
2. *Suppose $\lambda = 0$ and $(M,F)$ is forward geodesically complete. If*

$$\sup_{x \in \mathscr{B}_O(r)} \|A\|_x = o[r],$$

*then $L = 0$, and hence $F$ is a Landsberg metric. If additionally,*

$$\sup_{x \in \mathscr{B}_O(r)} \|\nabla_{\text{vert}}(A)\|_x = o[r],$$

*then $(M,F)$ is locally Minkowskian.*

We refer to [16] for the definition of $\nabla_{\text{vert}}$ in the above theorem. One can also find a detailed proof there. This theorem provides some insight as to why the Funk metric cannot be complete.

In the compact case, the norms of all the relevant tensors are bounded. Therefore we have the following corollary.

**Corollary 1.3 (Akbar-Zadeh [2]).** *Let $(M,F)$ be a compact Finsler space of constant flag curvature $\lambda$.*

1. *If $\lambda < 0$, then $F$ is a Riemannian metric.*
2. *If $\lambda = 0$, then $(M,F)$ is a locally Minkowskian space.*

Next we will recall the characterization of constant flag curvature for Randers spaces, in terms of the navigation data. The navigation data provide another method to express a Randers metric:

$$F(x,y) = \frac{\sqrt{h(y,W)^2 + \lambda h(y,y)}}{\lambda} - \frac{h(y,W)}{\lambda},$$

where $h$ is a Riemannian metric, $W$ is a vector field on $M$ with $h(W,W) < 1$, and $\lambda = 1 - h(W,W)$. The pair $(h,W)$ is called the navigation data of the Randers metric $F$. This version of Randers metric was introduced by Shen in [143]. If $F$ is a Randers metric with navigation data $(h,W)$, then we usually say that $F$ solves Zermelo's navigation problem of the Riemannian metric $h$ under the influence of an external vector field $W$. The navigation data prove to be convenient in handling problems concerning flag curvature and Ricci scalar.

In a local coordinate system, the transformation laws between the defining form and navigation data can be described as follows. If

$$F = \alpha + \beta = \sqrt{a_{ij}y^i y^j} + b_i y^i,$$

then the navigation data have the form

$$h_{ij} = (1 - \|\beta\|^2)(a_{ij} - b_i b_j), \quad W^i = -\frac{a^{ij} b_j}{1 - \|\beta\|_\alpha^2}. \tag{1.11}$$

Conversely, the defining form can also be expressed by the navigation data by the formula

$$a_{ij} = \frac{h_{ij}}{\lambda} + \frac{W_i}{\lambda} \frac{W_j}{\lambda}, \quad b_i = \frac{-W_i}{\lambda}, \tag{1.12}$$

where $W_i = h_{ij} W^j$ and $\lambda = 1 - W^i W_i = 1 - h(W, W)$. All these formulas can be found in [19].

**Theorem 1.15 (Bao–Robles–Shen [21]).** *Let $(M, F)$ be a Randers space that solves Zermelo's navigation problem on the Riemannina manifold $(M, h)$ under the influence of an external vector field $W$. Then $(M, F)$ has constant flag curvature $K$ if and only if there exists a constant $\sigma$ such that*

*1. $h$ is of constant flag curvature $(K + \frac{1}{16}\sigma^2)$,*
*2. $W$ is an infinitesimal homothety of $h$, namely,*

$$\mathscr{L}_W(h) = -\sigma h,$$

*where $\mathscr{L}$ is the Lie derivative of $M$.*

Moreover, $\sigma$ must vanish whenever $h$ is not flat.

For the proof, see [19, 21]. This theorem is applied in [21] to obtain a complete classification of Randers spaces of constant flag curvature. Bao–Robels–Shen also compute the dimension of the module spaces of such spaces.

Now we recall some results on Einstein–Finsler spaces. We first give the definition.

**Definition 1.8.** Let $(M, F)$ be a connected Finsler space. If there exists a smooth function $K(x)$ on $M$ such that $\mathrm{Ric}(x, y) = (n - 1)K(x)$, where Ric is the Ricci scalar of $F$, then $(M, F)$ is called an Einstein–Finsler space. If in addition the function $K(x)$ is a constant, then $(M, F)$ is said to be Ricci-constant.

If $F$ is an Einstein Riemannian manifold with dimension $\geq 3$, then it is necessarily Ricci-constant. This is the so-called Schur's lemma for Einstein metrics. For a general Finsler space, it is still unknown whether the above assertion holds. However, Robles proved that Schur's lemma holds for Einstein–Randers metrics [133]. In the Riemannian case, there are numerous results on Einstein metrics in the literature; see [28]. However, in the general case, there are very little research related to this topic; see [140] for some information. The following result is due to Bao and Robles.

**Theorem 1.16 (Bao–Robles [19]).** *Let $(M,F)$ be a Randers space with navigation data $(h,W)$. Then $F$ is Einstein with Ricci curvature $\mathrm{Ric}(x,y) = (n-1)K(x)$ if and only if there exists a real number $\sigma$ such that the following two conditions hold:*

1. *The Riemannian metric $h$ is Einstein with Ricci scalar $(n-1)(K(x) + \frac{1}{16}\sigma^2)$.*
2. *The vector field $W$ is an infinitesimal homothety of $h$, namely,*

$$\mathscr{L}_W h = -\sigma h.$$

*Furthermore, $\sigma$ must vanish whenever $h$ is not Ricci-flat.*

This theorem can be applied to construct many interesting examples of non-Riemannian Einstein metrics; see [19]. It also has the following important corollary:

**Proposition 1.10 (Bao–Robles [19]).** *Let $(M,F)$ be an Einstein–Randers space. Then $(M,F)$ has constant S-curvature.*

In Chap. 7, we will use Theorem 1.16 to deduce some interesting results on homogeneous Einstein–Randers metrics.

# Chapter 2
# Lie Groups and Homogeneous Spaces

In this chapter we will give a brief introduction to Lie groups and homogeneous spaces. Today, Lie theory is an important field of mathematics with so many topics that it is impossible to explain the details in such a short survey. So we will omit many proofs of the results in this chapter. Readers who are familiar with Lie theory may skip this chapter and go directly to the following chapters.

## 2.1 Lie Groups and Lie Algebras

We first give the definition of Lie groups.

**Definition 2.1.** Let $G$ be a smooth manifold with the structure of an abstract group. If the map

$$\rho : G \times G \to G, \quad \rho(g_1, g_2) = g_1 g_2^{-1}, \tag{2.1}$$

is smooth, then $G$ is called a Lie group.

In the literature, it is generally required that $G$ be an analytic manifold and the map in (2.1) be analytic. However, it can be proved that these two definitions are equivalent. More precisely, if $G$ is a Lie group in the sense of Definition 2.1, then there exists a unique analytic structure on $G$ such that the group operations are analytic. Setting $g_1 = e$ in (2.1), we see that the map $g \to g^{-1}$ is smooth. This then implies that the map $G \times G \to G : (g_1, g_2) \mapsto g_1 g_2$ is also smooth.

Let $G$ be a Lie group. Given $g \in G$, it is easily seen that the left translation $L_g : G \to G$, $L_g(h) = gh$, for $h \in G$, is a diffeomorphism of $G$ onto itself. Similarly, the right translation $R_g(h) = hg$, $h \in G$, is also a diffeomorphism. A (smooth) vector field $X$ on $G$ is called left-invariant if for any $g \in G$ we have $dL_g(X) = X$. Similarly, we can define right-invariant vector fields. Let $\mathfrak{g}$ denote the set of all left-invariant vector fields of $G$. Then the map $X \to X_e$, where $e$ is the identity element of $G$, is a linear isomorphism from $\mathfrak{g}$ onto $T_e(M)$. On the other hand, if $X, Y$ are two left-invariant vector fields, then the Lie bracket $[X, Y]$ is also left-invariant. This implies

S. Deng, *Homogeneous Finsler Spaces*, Springer Monographs in Mathematics, 31
DOI 10.1007/978-1-4614-4244-8_2, © Springer Science+Business Media New York 2012

that $\mathfrak{g}$ has the structure of a Lie algebra over the field of real numbers. Namely, there is a binary operation $[\,,]$ on $\mathfrak{g}$ that satisfies the following conditions

1. $[\,,]$ is bilinear in each of its two entries.
2. $[\,,]$ is skew-symmetric: $[X,X] = 0$, for any $X \in \mathfrak{g}$.
3. $[\,,]$ satisfies the Jacobi identity: for any $X,Y,Z \in \mathfrak{g}$,

$$[X,[Y,Z]] + [Y,[Z,X]] + [Z,[X,Y]] = 0.$$

In general, let $\mathbb{F}$ be a field and $V$ a vector space over $\mathbb{F}$. If there is a binary operation $[\,,]$ on $V$ satisfying the conditions (1)–(3) above, then $V$ is called a Lie algebra over $\mathbb{F}$. Hence in this sense $\mathfrak{g}$ is a real Lie algebra.

Since as real vector spaces, $\mathfrak{g} \simeq T_e(M)$, we have $\dim \mathfrak{g} = \dim G$. We call $\mathfrak{g}$ the Lie algebra of $G$.

Sometimes the Lie algebra of $G$ will be denoted by Lie $G$.

**Definition 2.2.** Let $G$ be a Lie group. A submanifold $H$ of $G$ is called a Lie subgroup of $G$ if

1. $H$ is an abstract subgroup of $G$.
2. $H$ is a topological group.

A Lie subgroup $H$ is called a topological Lie subgroup if the topology of $H$ is equal to the induced topology.

Note that there are examples of Lie subgroups that are not topological Lie subgroups; see for example [83].

**Definition 2.3.** Let $\mathfrak{g}$ be a Lie algebra over a field $\mathbb{F}$. A subspace $\mathfrak{h}$ is called a subalgebra of $\mathfrak{g}$ if for any $X,Y \in \mathfrak{h}$, we have $[X,Y] \in \mathfrak{h}$. If in addition $[X,Y] \in \mathfrak{h}$, for all $X \in \mathfrak{h}$ and $Y \in \mathfrak{g}$, then $\mathfrak{h}$ is called an ideal of $\mathfrak{g}$.

*Example 2.1.* Let $M$ be a vector subspace of a Lie algebra $\mathfrak{g}$. Define

$$Z_{\mathfrak{g}}(M) = \{X \in \mathfrak{g} \,|\, [X,Y] = 0, \forall Y \in M\}.$$

Then it is easily seen that $Z_{\mathfrak{g}}(M)$ is a subalgebra of $\mathfrak{g}$. The subalgebra $Z_{\mathfrak{g}}(M)$ is called the centralizer of $M$ in $\mathfrak{g}$. In the special case that $M = \mathfrak{g}$, $Z_{\mathfrak{g}}(\mathfrak{g})$ is an ideal of $\mathfrak{g}$. The ideal $Z_{\mathfrak{g}}(\mathfrak{g})$ is usually denoted simply by $Z(\mathfrak{g})$ and is called the center of $\mathfrak{g}$. Similarly, for a subalgebra $\mathfrak{h}$ of $\mathfrak{g}$, define

$$N_{\mathfrak{g}}(\mathfrak{h}) = \{X \in \mathfrak{g} \,|\, [X,Y] \in \mathfrak{h}, \forall Y \in \mathfrak{h}\}.$$

Then it is easy to check that $N_{\mathfrak{g}}(\mathfrak{h})$ is a subalgebra of $\mathfrak{g}$ containing $\mathfrak{h}$. This subalgebra is called the normalizer of $\mathfrak{h}$ in $\mathfrak{g}$.

Now we state an important theorem on Lie subgroups and Lie subalgebras.

**Theorem 2.1.** *Let G be a Lie group and H a Lie subgroup of G. Then the Lie algebra $\mathfrak{h}$ of H is a subalgebra of the Lie algebra $\mathfrak{g}$ of G. Conversely, given any subalgebra $\mathfrak{h}$ of $\mathfrak{g}$, there exists a unique connected Lie subgroup H of G whose Lie algebra is $\mathfrak{h}$.*

**Definition 2.4.** Let $G$ and $H$ be two Lie groups. A map $\varphi$ from $G$ to $H$ is called a homomorphism if $\varphi$ is an abstract group homomorphism and it is continuous with respect to the topology of the groups. A homomorphism $\varphi$ is called an isomorphism if $\varphi$ is a homeomorphism. In case there exists an isomorphism from $G$ onto $H$, we say that $G$ is isomorphic to $H$.

Similarly, we can define the notion of a homomorphism between Lie algebras. Let $\mathfrak{g}_1, \mathfrak{g}_2$ be two Lie algebras over a field $F$. A linear map $\phi$ from $\mathfrak{g}_1$ to $\mathfrak{g}_2$ is called a homomorphism if $\phi([X,Y]) = [\phi(X), \phi(Y)]$. A homomorphism between two Lie algebras is called an isomorphism of Lie algebras if it is also a linear isomorphism. In this case, we say that the two Lie algebras are isomorphic.

It is easy to prove that if $\varphi$ is a Lie group homomorphism, then its differential at the unit element, $d\varphi|_e$, is a homomorphism of the Lie algebras. Usually, we denote $d\varphi|_e$ simply by $d\varphi$.

Now we introduce the notion of the exponential map of a Lie group. For this we first define the one-parameter subgroups of a Lie group.

**Definition 2.5.** A one-parameter subgroup of a Lie group $G$ is a homomorphism from the additive group $\mathbb{R}$ to $G$.

The next theorem gives the existence and classification of one-parameter subgroups.

**Theorem 2.2.** *Let G be a Lie group with Lie algebra $\mathfrak{g}$. Then for any $X \in \mathfrak{g}$ there exists a unique one-parameter subgroup $\varphi$ with $\dot{\varphi}(0) = X_e$.*

*Proof.* We first prove the existence. Since $X$ is a vector field on $G$, there exists $\varepsilon > 0$ such that the maximal integral curve $\varphi(t)$ through $e$ is defined on $(-\varepsilon, \varepsilon)$, i.e., $\varphi(t)$ is a smooth curve with $\varphi(0) = e$ and $\dot{\varphi}(t) = X_{\varphi(t)}$. Moreover, such a curve is unique where it is defined. Now we show that if $(a,b)$ is an open interval containing 0 such that $\varphi$ is defined on $(a,b)$, and $s, t, s+t \in (a,b)$, then

$$\varphi(s)\varphi(t) = \varphi(s+t).$$

To see this, we just fix $s$ and consider $\varphi_1(t) = \varphi(s)\varphi(t)$ and $\varphi_2(t) = \varphi(s+t)$ as two curves in $t$. Then $\varphi_1(0) = \varphi_2(0) = \varphi(s)$. On the other hand, we have

$$\dot{\varphi}_1(t) = dL_{\varphi(s)}\dot{\varphi}(t) = dL_{\varphi(s)}X_{\varphi(t)}$$

and

$$\dot{\varphi}_2(t) = X_{\varphi(s+t)}. \tag{2.2}$$

Since $X$ is left-invariant, we have

$$dL_{\varphi(s)}X_{\varphi(t)} = X_{\varphi(s)\varphi(t)}.$$

This means that both the curves $\varphi_1$ and $\varphi_2$ are integral curves of the vector field $X$. Since they coincide at 0, they must coincide wherever they are defined. This proves the assertion.

Now we prove that the integral curve $\varphi$ is defined on the whole space $\mathbb{R}$. By the above argument, if we define $\gamma(t) = \varphi(\frac{\varepsilon}{2})\varphi(t - \frac{\varepsilon}{2})$, then $\gamma(t)$ is again an integral curve of $X$ and it is defined on $(-\frac{3}{2}\varepsilon, \frac{3}{2}\varepsilon)$. Therefore $\varphi$ can be extended to $(-\frac{3}{2}\varepsilon, \frac{3}{2}\varepsilon)$. This then implies that $\varphi$ can be defined on the whole real line $\mathbb{R}$. On the other hand, (2.2) shows that $\varphi$ is the required one-parameter subgroup.                     $\square$

Now we can define the notion of exponential map of a Lie group.

**Definition 2.6.** Let $G$ be a Lie group with Lie algebra $\mathfrak{g}$. The exponential map from $\mathfrak{g}$ to $G$ is defined by

$$\exp(X) = \varphi_X(1),$$

where $\varphi_X$ is the one-parameter subgroup determined by $X$.

From the definition, it is easily seen that the exponential map is smooth. In fact, it has better properties. Recall that the tangent space $T_e(G)$ is spanned by $X_e$, $X \in \mathfrak{g}$. Since the curve $\exp tX$ is the integral curve of $X$ through $e$, we easily see that $d\exp|_e = \mathrm{id}$. By the inverse function theorem, we have the following result.

**Theorem 2.3.** *There exists a neighborhood $U$ of the zero vector $0$ in $\mathfrak{g}$ such that $\exp$ is a diffeomorphism from $U$ onto $\exp(U)$.*

We now collect some properties of the exponential map.

**Proposition 2.1.** *For $X, Y \in \mathfrak{g}$, we have*

$$\exp(tX)\exp(tY) = \exp\left(t(X+Y) + \frac{t^2}{2}[X,Y] + \mathrm{O}(t^3)\right),$$

$$\exp(tX)\exp(tY)\exp(-tX) = \exp(tY + t^2[X,Y] + \mathrm{O}(t^3)),$$

$$\exp(tX)\exp(tY)\exp(-tX)\exp(-tY) = \exp(t^2[X,Y] + \mathrm{O}(t^3)),$$

*where $\mathrm{O}(t^3)$ means a vector in $\mathfrak{g}$ satisfying the condition that there exists $\varepsilon > 0$ such that $\frac{1}{t^3}\mathrm{O}(t^3)$ is bounded and smooth for $|t| < \varepsilon$.*

For the proof, see [83].

Using the exponential map, we can determine the Lie algebra of a Lie subgroup, especially when the subgroup is closed. We first recall the following result.

**Theorem 2.4.** *Let $G$ be a Lie group and $H$ an abstract subgroup of $G$. If $H$ is a closed subset, then there exists a unique differential structure on $H$ such that $H$ is a topological subgroup of $G$.*

We refer the proof of this theorem to [83]. The following theorem gives an effective method to compute the Lie algebra of a Lie subgroup.

**Theorem 2.5.** *Let G be a Lie group with Lie algebra $\mathfrak{g}$ and H a Lie subgroup of G with Lie algebra $\mathfrak{h}$. If H is a topological Lie subgroup, or H has at most countably many connected components, then*

$$\mathfrak{h} = \{X \in \mathfrak{g} \,|\, \exp(tX) \in H, \, \forall t \in \mathbb{R}\}.$$

*Example 2.2.* Let $G = \mathrm{GL}(n, \mathbb{R})$ be the set of all $n \times n$ nonsingular matrices. Obviously $G$ has the structure of an abstract group under the usual matrix multiplication. Note that $G$ can also be viewed as an open subset of the Euclidean space $\mathbb{R}^{n^2}$, and hence it has the structure of a smooth manifold. It can be easily checked that the group operations are smooth with respect to the manifold structure; hence $G$ is a Lie group, called the *real general linear group*. It is not hard to compute its Lie algebra:

$$\mathfrak{g} = \mathfrak{gl}(n, \mathbb{R}) = \mathbb{R}^{n \times n},$$

with Lie brackets

$$[X, Y] = XY - YX.$$

This Lie algebra is called the *real general linear Lie algebra*. The exponential map of $G$ is

$$\exp(X) = e^X = I_n + X + \frac{1}{2}X^2 + \cdots + \frac{1}{n!}X^n + \cdots.$$

In general, let $V$ be an $n$-dimensional real vector space. Then the set of all invertible linear transformations forms a Lie group isomorphic to $\mathrm{GL}(n, \mathbb{R})$. In this case, we usually denote the Lie group by $\mathrm{GL}(V)$.

Now we consider the following subgroups of $G$:

1. The special linear group $\mathrm{SL}(n, \mathbb{R})$ consisting of all elements in $G$ with determinant 1.
2. The orthogonal group $\mathrm{O}(n)$ consisting of the orthogonal matrices in $G$.
3. The special orthogonal group $\mathrm{SO}(n)$ consisting of the orthogonal matrices with determinant 1.

As in the case of general Lie groups, we sometimes write the special linear group as $\mathrm{SL}(V)$, where $V$ is a real vector space. Moreover, if $V$ is an $n$-dimensional Euclidean space, then the set of all orthogonal transformations of $V$ form a Lie group isomorphic to $\mathrm{O}(n)$. This group will usually be denoted by $\mathrm{O}(V)$. Similarly, we have the notation $\mathrm{SO}(V)$. The Lie algebras of these Lie groups will be denoted by corresponding notation.

By Theorem 2.5, we easily get the Lie algebras of these Lie subgroups:

1. The Lie algebra of $\mathrm{SL}(n, \mathbb{R})$ is $\mathfrak{sl}(n, \mathbb{R})$, consisting of all the traceless matrices in $\mathfrak{g}$. This Lie algebra is called the special linear Lie algebra.
2. The Lie algebras of $\mathrm{O}(n)$ and $\mathrm{SO}(n)$ are equal. It is $\mathfrak{so}(n)$, consisting of all the skew-symmetric matrices in $\mathfrak{g}$. This Lie algebra is called the orthogonal Lie algebra.

*Example 2.3.* The complex version of Example 2.2 is more interesting and useful. Let $GL(n, \mathbb{C})$ denote the set of all invertible complex $n \times n$ matrices. It is a complex manifold. Moreover, it is easily seen that the map $(g_1, g_2) \mapsto g_1 g_2^{-1}$ is holomorphic. In general, if $G$ is a complex manifold as well as an abstract group such that the map $G \times G \to G$, $(g_1, g_2) \mapsto g_1 g_2^{-1}$ is holomorphic, then $G$ is called a complex Lie group. A complex Lie group is automatically a (real) Lie group. Hence $GL(n, \mathbb{C})$ is a Lie group.

The Lie algebra of $GL(n, \mathbb{C})$ is $\mathfrak{gl}(n, \mathbb{C})$, the complex general linear Lie algebra, consisting of all complex $n \times n$ matrices, with Lie brackets $[A, B] = AB - BA$. One can also define the similar Lie subgroups and determine their Lie algebras as in the real case. We list some of the examples below:

1. The complex special linear group $SL(n, \mathbb{C})$. It is the subgroup of $GL(n, \mathbb{C})$ consisting of the matrices with determinant 1. Its Lie algebra is $\mathfrak{sl}(n, \mathbb{C})$, consisting of all the complex $n \times n$ matrices with zero trace.
2. The unitary group $U(n)$. It is the subgroup of $GL(n, \mathbb{C})$ consisting of all unitary matrices. Its Lie algebra is $\mathfrak{u}(n)$, consisting of all the skew-Hermitian matrices.
3. The special unitary group $SU(n) = SL(n, \mathbb{C}) \cap U(n)$. Its Lie algebra is $\mathfrak{su}(n)$, consisting of all the traceless skew-Hermitian matrices.
4. The complex symplectic group $Sp(n, \mathbb{C})$. It is the group of matrices $g$ in $GL(2n, \mathbb{C})$ satisfying the condition

$$g^t J_n g = J_n,$$

where

$$J_n = \begin{pmatrix} 0 & I_n \\ -I_n & 0 \end{pmatrix}.$$

Its Lie algebra is

$$\mathfrak{sp}(n, \mathbb{C}) = \left\{ \begin{pmatrix} Z_1 & Z_2 \\ Z_3 & -Z_1^t \end{pmatrix} \,\middle|\, Z_1, Z_2, Z_3 \text{ complex } n \times n \text{ matrices}, Z_2, Z_3 \text{ symmetric} \right\}.$$

We put $Sp(n) = Sp(n, \mathbb{C}) \cap U(2n)$ and call it the symplectic group. Note that in some books the definition of symplectic group is different from the above. We leave as an exercise for the reader to determine the Lie algebra of $Sp(n)$.

Note that the Lie groups $O(n)$ and $SO(n)$ are not simply connected. Their universal covering groups are denoted by $Pin(n)$ and $Spin(n)$ (called the spin group), respectively. We now recall briefly the construction of these two groups. For the details, we refer the reader to [34, Sect. 1.6].

*Example 2.4 (See [12]).* Let $V$ be a real inner product space. The Clifford algebra $Cliff(V)$ is the associative algebra freely generated by $V$ modulo the relations

$$vw + wv = -2(v, w), \quad v, w \in V.$$

If $V = \mathbb{R}^n$ is the standard Euclidean space, then $\mathrm{Cliff}(V)$ is generally denoted by $\mathrm{Cliff}(n)$. For example, $\mathrm{Cliff}(0) = \mathbb{R}$, $\mathrm{Cliff}(1) = \mathbb{C}$, $\mathrm{Cliff}(2) = \mathbb{H}$ (the quaternions). Now consider $\mathrm{Cliff}(n)$. Let $\mathrm{Pin}(n)$ be the group sitting inside $\mathrm{Cliff}(n)$ consisting of all the products of the unit elements in $\mathbb{R}^n$. Recall that each element in $O(n)$ can be written as a finite product of reflections. Since each unit element $v$ in $\mathbb{R}^n$ induces a reflection transformation $\tau_v$ of $\mathbb{R}^n$, and $\pm v$ induce the same transformation, the group $\mathrm{Pin}(V)$ is a double covering of $O(n)$. Let $\mathrm{Spin}(n) \subset \mathrm{Pin}(n)$ be the subgroup of the elements that are products of an even number of unit elements in $\mathbb{R}^n$. Then $\mathrm{Spin}(n)$ is a double covering of $SO(n)$. It is not difficult to prove that $\mathrm{Spin}(n)$ is simply connected.

In the final part of this section, we introduce some terminology in representation theory. Since in this book we will use only some results on finite-dimensional representations, we shall consider only the finite-dimensional cases. Let $G$ be a Lie group and $V$ a finite-dimensional real vector space. A representation of $G$ on $V$ is a continuous homomorphism $\rho$ from $G$ to the general linear group $\mathrm{GL}(V)$. We say that $(V, \rho)$ is a representation of $G$, or a $G$-module. A representation $(V, \rho)$ of $G$ is called faithful if the homomorphism $\rho$ has trivial kernel.

Similarly, let $\mathfrak{g}$ be a Lie algebra over the field $F$ and let $V$ be a vector space over $F$. A Lie algebra homomorphism from $\mathfrak{g}$ to $\mathfrak{gl}(V)$ is called a representation of $\mathfrak{g}$.

It is clear that if $\rho$ is a representation of a Lie group $G$ on the vector space $V$, then its differential $d\rho$ is a representation of its Lie algebra $\mathfrak{g}$. The converse is generically not true. But if $G$ is a connected simply connected Lie group, then every representation of $\mathfrak{g}$ can be lifted to a representation of $G$.

For a Lie group $G$ with Lie algebra $\mathfrak{g}$, we have a natural representation of $G$ on $\mathfrak{g}$. Note that for $X \in \mathfrak{g}$, $\exp(tX)$ is a one-parameter subgroup of $G$. Given $g \in G$, it is easy to check that $g(\exp(tX))g^{-1}$ is again a one-parameter subgroup. Hence there exists a unique $\widetilde{X}_g \in \mathfrak{g}$ such that $g(\exp(tX))g^{-1} = \exp(t\widetilde{X}_g)$. We denote $\widetilde{X}_g$ by $\mathrm{Ad}(g)(X)$. It is easy to check that the map $G \to \mathrm{GL}(\mathfrak{g})$, $g \to \mathrm{Ad}(g)$ is a representation of $G$ on $\mathfrak{g}$. This representation is called the adjoint representation of $G$. The differential of the adjoint representation of the Lie group is called the adjoint representation of the Lie algebra.

Let $G$ be a Lie group and $(V, \rho)$ a real representation of $G$. Suppose $V$ is endowed with an inner product $\langle\, , \rangle$. If

$$\langle \rho(g)(X), \rho(g)(Y) \rangle = \langle X, Y \rangle, \quad \forall X, Y \in V, \quad g \in G,$$

then we say that $\langle\, , \rangle$ is an invariant inner product, and $(V, \rho, \langle\, , \rangle)$ is an orthogonal representation of $G$.

A Lie group $G$ is called compact if $G$ is a compact manifold. The following theorem is called Weyl's unitary trick.

**Theorem 2.6.** *Let $G$ be a connected compact Lie group and $(V, \rho)$ a representation of $G$. Then there exists an invariant inner product on $V$.*

## 2.2  Lie Transformation Groups and Coset Spaces

In this section we will introduce the fundamental properties of Lie group actions on smooth manifolds. The general idea is to consider this problem originating from Felix Klein's Erlangen Program proposed in 1872. Klein's proposal was to categorize the new geometries by their characteristic groups of transformations. This program has been extremely successful. Today, it is the expectation of each geometer to find an effective method to study geometry by applying group theory.

We first define the notion of the action of a Lie group on a manifold. Let $G$ be a Lie group and $M$ a smooth manifold. A smooth action of $G$ on $M$ is a map $\rho$ from $G \times M$ onto $M$ satisfying the following two conditions:

1. $\rho$ is a smooth map.
2. $\rho(g_2, \rho(g_1, x)) = \rho(g_2 g_1, x)$, $\forall g_1, g_2 \in G$, $x \in M$.

If we write $\rho(g, x)$ as $g \cdot x$, then condition (2) can be rewritten as $g_2 \cdot (g_1 \cdot x) = (g_2 g_1) \cdot x$. Meanwhile, this condition implies that $e \cdot x = x$ (note that $\rho$ is assumed to be onto). This then implies that for each $g \in G$, the map $x \to g \cdot x$ is a diffeomorphism of $M$.

If $G$ has a smooth action on $M$, then $G$ is called a Lie transformation group of $M$. The action is called effective (resp. almost effective) if $e$ is the only element in $G$ such that $\rho(e)$ is the identity map (resp. if the subgroup $\rho^{-1}(\mathrm{id})$ of $G$ is discrete). It is called free if for any $g \neq e$ in $G$, the map $\rho(g)$ has no fixed point.

The most important action is the action of a Lie group on the coset spaces. We have the following theorem.

**Theorem 2.7.** *Let $G$ be a Lie group and $H$ a closed subgroup of $G$. Then there exists a unique differentiable structure on the left coset space $G/H$ with the induced topology that turns $G/H$ into a smooth manifold such that $G$ is a Lie transformation group of $G/H$.*

For the proof, see [83]. Note that the induced topology on $G/H$ is the unique topology such that the natural projection $\pi$ from $G$ onto $G/H$ is both continuous and open. More precisely, a subset $U$ in $G/H$ is open if and only if the preimage $\pi^{-1}(U)$ is an open subset of $G$.

Suppose $G$ is a Lie group acting transitively on a smooth manifold $M$. Given $p \in M$, denote by $G_p$ the set of elements of $G$ that keep $p$ fixed. Then $G_p$ is a closed subgroup of $G$. If we endow $G/G_p$ with the induced topology, then Theorem 2.7 asserts that there exists a unique differentiable structure such that $G$ is a Lie transformation group of $G/G_p$. Let $\alpha$ be the map from $G$ onto $M$ defined by $\alpha(g) = g \cdot p$. Then $\alpha$ is a diffeomorphism if it is a homeomorphism (see [83]). On the other hand, if $G$ has a countable base (this is the case if $G$ has countably many connected components), then $\alpha$ must be a homeomorphism. Hence in this case $G/G_p$ is diffeomorphic to $M$. The following proposition is very useful in practice.

**Proposition 2.2.** *Let $G$ be a Lie group acting transitively on a connected manifold $M$. Fix $p \in M$ and let $\alpha$ be the map from $G/G_p$ onto $M$ defined by $\alpha(g) = g \cdot p$. If $\alpha$*

*is a homeomorphism, then the unit connected component $G_0$ is also transitive on M.*
*In particular, if G has countably many connected components, then $G_0$ is transitive*
*on M.*

Let $G$ be a Lie group and $H$ a closed subgroup of $G$. In the following we will
always endow $G/H$ with the differentiable structure so that $G$ is a Lie transformation
group of $G/H$. In general, $H$ will be called the isotropy subgroup. The tangent space
of $G/H$ at the origin $eH = H$ of $G/H$ can be identified with the quotient vector space
$\mathfrak{g}/\mathfrak{h}$. Since the adjoint action of the group $H$ keeps $\mathfrak{h}$ invariant, $H$ has an action on
$\mathfrak{g}/\mathfrak{h}$ defined by

$$\text{Ad}_{\mathfrak{g}/\mathfrak{h}}(h)(X + \mathfrak{h}) = \text{Ad}(h)(X) + \mathfrak{h},$$

where Ad is the adjoint action of $G$. This action defines a representation of $H$ on
$\mathfrak{g}/\mathfrak{h}$ and is called the linear isotropy representation.

In some special cases there exists a subspace $\mathfrak{m}$ of $\mathfrak{g}$ such that

$$\mathfrak{g} = \mathfrak{h} + \mathfrak{m} \quad \text{(direct sum of subspaces)} \tag{2.3}$$

and

$$\text{Ad}(h)(\mathfrak{m}) \subset \mathfrak{m}, \quad \forall h \in H.$$

Then the coset space $G/H$ is called a reductive homogeneous manifold and (2.3)
is usually called a reductive decomposition of $\mathfrak{g}$. In this case, the tangent space
$T_{eH}(G/H)$ of $G/H$ at the origin $eH$ can be identified with $\mathfrak{m}$ through the map

$$X \mapsto \frac{d}{dt} \exp(tX)H\Big|_{t=0}, \quad X \in \mathfrak{m}.$$

Then the linear isotropy representation corresponds to the adjoint action of $H$ on $\mathfrak{m}$.
For the geometry of reductive homogeneous space, we refer the reader to [102].

Next we introduce some notions about the adjoint group of a Lie algebra. Let $\mathfrak{g}$
be a real Lie algebra. Since $\mathfrak{g}$ is a real vector space, we have the general linear group
$\text{GL}(\mathfrak{g})$. Note that a priori, $\text{GL}(\mathfrak{g})$ has no relation with the Lie algebra structure of
$\mathfrak{g}$. As pointed out in Sect. 2.1, the Lie algebra of the Lie group $\text{GL}(\mathfrak{g})$ is $\mathfrak{gl}(\mathfrak{g})$,
consisting of all the linear endomorphisms of $\mathfrak{g}$ with Lie brackets $[A, B] = AB - BA$.
Now for each $X \in \mathfrak{g}$, we can define a linear endomorphism $\text{ad}(X)$ of $\mathfrak{g}$ by

$$\text{ad}(X)(Y) = [X, Y], \quad Y \in \mathfrak{g}.$$

It is obvious that the endomorphisms $\text{ad}(X)$, $X \in \mathfrak{g}$, form a subalgebra of $\mathfrak{gl}(\mathfrak{g})$. We
denote this subalgebra by $\text{ad}(\mathfrak{g})$. By Theorem 2.1, there exists a unique connected
Lie subgroup of $\text{GL}(\mathfrak{g})$ with Lie algebra $\text{ad}(\mathfrak{g})$. This Lie group is called the adjoint
group of $\mathfrak{g}$ and will be denoted by $\text{Int}\,\mathfrak{g}$.

There is another Lie group that is closely related to the Lie algebra structure of
$\mathfrak{g}$. A linear isomorphism $\sigma$ of $\mathfrak{g}$ is called an automorphism of $\mathfrak{g}$ if

$$\sigma([X, Y]) = [\sigma(X), \sigma(Y)], \quad \forall X, Y \in \mathfrak{g}.$$

It is clear that the set of all the automorphisms of $\mathfrak{g}$ is a closed subgroup of $GL(\mathfrak{g})$. Hence it is a topological Lie subgroup of $GL(\mathfrak{g})$. This group is called the automorphism group of $\mathfrak{g}$ and will be denoted by $\mathrm{Aut}(\mathfrak{g})$. The Lie algebra of $\mathrm{Aut}(\mathfrak{g})$ is denoted by $\partial(\mathfrak{g})$. It is easy to prove that $\partial(\mathfrak{g})$ consists of all the linear endomorphisms of $\mathfrak{g}$ satisfying

$$D([X,Y]) = [D(X),Y] + [X,D(Y)], \quad \forall X,Y \in \mathfrak{g}. \tag{2.4}$$

A linear endomorphism satisfying (2.4) is called a derivation. It is easily seen that for a derivation $D$, $e^{tD}$ is an automorphism of $\mathfrak{g}$. Hence $\partial(\mathfrak{g})$ consists of all the derivations of $\mathfrak{g}$. By the Jacobi identity we see that for $X \in \mathfrak{g}$, $\mathrm{ad}(X)$ is a derivation. Thus $\mathrm{ad}(\mathfrak{g})$ is a subalgebra of $\partial(\mathfrak{g})$ and $\mathrm{Int}(\mathfrak{g})$ is a subgroup of $\mathrm{Aut}\,\mathfrak{g}$. Further, one easily checks that for any automorphism $\sigma$ of $\mathfrak{g}$ and any $X \in \mathfrak{g}$ we have, as endomorphisms of $\mathfrak{g}$,

$$\sigma e^{\mathrm{ad}(X)} \sigma^{-1} = e^{\mathrm{ad}(\sigma(X))}.$$

Since $\mathrm{Int}\,\mathfrak{g}$ is connected, it is generated by the elements of the form $e^{\mathrm{ad}(X)}$, $X \in \mathfrak{g}$. The above identity then implies that $\mathrm{Int}(\mathfrak{g})$ is a normal subgroup of $\mathrm{Aut}(\mathfrak{g})$.

Let $G$ be a connected Lie group with Lie algebra $\mathfrak{g}$. For $g \in G$, $\mathrm{Ad}(g)$ is an automorphism of $\mathfrak{g}$. We assert that $\mathrm{Ad}(g)$ lies in $\mathrm{Int}(\mathfrak{g})$. In fact, this follows from the facts that $G$ is generated by the elements of the form $\exp(X)$, $X \in \mathfrak{g}$, and that

$$\mathrm{Ad}(\exp(X)) = e^{\mathrm{ad}(X)}.$$

(The proof of the above identity is left to the reader). This means that $\mathrm{Ad}$ defines a map from $G$ onto $\mathrm{Int}(\mathfrak{g})$. In fact we have the following.

**Proposition 2.3.** *Let $G$ be a connected Lie group with Lie algebra $\mathfrak{g}$. Then*

1. *The map $g \rightarrow \mathrm{Ad}(g)$ is a homomorphism from $G$ onto $\mathrm{Int}(\mathfrak{g})$ with kernel $Z(G)$.*
2. *The map $gZ(G) \rightarrow \mathrm{Ad}(g)$ is an isomorphism from $G/Z(G)$ onto $\mathrm{Int}(\mathfrak{g})$.*

Now we define the notions of compactly embedded Lie algebras and compact Lie algebras. This notion is very useful in the study of homogeneous Riemannian and Finsler manifolds.

**Definition 2.7.** Let $\mathfrak{g}$ be a real Lie algebra. A subalgebra $\mathfrak{k}$ is called a compactly embedded subalgebra of $\mathfrak{g}$ if the connected Lie subgroup $K^*$ of $\mathrm{Int}(\mathfrak{g})$, corresponding to the subalgebra $\mathrm{ad}_{\mathfrak{g}}(\mathfrak{k})$ of $\mathrm{ad}(\mathfrak{g})$, is a compact Lie group. A real Lie algebra is called compact if it is a compactly embedded subalgebra of itself.

As an example, any abelian real Lie algebra $\mathfrak{g}$ is compact, since in this case its adjoint Lie group $\mathrm{Int}(\mathfrak{g})$ consists of the single element $\{e\}$, which is compact. It can be proved that a real Lie algebra $\mathfrak{g}$ is compact if and only if there is a compact Lie group $G$ whose Lie algebra is $\mathfrak{g}$; see [83].

At the final part of this section, we introduce some terminology concerning the orbits of general Lie group actions on smooth manifolds, which need not be transitive. In particular, we will recall the definition of principal orbits. Let $M$ be a (connected) smooth $n$-dimensional manifold and $G$ a Lie group acting smoothly on $M$. For $x \in M$, the orbit $G \cdot x = \{g(x) \mid g \in G\}$ is a submanifold of $M$ and if the action is proper, namely, the inverse image of every compact subset of $M \times M$ under the map

$$G \times M \to M \times M : \quad (g, p) \mapsto (p, g(p))$$

is compact, then every orbit is a closed submanifold of $M$. It is easily seen that if $G$ is a compact Lie group, then the action is proper. If the action is proper, then the space of orbits is a Hausdorff space and can be endowed with a differentiable structure.

If $G$ acts transitively on $M$, then $M$ is diffeomorphic to $G/G_x$, $\forall x \in M$. In this case, the action has only one orbit. When the action is nontransitive, the orbit structure is in general very complicated. One important tool for studying the orbit structure is the introduction of the notion of a slice of a proper action.

**Definition 2.8 (see [122]).** Let $G$ be a Lie group acting smoothly and properly on the manifold $M$ and $p \in M$. Then a slice of the action at $p$ is a submanifold $\Sigma$ satisfying the conditions

1. $p \in \Sigma$.
2. $G \cdot \Sigma = \{g(q) \mid g \in G, q \in \Sigma\}$ is an open submanifold of $M$.
3. $G_p \cdot \Sigma = \Sigma$.
4. The action of $G_p$ on $\Sigma$ is isomorphic to an orthogonal linear action of $G_p$ on an open ball of a Euclidean space.
5. Let $(G \times \Sigma)/G_p$ be the orbit space of the action of $G_p$ on $G \times \Sigma$ given by $k((g, q)) = (gk^{-1}, k(q))$, $k \in G_p$, $g \in G$, $q \in \Sigma$. Then the map

$$(G \times \Sigma)/G_p \to M : \quad G_p \cdot (g, q) \mapsto g(q)$$

is a diffeomorphism onto $G \cdot \Sigma$.

It was proved by Montgomery and Yang that every proper action admits a slice at each point [122]. This fact enables us to define a partial ordering on the orbit space. Given $p, q \in M$, we say that the two orbits $G \cdot p$ and $G \cdot q$ have the same orbit type if the isotropic subgroups $G_p$ and $G_q$ are conjugate in $G$. This defines an equivalence relation among the orbits of $G$. Denote the orbit type of $G \cdot p$ by $[G \cdot p]$. Then we can introduce a partial ordering on the set of orbit types by saying that $[G \cdot p] \leq [G \cdot q]$ if and only if $G_q$ is conjugate in $G$ to some subgroup of $G_q$. If the orbit space is connected, then there exists an orbit type that is the largest among all the orbit types. Each representative of the largest orbit type will be called a principal orbit. Each principal orbit has the maximal dimension among all the orbits. Note that there may be some orbit that is of maximal dimension but not principal. The following result is useful.

**Proposition 2.4 (see [122]).** *Let $G$ be a connected Lie group acting smoothly and properly on a connected manifold $M$. Then the union of the principal orbits is an open and dense subset of $M$.*

## 2.3   Semisimple Lie Algebras

In this section we will present a survey of the main results on the structure and classification of real and complex semisimple Lie algebras. This is the basis for the classification of Riemannian symmetric spaces. We first recall some notions on the solvability and nilpotency of Lie algebras.

**Definition 2.9.** A Lie algebra $\mathfrak{g}$ over a field $\mathbb{F}$ is called solvable if there exists a positive integer $m$ such that $\mathfrak{g}^{(m)} = 0$, where $\mathfrak{g}^{(1)} = \mathfrak{g}$ and $\mathfrak{g}^{(l+1)} = [\mathfrak{g}^{(l)}, \mathfrak{g}^{(l)}]$, for $l \geq 1$.

**Definition 2.10.** A Lie algebra $\mathfrak{g}$ over a field $\mathbb{F}$ is called nilpotent if there exists a positive integer $m$ such that $\mathfrak{g}^m = 0$, where $\mathfrak{g}^1 = \mathfrak{g}$ and $\mathfrak{g}^{l+1} = [\mathfrak{g}^l, \mathfrak{g}]$, for $l \geq 1$. The minimal integer $m$ satisfying the condition $\mathfrak{g}^m = 0$ is called the nilpotent index of $\mathfrak{g}$.

Note that here $\mathbb{F}$ can be any field. It is obvious that any nilpotent Lie algebra must be solvable. The converse is not true, as can be seen from the 2-dimensional solvable Lie algebra. The main tools to study solvable and nilpotent Lie algebras are the following Engel's theorem and Lie's theorem.

Recall that an element $x$ in a Lie algebra $\mathfrak{g}$ is called ad-nilpotent if there exists a natural number $n$ such that $[\mathrm{ad}(x)]^n = 0$. The Lie algebra $\mathfrak{g}$ is called ad-nilpotent if all its elements are ad-nilpotent.

**Theorem 2.8 (Engel's theorem).** *A finite-dimensional Lie algebra is nilpotent if and only if it is* ad-*nilpotent.*

The following is an equivalent version of this result.

**Theorem 2.9 (Engel's theorem).** *Let $V$ be a nonzero finite-dimensional vector space and $L$ a subalgebra of $\mathfrak{gl}(V)$. If $L$ consists of nilpotent endomorphisms of $V$, then there exists a nonzero element $v$ in $V$ such that $A(v) = 0$, for all $A \in L$.*

**Theorem 2.10 (Lie's theorem).** *Let $V$ be a nonzero finite-dimensional vector space over an algebraic closed field $\mathbb{F}$ of characteristic $0$. Suppose $L$ is a solvable subalgebra of $\mathfrak{gl}(V)$. Then there exists a common eigenvector for all the endomorphisms in $L$.*

**Theorem 2.11 (Cartan's criterion).** *Let $\mathfrak{g}$ be a finite-dimensional Lie algebra over an algebraically closed field of characteristic $0$. Then $\mathfrak{g}$ is solvable if and only if for every $x \in [\mathfrak{g}, \mathfrak{g}]$ and $y \in \mathfrak{g}$, $\mathrm{tr}(\mathrm{ad}(x)\mathrm{ad}(y)) = 0$.*

The Killing form of the Lie algebra $\mathfrak{g}$ is the bilinear function

$$B(x, y) = \mathrm{tr}(\mathrm{ad}(x)\mathrm{ad}(y)), \quad x, y \in \mathfrak{g}.$$

Then Cartan's criterion can be restated that $\mathfrak{g}$ is solvable if and only if

$$B([\mathfrak{g}, \mathfrak{g}], \mathfrak{g}) = 0.$$

The Killing form is symmetric with respect to its two entries and it is easy to check (using the Jacobi identity) that the following identity holds:

$$B([x,y],z) + B(y,[x,z]) = 0, \quad \forall x,y,z \in \mathfrak{g}.$$

**Definition 2.11.** A Lie algebra $\mathfrak{g}$ is called semisimple if it has no nonzero abelian ideal.

We can also define a semisimple Lie algebra as the following. Note that the sum of two solvable ideals of a Lie algebra is still a solvable ideal. This fact enables us to define the radical of a Lie algebra as the maximal solvable ideal of it. Then a Lie algebra is semisimple if and only if its radical is zero.

From now on we will consider only real or complex Lie algebras.

**Theorem 2.12.** *Let $\mathfrak{g}$ be a real or complex Lie algebra. Then $\mathfrak{g}$ is semisimple if and only if the Killing form is nondegenerate.*

**Theorem 2.13.** *Let $\mathfrak{g}$ be a real or complex semisimple Lie algebra. Then each derivation of $\mathfrak{g}$ is inner.*

Recall that a derivation of a Lie algebra $\mathfrak{g}$ is a linear endomorphism $D$ of $\mathfrak{g}$ such that

$$D([x,y]) = [D(x),y] + [x,D(y)], \quad \forall x,y \in \mathfrak{g}.$$

It is called inner if there exists $z \in \mathfrak{g}$ such that $D = \mathrm{ad}(z)$.

A Lie algebra $\mathfrak{g}$ is called simple if $\mathfrak{g}$ has no ideal other than $\{0\}$ and $\mathfrak{g}$ itself. It can be easily proved that a real or complex simple Lie algebra must be semisimple. We have the following result.

**Theorem 2.14.** *Let $\mathfrak{g}$ be a real or complex semisimple Lie algebra. Then $\mathfrak{g}$ has a decomposition*

$$\mathfrak{g} = \mathfrak{g}_1 + \mathfrak{g}_2 + \cdots + \mathfrak{g}_k,$$

*where $\mathfrak{g}_i$, $1 \le i \le k$, are simple ideals of $\mathfrak{g}$. Moreover, the decomposition is unique up to an adjustment of the order of the simple ideals.*

This theorem reduces the classification of semisimple Lie algebras to that of the simple ones. In the following we will consider the classification of complex simple Lie algebras. We begin with the structure theorems for semisimple Lie algebras.

Let $\mathfrak{g}$ be a complex semisimple Lie algebra. An element $x \in \mathfrak{g}$ is called semisimple if $\mathrm{ad}(x)$ is a semisimple endomorphism of $\mathfrak{g}$ (i.e., it has a diagonal matrix under a certain basis of $\mathfrak{g}$). It is called nilpotent if $\mathrm{ad}(x)$ is a nilpotent endomorphism. The Jordan–Chevalley decomposition theorem asserts that each endomorphism can be uniquely written as the sum of a semisimple and a nilpotent endomorphism that commute with each other. Now given $x \in \mathfrak{g}$, the endomorphism $\mathrm{ad}(x)$ can be decomposed into the sum of two endomorphisms: $\mathrm{ad}(x) = s + n$, where $s$ is semisimple and $n$ is nilpotent with $[s,n] = sn - ns = 0$. It can be easily checked that $s$ and $n$ are also derivations of the Lie algebra $\mathfrak{g}$. Since $\mathfrak{g}$ is semisimple, each derivative is inner. Thus there exist a semisimple element $x_s$ and a nilpotent element $x_n$ such that $s = \mathrm{ad}(x_s)$ and $n = \mathrm{ad}(x_n)$, with $[\mathrm{ad}(x_s), \mathrm{ad}(x_n)] = 0$. Then we have $x = x_s + x_n$ and $[x_s, x_n] = 0$. We usually call this decomposition the Jordan–Chevalley decomposition of $x$.

A subalgebra $\mathfrak{t}$ of $\mathfrak{g}$ is called a toral subalgebra of $\mathfrak{g}$ if $\mathfrak{t}$ consists of semisimple elements. It is easily seen that a toral subalgebra of a complex semisimple Lie algebra must be abelian.

**Theorem 2.15.** *Let $\mathfrak{g}$ be a complex semisimple Lie algebra and $\mathfrak{h}$ a maximal toral subalgebra of $\mathfrak{g}$. Then $\mathfrak{g}$ has a decomposition*

$$\mathfrak{g} = \mathfrak{h} + \sum_{\alpha \in \Delta} \mathfrak{g}_\alpha,$$

*where $\Delta$ is a subset of $\mathfrak{h}^* \setminus \{0\}$ and*

$$\mathfrak{g}_\alpha = \{x \in \mathfrak{g} | [h, x] = \alpha(h)x, \; \forall h \in \mathfrak{h}\}.$$

*The decomposition has the following properties:*

1. *The span of $\Delta$ is $\mathfrak{h}^*$.*
2. *For any $\alpha \in \Delta$, $\dim \mathfrak{g}_\alpha = 1$.*
3. *If $\alpha \in \Delta$, then $-\alpha \in \Delta$. Moreover, if $\alpha \in \Delta$ and $c$ is a nonzero number such that $c\alpha \in \Delta$, then $c = \pm 1$.*
4. *If $\alpha, \beta \in \Delta$, then $[\mathfrak{g}_\alpha, \mathfrak{g}_\beta] \subseteq \mathfrak{g}_{\alpha+\beta}$ (here $\mathfrak{g}_\gamma = 0$ if $\gamma \notin \Delta \cup \{0\}$).*
5. *The restriction of the Killing form $B$ to $\mathfrak{h}$ is nondegenerate.*
6. *Fix $\alpha \in \Delta$ and let $t_\alpha \in \mathfrak{h}$ be the unique element satisfying $\alpha(h) = B(t_\alpha, h)$, $\forall h \in \mathfrak{h}$. Then $B(t_\alpha, t_\alpha) \neq 0$, and for $x_\alpha \in \mathfrak{g}_\alpha$, $y_\alpha \in \mathfrak{g}_{-\alpha}$, $[x_\alpha, y_\alpha] = B(x_\alpha, y_\alpha)t_\alpha$.*
7. *Fix $\alpha \in \Delta$. For any nonzero $x_\alpha \in \mathfrak{g}_\alpha$, there exists $y_\alpha \in \mathfrak{g}_{-\alpha}$ such that $x_\alpha, y_\alpha, h_\alpha = [x_\alpha, y_\alpha]$ span a 3-dimensional Lie algebra isomorphic to $\mathfrak{sl}(2, \mathbb{C})$. Moreover, $h_\alpha$ is independent of the choice of $x_\alpha$ and $y_\alpha$, and $h_\alpha = \frac{2t_\alpha}{B(t_\alpha, t_\alpha)}$, $h_\alpha = -h_{-\alpha}$.*

The subset $\Delta$ of $\mathfrak{h}^*$ in the above theorem is called the root system of $\mathfrak{g}$ with respect to $\mathfrak{h}$. An element $\alpha \in \Delta$ is called a root. For $\alpha, \beta \in \Delta$, we define

$$(\alpha, \beta) = B(t_\alpha, t_\beta).$$

Then we have the following theorem.

**Theorem 2.16.** *Let $\alpha, \beta \in \Delta$. Then*

1. *The number $\frac{2(\beta, \alpha)}{(\alpha, \alpha)}$ is an integer.*
2. *$\beta - \frac{2(\beta, \alpha)}{(\alpha, \alpha)}\alpha \in \Delta$.*
3. *The elements of the form $\beta + l\alpha$, $l \in \mathbb{Z}$, contained in $\Delta$ constitute a continuous string. Let $q, r$ be the largest integers such that $\beta + q\alpha \in \Delta$ and $\beta - r\alpha \in \Delta$. Then $q - r = -\frac{2(\beta, \alpha)}{(\alpha, \alpha)}$.*

The set of the roots $\beta + l\alpha$, $-r \le l \le q$, is called the $\alpha$-string through $\beta$. Select a basis $\alpha_1, \alpha_2, \ldots, \alpha_m$ of $\mathfrak{h}^*$ such that $\alpha_i \in \Delta$. Then the bilinear form $(\alpha, \beta) = B(t_\alpha, t_\beta)$ defined above can be extended to an inner product of the real vector space

$$E = \sum_{i=1}^{m} \mathbb{R}\alpha_i.$$

Then $\Delta$ is a root system in the Euclidean space $E$ in the following sense.

**Definition 2.12.** Let $E$ be a finite-dimensional Euclidean space, with inner product $(\,,\,)$. A subset $\Phi$ of $E$ is called a root system if the following axioms are satisfied:

1. $\Phi$ is a finite subset spanning $E$ and it does not contain 0.
2. $\alpha \in \Phi$ and $c\alpha \in \Phi$, $c \in \mathbb{R}$, implies that $c = \pm 1$.
3. For any $\alpha \in \Phi$, the reflection $\sigma_\alpha$ defined by $\alpha$ leaves the set $\Phi$ invariant.
4. For any $\alpha, \beta \in \Phi$, the number

$$\langle \alpha, \beta \rangle = \frac{2(\beta, \alpha)}{(\alpha, \alpha)}$$

is an integer.

Let $\Phi$ be a root system in $E$. The group generated by all the reflections $\sigma_\alpha$, $\alpha \in \Phi$, is a finite group of isometric transformations of $E$. This group is called the Weyl group of $\Phi$ and will be denoted by $W$. Using the axioms of a root system, we can easily deduce the following.

**Proposition 2.5.** *Let $\Phi$ be a root system in $E$. Then*

1. *Suppose $\alpha, \beta \in \Phi$, $\alpha \ne \pm\beta$. If $(\alpha, \beta) > 0$, then $\alpha - \beta \in \Phi$. If $(\alpha, \beta) < 0$, then $\alpha + \beta \in \Phi$.*
2. *For $\alpha, \beta \in \Phi$, the set of roots of the form $\beta + i\alpha$ is a unbroken string, called the $\alpha$-string through $\beta$.*
3. *Let $q, r$ be respectively the largest integers such that $\beta + q\alpha \in \Phi$ and $\beta - r\alpha \in \Phi$. Then $r - q = -\langle \beta, \alpha \rangle = -\frac{2(\beta, \alpha)}{(\alpha, \alpha)}$.*

A subset $\Pi$ of $\Phi$ is called a base if $\Pi$ is a basis of the vector space $E$ and each root $\beta$ of $\Phi$ can be uniquely written as the sum $\beta = \sum_{\alpha \in \pi} k_\alpha \alpha$, where $k_\alpha$ are integers that are either all nonnegative or all nonpositive. An element in $\Pi$ is called a simple root. It is a fundamental result that every root system has a base. Now we introduce a method to construct a base. Given $\alpha \in \Phi$, set

$$P_\alpha = \{x \in E \,|\, \alpha(x) = 0\}.$$

An element $\gamma \in E \setminus \cup_{\alpha \in \Phi} P_\alpha$ is called regular. For a regular element $\gamma$, define

$$\Phi^+(\gamma) = \{\alpha \in \Phi \mid (\gamma, \alpha) > 0\}, \ \Phi^-(\gamma) = \{\alpha | (\gamma, \alpha) < 0\}.$$

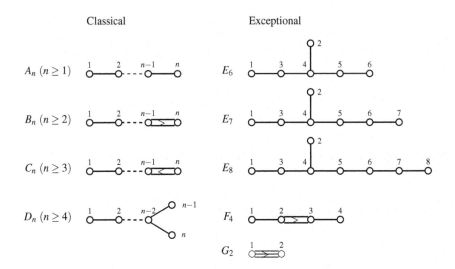

**Fig. 2.1** Dynkin diagrams

A root in $\Phi^+(\gamma)$ is called indecomposable if it cannot be written as the sum of two roots in $\Phi^+(\gamma)$.

**Theorem 2.17.** *Let $\gamma$ be a regular element in $E$. Then the set of all indecomposable roots in $\Phi^+(\gamma)$ is a base of $\Phi$. Moreover, every base of $\Phi$ can be obtained in this manner.*

A connected component of $E \setminus \cup_{\alpha \in \Phi} P_\alpha$ is called a Weyl chamber. It is clear that each regular element of $E$ lies in exactly one Weyl chamber. There is a one-to-one correspondence between the Weyl chambers and the bases of $\Phi$. The Weyl group can be viewed as a permutation group of the set of Weyl chambers. In fact, the action of the Weyl group on the set of Weyl chambers is simply transitive.

Let $\Phi$ be a root system and $\Pi = \{\alpha_1, \alpha_2, \ldots, \alpha_l\}$ a base of $\Phi$. The Dynkin diagram of $\Pi$ is a diagram with $l$ vertices such that the $i$th is joined to the $j$th by $\langle \alpha_j, \alpha_i \rangle \langle \alpha_i, \alpha_j \rangle$ edges. Moreover, if the lengths $|\alpha_i|$ and $|\alpha_j|$ are not equal, then we add an arrow pointing to the shorter one.

A root system $\Phi$ is called irreducible if it cannot be partitioned into two proper and orthogonal subsets. It is easily seen that $\Phi$ is irreducible if and only if its Dynkin diagram is connected. Two root systems $\Phi_1, \Phi_2$ in $E_1, E_2$ respectively are called isomorphic if there exists a linear isomorphism $\sigma$ from $E_1$ onto $E_2$ such that $\sigma(\Phi_1) = \Phi_2$ and for any $\alpha, \beta \in \Phi_1$ we have $\langle \alpha, \beta \rangle_1 = \langle \sigma(\alpha), \sigma(\beta) \rangle_2$. It is clear that two irreducible root systems are isomorphic if and only if they have the same Dynkin diagram. The following theorem gives a complete classification of irreducible root systems.

**Theorem 2.18.** *Let $\Phi$ be an irreducible root system. Then the Dynkin diagram of $\Phi$ must be one of the diagrams in Fig. 2.1. Moreover, any of the diagrams in Fig. 2.1 is the Dynkin diagram of a root system.*

As we have seen above, for a complex semisimple Lie algebra $\mathfrak{g}$ and a maximal toral subalgebra $\mathfrak{h}$ of $\mathfrak{g}$, the root system $\Delta$ of $\mathfrak{g}$ with respect to $\mathfrak{h}$ is a root system in the sense of Definition 2.12. If $\mathfrak{h}_1$ is another maximal toral subalgebra of $\mathfrak{g}$, then we have another root system $\Delta_1$, the root system of $\mathfrak{g}$ with respect to $\mathfrak{h}_1$. It is a very nice fact that different maximal toral subalgebras are conjugate to each other. In other words, there exists an automorphism $\tau$ of $\mathfrak{g}$ such that $\tau(\mathfrak{h}) = \mathfrak{h}_1$ (see Sect. 2.5 below). It is easily seen that this automorphism induces an isomorphism between the root systems $\Delta$ and $\Delta_1$. On the other hand, it is clear that $\mathfrak{g}$ is simple if and only if $\Delta$ is irreducible. Moreover, it can be proved that for each Dynkin diagram in Fig. 2.1, there exists a complex simple Lie algebra whose Dynkin diagram is exactly the given one (see [83]). Thus we have the following theorem.

**Theorem 2.19.** *Up to isomorphism, there are nine types of complex simple Lie algebras, whose Dynkin diagram are respectively* $A_l$, $B_l$, $C_l$, $D_l$, $E_6$, $E_7$, $E_8$, $F_4$, $G_2$.

This theorem gives a complete classification of complex simple Lie algebras, as well as semisimple Lie algebras. Usually, we use the symbols of the Dynkin diagram to represent the corresponding Lie algebra. The first four classes are called classical simple Lie algebras and the other five types are called exceptional ones.

In the final part of this section we introduce some results on real semisimple Lie algebras. We first recall the notion of the complexification of a real Lie algebra.

Suppose $\mathfrak{g}_0$ is a real Lie algebra. The complexification of $\mathfrak{g}_0$ is the complex vector space $\mathfrak{g}_0 \otimes \mathbb{C}$ endowed with the Lie brackets

$$\left[x_1 + \sqrt{-1}y_1, x_2 + \sqrt{-1}y_2\right] = [x_1, x_2] - [y_1, y_2] + \sqrt{-1}\left([x_1, y_2] + [y_1, x_2]\right),$$

where $x_i, y_i \in \mathfrak{g}_0$, $i = 1, 2$. It is easy to check that the above brackets make $\mathfrak{g}_0 \otimes \mathbb{C}$ into a complex Lie algebra. We will denote this Lie algebra by $\mathfrak{g}_0^C$. For convenience, we sometimes use $\mathfrak{g}$ to denote the complexification of $\mathfrak{g}_0$.

Conversely, suppose $\mathfrak{g}$ is a complex Lie algebra. Then $\mathfrak{g}$ can also be viewed as a real Lie algebra: just view the complex vector space $\mathfrak{g}$ as a real vector space and define the same Lie brackets. This real Lie algebra will be denoted by $\mathfrak{g}^R$. Note that set theoretically $\mathfrak{g}^R$ and $\mathfrak{g}$ are the same thing and that $\dim_{\mathbb{R}} \mathfrak{g}^R = 2\dim_{\mathbb{C}} \mathfrak{g}$. The following proposition can be proved by a direct computation.

**Proposition 2.6.** *Let* $\mathfrak{g}_0$ *be a real lie algebra and* $\mathfrak{g}$ *its complexification. Denote the Killing forms of* $\mathfrak{g}_0$, $\mathfrak{g}$, *and* $\mathfrak{g}^R$ *by* $K_0$, $K$, *and* $K_R$, *respectively. Then we have*

$$K_0(X,Y) = K(X,Y), \quad \text{for } X, Y \in \mathfrak{g}_0,$$
$$K_R(X,Y) = 2\text{Re}(K(X,Y)), \quad \text{for } X, Y \in \mathfrak{g}^R,$$

In view of Theorem 2.12, we have the following.

**Proposition 2.7.** *A real Lie algebra is semisimple if and only if its complexification is semisimple.*

Let $\mathfrak{g}$ be a complex Lie algebra. A real form of $\mathfrak{g}$ is a subalgebra $\mathfrak{g}_0$ of the real Lie algebra $\mathfrak{g}^R$ such that

$$\mathfrak{g}^R = \mathfrak{g}_0 + \sqrt{-1}\mathfrak{g}_0 \quad \text{(direct sum of vector spaces)}.$$

In this case $\mathfrak{g}$ is isomorphic to the complexification of $\mathfrak{g}_0$. The following theorem is very important for the classification of real semisimple Lie algebras.

**Theorem 2.20.** *Let $\mathfrak{g}$ be a complex semisimple Lie algebra. Then there exists a real form $\mathfrak{g}_0$ of $\mathfrak{g}$ such that $\mathfrak{g}_0$ is a compact Lie algebra.*

**Definition 2.13.** Let $\mathfrak{g}_0$ be a real semisimple Lie algebra and $\mathfrak{g}$ its complexification. Denote the conjugation of $\mathfrak{g}$ with respect to $\mathfrak{g}_0$ by $\sigma$. A decomposition $\mathfrak{g}_0 = \mathfrak{k}_0 + \mathfrak{p}_0$ of $\mathfrak{g}_0$ into the direct sum of a subalgebra $\mathfrak{k}_0$ and a vector subspace $\mathfrak{p}_0$ is called a Cartan decomposition if there exists a compact real form $\mathfrak{g}_k$ of $\mathfrak{g}$ such that

$$\sigma(\mathfrak{g}_k) \subset \mathfrak{g}_k, \quad \mathfrak{k}_0 = \mathfrak{g}_0 \cap \mathfrak{g}_k, \quad \mathfrak{p}_0 = \mathfrak{g}_0 \cap (\sqrt{-1}\mathfrak{g}_k).$$

**Theorem 2.21.** *Let $\mathfrak{g}_0$ be a real semisimple Lie algebra. Then there exists a Cartan decomposition of $\mathfrak{g}_0$. Moreover, if $\mathfrak{g}_0 = \mathfrak{k}_1 + \mathfrak{p}_1$ and $\mathfrak{g}_0 = \mathfrak{k}_2 + \mathfrak{p}_2$ are two Cartan decompositions of $\mathfrak{g}_0$, then there exists $g \in \mathrm{Int}(\mathfrak{g}_0)$ such that $g(\mathfrak{k}_1) = \mathfrak{k}_2$ and $g(\mathfrak{p}_1) = \mathfrak{p}_2$.*

**Theorem 2.22.** *Let $\mathfrak{g}_0$ be a real semisimple Lie algebra. Then a decomposition $\mathfrak{g}_0 = \mathfrak{k}_0 + \mathfrak{p}_0$ of $\mathfrak{g}_0$ into a subalgebra $\mathfrak{k}_0$ and a vector space $\mathfrak{p}_0$ is a Cartan decomposition if and only if the map $s : X + T \to X - T$, $X \in \mathfrak{k}_0$, $T \in \mathfrak{p}_0$, of $\mathfrak{g}_0$ is an automorphism and the Killing form $B$ is positive definite on $\mathfrak{p}_0$ and negative definite on $\mathfrak{k}_0$.*

## 2.4  Homogeneous Riemannian Manifolds

In this section we will collect some results on homogeneous Riemannian manifolds. For a connected Riemannian manifold $(M, Q)$, we have two ways to define an isometry. On the one hand, we call a diffeomorphism $\sigma$ of $M$ onto itself an isometry if for $x \in M$ and any tangent vectors $X, Y \in T_x(M)$, we have $Q(\mathrm{d}\sigma|_x(X), \mathrm{d}\sigma|_x(Y)) = Q(X, Y)$. On the other hand, we can also define an isometry of $(M, Q)$ to be a map $\tau$ of $M$ onto itself such that $d(\tau(x), \tau(y)) = d(x, y)$, for any $x, y \in M$. The essence of the celebrated Myers–Steenrod theorem is that the above two definitions are equivalent. In the literature, the following corollary of the above result is often referred to as the Myers–Steenord theorem [123].

**Theorem 2.23 (Myers–Steenrod).** *Let $(M, Q)$ be a connected Riemannian manifold. Then the group of isometries $I(M, Q)$ of $(M, Q)$ admits a differentiable structure such that $I(M, Q)$ is a Lie transformation group of $(M, Q)$.*

**Definition 2.14.** A connected Riemannian manifold $(M, Q)$ is called homogeneous if the group of isometries of $(M, Q)$ is transitive on $M$.

If $(M, Q)$ is a homogeneous Riemannian manifold, then $I(M, Q)$ has at most countably many connected components. Therefore by Proposition 2.2, the identity component $I_0(M, G)$ is also transitive on $M$. Fix $x \in M$ and denote by $K$ the isotropy subgroup of $I_0(M, Q)$ at $x$, Then $K$ is a compact subgroup of $I_0(M, Q)$ and $M$ is diffeomorphic to $I_0(M, Q)/K$. The Riemannian metric $Q$ can be viewed as an $I_0(M, Q)$-invariant Riemannian metric on $M$. This means that any homogeneous Riemannian manifold can be written as a coset space of a connected Lie group with an invariant Riemannian metric.

Now we consider the more general case. Suppose $G$ is a connected Lie group and $H$ is a closed subgroup of $G$ such that $G$ acts almost effectively on $G/H$. We study the invariant Riemannian metrics on the coset space $G/H$. Let $\mathfrak{g}, \mathfrak{h}$ be respectively the Lie algebras of $G$ and $H$. Denote the adjoint representation of $G$ on $\mathfrak{g}$ by Ad. Note that for $h \in H$, $\mathrm{Ad}(h)$ keeps the subalgebra $\mathfrak{h}$ invariant. Hence $\mathrm{Ad}(h)$ induces a linear map on the quotient space $\mathfrak{g}/\mathfrak{h}$. We denote this map by $\mathrm{Ad}_{\mathfrak{g}/\mathfrak{h}}(h)$. The differential of this action, which is a representation of $\mathfrak{h}$ on $\mathfrak{g}/\mathfrak{h}$, will be denoted by $\mathrm{ad}_{\mathfrak{g}/\mathfrak{h}}$.

**Proposition 2.8.** *There is a one-to-one correspondence between the $G$-invariant Riemannian metrics on $G/H$ and the $\mathrm{Ad}_{\mathfrak{g}/\mathfrak{h}}(H)$-invariant inner products on $\mathfrak{g}/\mathfrak{h}$.*

For the proof, see [102].

Given $X \in \mathfrak{h}$, $\exp tX$ is a one-parameter subgroup of $H$. With respect to an $\mathrm{Ad}_{\mathfrak{g}/\mathfrak{h}}(H)$-invariant inner product $\langle , \rangle$ on $\mathfrak{g}/\mathfrak{h}$, $\exp(tX)$ are orthogonal transformations. Taking the derivation with respect to $t$, we see that $\mathrm{ad}_{\mathfrak{g}/\mathfrak{h}}(X)$ is skew-symmetric with respect to $\langle , \rangle$. This observation leads to the following assertion.

**Proposition 2.9.** *Let $G/H$ be a coset space. Suppose $G/H$ admits a $G$-invariant Riemannian metric. If the action of $G$ on $G/H$ is almost effective, i.e., if $\mathfrak{h}$ contains no ideal of $\mathfrak{g}$ other than $\{0\}$, then the following assertions hold:*

*1. The Killing form of $\mathfrak{h}$ is negative semidefinite.*
*2. The restriction of the Killing form of $\mathfrak{g}$ to $\mathfrak{h}$ is negative definite.*

The following proposition gives a necessary and sufficient condition for a coset space to admit an invariant Riemannian metric.

**Proposition 2.10.** *Suppose the action of $G$ on the coset space $G/H$ is almost effective. Then $G/H$ admits an invariant Riemannian metric if and only if there is an inner product $\langle , \rangle$ on the Lie algebra $\mathfrak{g}$ of $G$ such that for every $X, Y \in \mathfrak{g}$ and $h \in H$,*

$$\langle \mathrm{Ad}(h)(X), \mathrm{Ad}(h)(Y) \rangle = \langle X, Y \rangle.$$

Next we will consider the Levi-Civita connection and the curvatures of homogeneous Riemannian manifolds. Let $(M, Q)$ be a homogeneous Riemannian manifold. A vector field $X$ on $M$ is called a Killing vector field if every local one-parameter transformation group generated by $X$ consists of local isometries of $(M, Q)$.

**Theorem 2.24.** *Let $(M, Q)$ be a homogeneous Riemannian manifold with the Levi-Civita connection $\nabla$. Suppose $X, Y, Z$ are Killing vector fields on $M$. Then we have*

$$\nabla_X Y = \frac{1}{2}([X, Y] + U(X, Y)),  \tag{2.5}$$

*where $U(X, Y)$ is a symmetric tensor field of type $(2, 1)$ on $M$ determined by*

$$Q(U(X, Y), Z) = \frac{1}{2}(Q([X, Z], Y) + Q(X, [Y, Z])).$$

If we write $(M, Q)$ as a coset space $G/H$ with $H$ compact, then the Lie algebra $\mathfrak{g}$ has a decomposition

$$\mathfrak{g} = \mathfrak{h} + \mathfrak{m} \quad \text{(direct sum of subspaces)},  \tag{2.6}$$

where $\mathfrak{m}$ is a subspace of $\mathfrak{g}$ such that $\text{Ad}(h)(\mathfrak{m}) \subset \mathfrak{m}$, for every $h \in H$. In fact, we just need to fix an inner product as in Proposition 2.10 and take $\mathfrak{m}$ to be the orthogonal complement of $\mathfrak{h}$ with respect to this inner product. Thus in this case, the coset space $G/H$ is reductive.

The tangent space $T_o(G/H)$, where $o = H$ is the origin, can be identified with $\mathfrak{m}$ through the map

$$X \mapsto \frac{d}{dt}(\exp(tX) \cdot o)|_{t=0}.$$

Note that for any $Y \in \mathfrak{g}$, the vector field $\widetilde{Y}|_{gH} = \frac{d}{dt}(g \exp(tY)H)|_{t=0}$ is a Killing vector field (the fundamental Killing vector field generated by $Y$). If $Y \in \mathfrak{m}$, then $\widetilde{Y}_o = Y$. This fact combined with the homogeneity of $M$ implies that formula (2.5) completely determines the Levi-Civita connection of $(M, Q)$. This also leads to the following formulas of the sectional curvature and Ricci curvature of $(M, Q)$.

**Theorem 2.25.** *Let $(G/H, Q)$ be a homogeneous Riemannian manifold with a reductive decomposition (2.6). Suppose $X, Y$ are two orthogonal vectors in $\mathfrak{m}$ of unit length with respect to $Q$. Then the sectional curvature of the tangent plane spanned by $X, Y$ is given by*

$$K(X, Y) = -\frac{3}{4}|[X, Y]_{\mathfrak{m}}|^2 - \frac{1}{2}Q([X, [X, Y]]_{\mathfrak{m}}, Y)$$

$$-\frac{1}{2}Q([Y, [Y, X]]_{\mathfrak{m}}, X) + |U(X, Y)|^2 - Q(U(X, X), U(Y, Y)),$$

*where $Z_{\mathfrak{m}}$ denotes the $\mathfrak{m}$-component of $Z$ and $|\cdot|$ is the length with respect to $Q$.*

The proof can be obtained through a direct computation using (2.5); see [28]. Let $X_1, X_2, \ldots, X_n$ be an orthonormal basis of $\mathfrak{m}$ with respect to $Q$. Set

$$Z = \sum_{i=1}^{n} U(X_i, X_i).$$

It is easy to check that $Z$ is the unique element in $\mathfrak{m}$ such that

$$Q(Z,X) = \mathrm{tr}(\mathrm{ad}X), \quad \forall X \in \mathfrak{m}.$$

**Corollary 2.1.** *The Ricci curvature is given by*

$$\mathrm{Ric}(X,X) = -\frac{1}{2}\sum_i |[X,X_i]_\mathfrak{m}|^2 - \frac{1}{2}\sum_i Q([X,[X,X_i]_\mathfrak{m}]_\mathfrak{m},X_i)$$

$$-\sum_i Q([X,[X,X_i]_\mathfrak{h}]_\mathfrak{m}) + \frac{1}{4}\sum_{i,j}(Q([X_i,X_j]_\mathfrak{m},X))^2$$

$$-Q([Z,X]_\mathfrak{m},X).$$

Finally, we recall some results on homogeneous Einstein manifolds. The following definition is a special case of the Finslerian notion of Einstein metrics.

**Definition 2.15.** An $n$-dimensional connected Riemannian manifold $(M,Q)$ is called an Einstein manifold if its Ricci curvature is a multiple of the metric. More precisely, it is called an Einstein manifold if

$$\mathrm{Ric}(X,Y) = \lambda Q(X,Y), \quad \forall X,Y \in T_x(M), x \in M.$$

The number $c = \frac{1}{n}\lambda$ is called the Einstein constant in the literature. If an Einstein manifold is homogeneous as a Riemannian manifold, then it is called a homogeneous Einstein manifold.

There is a great difference between homogeneous Einstein manifolds with different signs of the Einstein constant. For $\lambda = 0$, we have the following result of Alekseevskii and Kimel'fel'd (see [5, 28]):

**Theorem 2.26.** *A homogeneous Einstein manifold with $c = 0$ is flat; hence it is isometric either to a Euclidean space or to a flat torus.*

By the Bonnect–Myers theorem, a homogeneous Einstein manifold with positive Einstein constant $c$ must be compact. A simple observation leads to the following theorem (see [28]).

**Theorem 2.27.** *A homogeneous Einstein manifold with negative Einstein constant must be noncompact.*

In recent years, homogeneous Einstein manifolds have been studied extensively. We now survey some important results in this field. It is obvious that the standard Riemannian metric on spheres are Einstein metrics. The first nonstandard homogeneous Einstein metric on spheres was found by Jensen in 1973 [94]. Then Ziller obtained a complete classification of all the homogeneous Einstein Riemannian metrics on spheres in 1982 [182]. D'Atri, Wang, and Ziller made a systematic study of compact homogeneous Einstein manifolds, including the normal homogeneous manifolds and naturally reductive left-invariant metrics on compact Lie groups; see [45, 165]. However, up to now, necessary and sufficient conditions for a compact homogeneous manifold to have invariant Einstein metrics are still unknown; see [166] for some partial results. Meanwhile, there has appeared a series of results on low-dimensional spaces.

In the noncompact case there are also many interesting results. We mention the following conjecture; see [4].

*Conjecture 2.1 (D.V. Alekseevskii).* Let $M = G/H$ be a noncompact homogeneous Einstein manifold. Then $H$ is a maximal compact subgroup of $G$.

This conjecture is still open. If it is true, then every noncompact homogeneous Einstein metric can be realized as a left-invariant metric on a solvable Lie group. Based on this point of view, many researchers have studied left-invariant Einstein metrics on solvable Lie groups; see, for example, [75, 80, 160].

## 2.5 Symmetric Spaces

In this section we recall the main results on the structure and classification of Riemannian symmetric spaces. A connected Riemannian manifold $(M, Q)$ is called locally symmetric if for every $x \in M$ there is a neighborhood $U$ of $x$ such that the geodesic symmetry

$$\exp(tX) \to \exp(-tX), \quad X \in T_x(M),$$

defines a local isometry on $U$. It is called globally symmetric if for every $x \in M$, there exists an involutive isometry $\sigma$ (i.e., $\sigma^2 = \mathrm{id}$) such that $x$ is an isolated fixed point of $\sigma$. It is obvious that a globally symmetric Riemannian space must be locally symmetric. The results of this section are mainly due to Cartan.

**Theorem 2.28.** *A connected Riemannian manifold $(M, Q)$ is locally symmetric if and only if its sectional curvature is invariant under parallel displacements. A complete, connected, and simply connected locally symmetric Riemannian manifold must be globally symmetric.*

Let $(M, Q)$ be a globally symmetric Riemannian space and $x \in M$. Then there is an involutive isometry $\tau_x$ with $x$ as its isolated fixed point. It is easily seen that $d\tau_x|_x = -\mathrm{id}|_{T_x(M)}$. Thus $\tau_x$ is actually a geodesic symmetry at $x$. Given $x_1, x_2 \in M$, we can connect $x_1$ to $x_2$ using a family of geodesic segments $\gamma_i$, $i = 1, 2, \dots, l$. Let $z_i$ be the midpoint of the geodesic segment $\gamma_i$. It is easily seen that the isometry $\tau_{z_l} \circ \tau_{z_{l-1}} \circ \cdots \circ \tau_{z_1}$ sends $x_1$ to $x_2$. Thus $(M, Q)$ is homogeneous. Since a homogeneous Riemannian manifold must be complete, $(M, Q)$ is complete. Moreover, it is easy to see that the universal covering Riemannian manifold of $(M, Q)$ is also globally symmetric.

Let $I(M,Q)$ be the full group of isometries of $(M,Q)$. Then $I(M,Q)$ is a Lie transformation group that acts transitively on $M$. Moreover, $I(M,Q)$ has at most countably many connected components. In view of Proposition 2.2, the unity component of $I(M,Q)$, denoted by $G$, is also transitive on $M$. Fix $x \in M$ and denote by $K$ the isotropy subgroup of $G$ at $x$. Then as a manifold, $M = G/K$ and the Riemannian metric $Q$ can be viewed as a $G$-invariant metric on the coset space $G/K$. Now we give a definition:

**Definition 2.16.** Let $G$ be a connected Lie group and $H$ a closed subgroup of $G$. If there exists an involutive automorphism $\sigma$ of $G$ such that $(G_\sigma)_0 \subset H \subset G_\sigma$, where $G_\sigma$ denotes the set of fixed points of $\sigma$ and $(G_\sigma)_0$ the unity component of $G_\sigma$, then the pair $(G,H)$ is called a symmetric pair. If in addition, the group $\mathrm{Ad}_G(H)$ is a compact Lie group, then $(G,H)$ is called a Riemannian symmetric pair.

Note that $\mathrm{Ad}_G(H)$ is the image of $H$ under the adjoint representation $\mathrm{Ad}_G(G)$ of $G$. This leads us to our next result.

**Theorem 2.29.** *If $(G,K)$ is a Riemannian symmetric pair, then there exists a $G$-invariant Riemannian metric $Q$ on $G/K$ such that $(G/K,Q)$ is a Riemannian globally symmetric space. Conversely, if $(M,Q)$ is a Riemannian globally symmetric space, then there exists a connected Lie group $G$ that acts isometrically and transitively on $M$ such that $(G,K)$ is a Riemannian symmetric pair, where $K$ is the isotropy subgroup of $G$ at certain $x \in M$.*

Note that in the above theorem, $G$ can be taken as the unity component of the full group of isometries. This theorem reduces the research of Riemannian symmetric spaces to coset spaces endowed with an invariant Riemannian metric. In fact, we can reduce the problem further.

**Definition 2.17.** Let $\mathfrak{g}$ be a real Lie algebra and $s$ an involutive automorphism of $\mathfrak{g}$. If the set of fixed points of $s$, denoted by $\mathfrak{k}$, is a compactly embedded subalgebra of $\mathfrak{g}$, then $(\mathfrak{g},s)$ is called an orthogonal symmetric Lie algebra. If in addition, $\mathfrak{k} \cap \mathfrak{z} = \{0\}$, where $\mathfrak{z}$ is the center of $\mathfrak{g}$, then the pair is called effective.

**Theorem 2.30.** *Let $(G,K)$ be a Riemannian symmetric pair and $\sigma$ an involutive automorphism of $G$ such that $(G_\sigma)_0 \subset K \subset G_\sigma$. Let $\mathfrak{g}$ be the Lie algebra of $G$ and $s$ the differential of $\sigma$ at $e \in G$. Then $(\mathfrak{g},s)$ is an orthogonal symmetric Lie algebra. If the action of $G$ on $G/K$ is effective, then $(\mathfrak{g},s)$ is effective.*

**Theorem 2.31.** *Let $(\mathfrak{g},s)$ be an orthogonal symmetric Lie algebra and $\mathfrak{k}$ the set of fixed points of $s$. Suppose $(G,K)$ is a pair associated with $(\mathfrak{g},s)$ (i.e., $G$ is a connected Lie group with Lie algebra $\mathfrak{g}$ and $K$ is a Lie subgroup of $G$ with Lie algebra $\mathfrak{k}$) such that $G$ is simply connected and $K$ is connected. Then $K$ is closed in $G$ and the pair $(G,K)$ is a Riemannian symmetric pair.*

These theorems reduce the study of symmetric spaces to that of the orthogonal symmetric Lie algebras. Now we consider the structure of such algebras.

**Definition 2.18.** Let $(\mathfrak{g}, s)$ be an orthogonal symmetric Lie algebra and $\mathfrak{g} = \mathfrak{k} + \mathfrak{p}$ the decomposition of $\mathfrak{g}$ into the eigenspaces of $s$ for the eigenvalues $+1$ and $-1$, respectively. If $\mathfrak{g}$ is a compact semisimple Lie algebra, then $(\mathfrak{g}, s)$ is said to be of compact type; if $\mathfrak{g}$ is noncompact and semisimple, then $(\mathfrak{g}, s)$ is said to be of noncompact type; if $\mathfrak{p}$ is an abelian ideal of $\mathfrak{g}$, then $(\mathfrak{g}, s)$ is said to be of Euclidean type. A pair $(G, K)$ associated with $(\mathfrak{g}, s)$ is said to be of compact, noncompact, or Euclidean type according to the type of $(\mathfrak{g}, s)$. A Riemannian globally symmetric space $(M, Q)$ is said to be of Euclidean, compact, or noncompact type according to the type of the corresponding symmetric pair.

Note that if a connected simply connected globally symmetric space $(M, Q)$ is of Euclidean type, then $(M, Q)$ is the standard Euclidean space.

**Theorem 2.32.** *Let $(\mathfrak{g}, s)$ be an effective orthogonal symmetric Lie algebra. Then $\mathfrak{g}$ has a decomposition into the direct sum of ideals $\mathfrak{g} = \mathfrak{g}_0 + \mathfrak{g}_c + \mathfrak{g}_n$, where $\mathfrak{g}_0$, $\mathfrak{g}_c$, and $\mathfrak{g}_n$ are all invariant under $s$ and orthogonal with respect to the Killing form of $\mathfrak{g}$. Moreover, let $s_0$, $s_c$, and $s_n$ be respectively the restrictions of $s$ to $\mathfrak{g}_0$, $\mathfrak{g}_c$, and $\mathfrak{g}_n$. Then $(\mathfrak{g}_0, s_0)$, $(\mathfrak{g}_c, s_c)$, and $(\mathfrak{g}_n, s_n)$ are orthogonal symmetric Lie algebras of Euclidean, compact, and noncompact type, respectively.*

The global version of the above theorem is the following.

**Theorem 2.33.** *Let $(M, Q)$ be a connected simply connected Riemannian globally symmetric space. Then $(M, Q)$ can be decomposed as the product $M = M_0 \times M_c \times M_n$, where $M_0$ is a Euclidean space, $M_c$ is a Riemannian globally symmetric space of compact type, and $M_n$ is a Riemannian globally symmetric space of noncompact type.*

There is a remarkable duality between Riemannian globally symmetric spaces of compact and noncompact type. We start with orthogonal symmetric Lie algebras. Let $(\mathfrak{g}, s)$ be an orthogonal Lie algebra and $\mathfrak{g} = \mathfrak{k} + \mathfrak{p}$ the canonical decomposition. Consider the subalgebra

$$\mathfrak{g}^* = \mathfrak{k} + \sqrt{-1}\mathfrak{p}$$

of the complexification $\mathfrak{g}^{\mathbb{C}}$ and define a linear endomorphism $s^*$ on $\mathfrak{g}^*$ by

$$s^*(X + \sqrt{-1}Y) = X - \sqrt{-1}Y, \quad X \in \mathfrak{k}, Y \in \mathfrak{p}.$$

Then it is easy to check that $(\mathfrak{g}^*, s^*)$ is again an orthogonal Lie algebra. The pair $(\mathfrak{g}^*, s^*)$ is called the dual of $(\mathfrak{g}, s)$. Two orthogonal symmetric Lie algebras $(\mathfrak{g}_1, s_1)$ and $(\mathfrak{g}_2, s_2)$ are called isomorphic if there is a Lie algebra isomorphism $\varphi$ from $\mathfrak{g}_1$ onto $\mathfrak{g}_2$ such that $\varphi \circ s_1 = s_2 \circ \varphi$.

**Proposition 2.11.** *The orthogonal symmetric Lie algebra $(\mathfrak{g}, s)$ is of compact type if and only if $(\mathfrak{g}^*, s^*)$ is of noncompact type. If $(\mathfrak{g}_1, s_1)$ is isomorphic to $(\mathfrak{g}_2, s_2)$, then $(\mathfrak{g}_1^*, s_1^*)$ is isomorphic to $(\mathfrak{g}_2^*, s_2^*)$.*

An orthogonal symmetric Lie algebra $(\mathfrak{g}, s)$ is called irreducible if in its canonical decomposition $\mathfrak{g} = \mathfrak{k} + \mathfrak{p}$, the adjoint representation of $\mathfrak{k}$ on $\mathfrak{p}$ is irreducible. It can be proved that every effective orthogonal Lie algebra of compact or noncompact type can be decomposed into a direct sum of irreducible orthogonal Lie algebras.

**Theorem 2.34.** *Let $(\mathfrak{g}, s)$ be an irreducible effective orthogonal Lie algebra. Then $(\mathfrak{g}, s)$ must be isomorphic to one of the following types:*

(CI) *$\mathfrak{g}$ is compact simple and $s$ is an involutive automorphism of $\mathfrak{g}$.*
(CII) *$\mathfrak{g} = \mathfrak{g}_1 \oplus \mathfrak{g}_1$, where $\mathfrak{g}_1$ is a compact simple Lie algebra, and $s$ is defined by $s(X, Y) = (Y, X)$, $X, Y \in \mathfrak{g}_1$.*
(NI) *$\mathfrak{g}$ is a noncompact simple Lie algebra whose complexification $\mathfrak{g}^{\mathbb{C}}$ is a complex simple Lie algebra, and $s$ is a Cartan involution of $\mathfrak{g}$.*
(NII) *$\mathfrak{g}$ is a noncompact simple Lie algebra whose complexification $\mathfrak{g}^{\mathbb{C}}$ is a complex nonsimple Lie algebra, and $s$ is a Cartan involution of $\mathfrak{g}$.*

*Among the above four types of orthogonal symmetric Lie algebras, (CI) is dual to (NI), and (CII) is dual to (NII).*

A connected simply connected Riemannian globally symmetric space is said to be of type (CI), (CII), (NI), or (NII) according to the type of its corresponding orthogonal symmetric Lie algebra. Note that the symmetric spaces of type (CII) are exactly all the compact simple Lie groups endowed with bi-invariant Riemannian metrics. Since each complex simple Lie algebra has a unique (up to an isomorphism) compact real form, each complex simple Lie algebra in Theorem 2.18 corresponds to exactly one symmetric space of type (CII). We will use the same notation as in Theorem 2.18 to denote the corresponding symmetric space.

A complete list of connected simply connected Riemannian globally symmetric spaces of type (CI) can be found in [83].

There is a great difference between Riemannian globally symmetric spaces of compact type and those of noncompact type. In the noncompact case, the structure is very simple:

**Theorem 2.35.** *Let $(\mathfrak{g}_0, s)$ be an effective orthogonal symmetric Lie algebra of noncompact type and let $\mathfrak{g}_0 = \mathfrak{h}_0 + \mathfrak{m}_0$ be the associated decomposition. Then $\mathfrak{g}_0$ is a noncompact semisimple Lie algebra and the above decomposition is a Cartan decomposition of $\mathfrak{g}_0$. Suppose $(G, H)$ is any pair associated with $(\mathfrak{g}_0, \mathfrak{h}_0)$. Then we have the following:*

1. *$H$ is connected, closed, and contains the center $Z$ of $G$. Moreover, $H$ is compact if and only if $Z$ is finite. In this case, $H$ is a maximal compact subgroup of $G$.*
2. *The pair $(G, H)$ is a Riemannian symmetric pair.*
3. *The map $\varphi : (X, h) \to (\exp X)h$ is a diffeomorphism from $\mathfrak{m}_0 \times H$ onto $G$ and the exponential map of the Riemannian manifold $M = G/H$ is a diffeomorphism from $\mathfrak{m}_0$ onto the manifold $M$.*

This theorem has the following corollary:

**Corollary 2.2.** *Suppose M and M' are two irreducible Riemannian globally symmetric spaces of noncompact type. If the full groups I(M) and I(M') have the same Lie algebra, then up to a positive scalar, M and M' are isometric.*

From Theorem 2.35 we can also deduce the conjugacy of maximal compact subgroups of noncompact semisimple Lie groups:

**Theorem 2.36.** *Let $(G, H)$ be a Riemannian symmetric pair of noncompact type. Then we have the following:*

(a) *For any compact subgroup $H_1$ of G, there exists $x \in G$ such that $x^{-1}H_1 \subset H$.*
(b) *H has a unique compact subgroup $H_0$ that is a maximal subgroup of G.*
(c) *All maximal compact subgroups of a connected semisimple Lie group are conjugate under an inner automorphism.*

From the above results we see that the structure of noncompact Riemannian symmetric spaces is usually very simple. In particular, if $G/H$ is a connected Riemannian globally symmetric space of noncompact type, and $G$ is a connected Lie group, then $H$ must be a connected compact subgroup of $G$. However, the above results are usually not true for a Riemannian globally symmetric space of compact type. In fact, one can construct some examples to show the following (see [83, Chap. VII]):

**Proposition 2.12.** *Let $(\mathfrak{u}, s)$ be an orthogonal symmetric Lie algebra of compact type. Let $(U, K)$ be any pair associated with $(\mathfrak{u}, s)$. Then*

1. *The center of U need not be contained in K.*
2. *Even if $U/K$ is Riemannian globally symmetric, K is not necessarily connected.*
3. *Even if $U/K$ is Riemannian globally symmetric, s does not necessarily correspond to an automorphism of U.*

This difference causes much difficulty in classifying Riemannian globally symmetric spaces of compact type. We omit the details here; see [83].

Now we recall some results on Hermitian symmetric spaces. Let $(M, J)$ be a complex manifold with complex structure $J$. A Riemannian metric $Q$ on $M$ is called Hermitian if $Q(J(X), J(Y)) = Q(X, Y)$ for all tangent vectors $X, Y$ on $M$. If in addition $\nabla_X J = 0$, then the Hermitian metric is called Kählerian.

**Definition 2.19.** Let $M$ be a connected complex manifold with a Hermitian metric $Q$. Then $(M, Q)$ is said to be a Hermitian symmetric space if each point $p \in M$ is an isolated fixed point of an involutive holomorphic isometry $s_p$ of $M$.

It is not hard to prove that a Hermitian symmetric space must be Kählerian. Moreover, a Hermitian symmetric space must be a Riemannian globally symmetric space. Therefore, the classification of Hermitian symmetric spaces is reduced to finding out which Riemannian globally symmetric spaces admit a complex structure that is admissible with the metric structure. We say that a Hermitian symmetric space is of compact or noncompact type according to the type of the corresponding Riemannian symmetric space. It is called irreducible if the corresponding Riemannian symmetric space is irreducible. It can be proved that a Hermitian symmetric space of compact or noncompact type is simply connected.

**Theorem 2.37.** *Let M be a simply connected Hermitian symmetric space. Then M is a product*

$$M = M_0 \times M_c \times M_n,$$

*where $M_0, M_c, M_n$ are simply connected Hermitian symmetric spaces, $M_0 = \mathbb{C} \times \mathbb{C} \times \cdots \times \mathbb{C}$, and $M_c$ and $M_n$ are of compact type and noncompact type, respectively.*

**Theorem 2.38.** *Let M be an irreducible Hermitian symmetric space.*

(i) *If M is of noncompact type, then M is the manifold $G/K$, where G is a connected noncompact simple Lie group with center $\{e\}$ and K is a maximal compact subgroup of G with nondiscrete center.*

(ii) *If M is of compact type, then M is the manifold $U/K$, where U is a connected compact simple Lie group with center $\{e\}$ and K is a maximal connected proper subgroup with nondiscrete center.*

Finally we recall some results on the rank of symmetric spaces. Let $M$ be a Riemannian globally symmetric space. The rank of $M$ is the maximal dimension of flat, totally geodesic submanifolds of $M$. Suppose $M$ is of compact or noncompact type and write $M$ as a coset space $G/H$, where $G$ is the identity component of the full group. Let $\mathfrak{g} = \mathfrak{h} + \mathfrak{p}$ be the canonical decomposition of the Lie algebra of $G$. A subspace $\mathfrak{s}$ of $\mathfrak{p}$ is called a Lie triple system if $[[\mathfrak{s}, \mathfrak{s}], \mathfrak{s}] \subset \mathfrak{s}$. A result of Cartan says that there is a one-to-one correspondence between totally geodesic submanifolds of $M$ and Lie triple systems in $\mathfrak{p}$ through the map $\mathfrak{s} \to \mathrm{Exp}(\mathfrak{s})$.

**Theorem 2.39.** *The totally geodesic submanifold $\mathrm{Exp}(\mathfrak{s})$ is flat if and only if $\mathfrak{s}$ is abelian.*

Therefore, the rank of a Riemannian globally symmetric space of compact type or noncompact type is equal to the dimension of the maximal abelian subspace of $\mathfrak{p}$.

**Theorem 2.40.** *Let $\mathfrak{a}$ and $\mathfrak{a}'$ be two maximal abelian subspaces of $\mathfrak{p}$. Then*

1. *There exists an element $u \in \mathfrak{p}$ whose centralizer in $\mathfrak{p}$ is $\mathfrak{a}$.*
2. *There exists an element $k \in H$ such that $\mathrm{Ad}(k)\mathfrak{a} = \mathfrak{a}'$.*
3. $\mathfrak{p} = \bigcup_{k \in H} \mathrm{Ad}(k)\mathfrak{a}.$

This theorem has an important corollary that any two maximal flat totally geodesic submanifolds of $M$ must be congruent under the group of isometries of $M$. Applying the above results to connected compact Lie groups, we get the following theorem.

**Theorem 2.41.** *Let $G$ be a connected compact Lie group with Lie algebra $\mathfrak{g}$. Suppose $\mathfrak{t}$ and $\mathfrak{t}'$ are two maximal abelian subalgebras of $\mathfrak{g}$. Then*

1. *There exists an element $u \in \mathfrak{t}$ whose centralizer in $\mathfrak{g}$ is $\mathfrak{t}$.*
2. *There exists an element $g \in G$ such that $\mathrm{Ad}(g)\mathfrak{t} = \mathfrak{t}'$.*
3. $\mathfrak{g} = \bigcup_{g \in G} \mathrm{Ad}(g)\mathfrak{t}$.

From this theorem it follows that two Cartan subalgebras of a complex semisimple Lie algebra are congruent under the group of automorphisms.

The material of this section is taken mainly from [83].

# Chapter 3
# The Group of Isometries

In this chapter we deal with isometries of a Finsler space. The first result on this topic is the theorem of Van Danzig and van der Waerden, published in 1928 (see [156] and Sect. 3.2 below), asserting that the group of isometries of a locally compact metric space is a topological transformation group of the underlying space. Cartan proved in [37] that the group of isometries of a globally symmetric Riemannian manifold is a Lie transformation group of the underlying manifold. This result plays a fundamental role in the theory of symmetric Riemannian spaces. In 1939, this result of Cartan was generalized to all Riemannian manifolds by the Myers–Steenrod theorem (Theorem 2.23). This theorem is the foundation of the theory of homogeneous Riemannian manifolds. The proof of the Myers–Steenrod theorem has been simplified by the work of several people; see, for example, Palais [127,128], Kobayashi [101], Kobayashi–Nomizu [102]. There are also several versions of the generalizations of this result. For example, in [102], it is proved that the group of all affine transformations of an affine manifold is a Lie transformation group. This result has been applied to the case of pseudo-Riemannian manifolds and has resulted in many powerful results in the study of pseudo-Riemannian metrics.

The first two sections of this chapter are devoted to generalizing the Myers–Steenrod theorem from Riemannian geometry to the Finslerian setting, through a thorough study of isometries and Killing vector fields. In Sect. 3.3, we re-prove a classical theorem of Wang concerning the maximal dimension of the group of isometries of non-Riemannian Finsler metrics on a manifold. Section 3.4 is devoted to studying two-point homogeneous Finsler spaces. The main result is that a two-point homogeneous Finsler manifold must be Riemannian. Finally, in Sect. 3.5, we consider the set of fixed points of isometries and generalize an interesting result of Kobayashi from Riemannian manifolds to Finsler spaces.

We should mention here that in the literature there has yet appeared no result on the group of isometries of a pseudo-Finsler space (that is, the Hessian matrix of the Finsler function is assumed only to be nondegenerate on the tangent spaces; see [139]). It would be interesting to consider this problem and to generalize some of the results of this chapter to this more generalized case.

S. Deng, *Homogeneous Finsler Spaces*, Springer Monographs in Mathematics, DOI 10.1007/978-1-4614-4244-8_3, © Springer Science+Business Media New York 2012

The standard references of this chapter are [39, 47, 55]; see [68, 69] for more information.

## 3.1  Isometries and Killing Vector Fields

Let $(M,F)$ be a Finsler space. As in the Riemannian case, we have two ways to define an isometry on $(M,F)$. On the one hand, we can define an isometry to be a diffeomorphism of $M$ onto itself that preserves the Finsler function. On the other hand, since on $M$ we still have the definition of distance function (although generically it is not a distance in the strict sense), we can define an isometry of $(M,F)$ to be a map of $M$ onto $M$ that keeps the *distance* of each pair of points of $M$. The Myers–Steenrod theorem asserts that the above two conditions are equivalent for a Riemannian manifold.

In this section we will generalize the Myers–Steenrod theorem to Finsler spaces. We first prove a result on the distance function.

**Lemma 3.1.** *Let $(M,F)$ be a Finsler space with distance function d and $x \in M$. Then for any $\varepsilon > 0$, there exists a neighborhood N of the origin of $T_x(M)$ such that $\exp_x$ is a $C^1$-diffeomorphism from N onto its image and for any $A, B \in U$, $A \neq B$, and any $C^1$ curve $\sigma_0(s), 0 \leq s \leq 1$, connecting A and B that satisfies $\sigma_0(s) \in N$ and $\dot{\sigma}_0(s) \neq 0$, $s \in [0,1]$, we have*

$$\left| \frac{L(\sigma)}{L(\sigma_0)} - 1 \right| \leq \varepsilon,$$

*where $L(\cdot)$ denotes the arc length of a curve and $\sigma(s) = \exp_x \sigma_0(s)$.*

*Proof.* By Theorem 1.3, there exists a tangent ball

$$B_x(r) = \{A \in T_x(M) | F(x,A) < r\}$$

of the origin in $T_x(M)$ such that $\exp = \exp_x$ is a $C^1$-diffeomorphism from $B_x(r)$ onto $\mathscr{B}_x^+(r) = \{w \in M | d(x,w) < r\}$. Assume that $A, B \in B_x(r)$, $A \neq B$. Let $\sigma_0(s)$, $0 \leq s \leq 1$, be a $C^1$ curve connecting $A$ and $B$ such that $\forall s, \sigma_0(s) \in B_x(r)$ and $\dot{\sigma}_0(s) \neq 0$. Then we can write the velocity vector of $\sigma_0(s)$ as $\dot{\sigma}_0(s) = t(s)X(s)$, where $X(s)$ satisfies $F(x, X(s)) = \frac{r}{2}, \forall s$, and $t(s) \geq 0$ is a continuous function on $[0,1]$. Therefore the arc length of $\sigma_0$ is

$$L(\sigma_0) = \int_0^1 t(s)F(x,X(s))\, ds.$$

Set $X_1(s) = d(\exp_x)|_{\sigma_0(s)} X(s)$. Then the velocity vector of the curve $\sigma(s) = \exp_x(\sigma_0(s))$, $0 \leq s \leq 1$, is

$$\dot{\sigma}(s) = d(\exp_x)|_{\sigma_0(s)}(t(s)X(s)) = t(s)d(\exp_x)|_{\sigma_0(s)}(X(s)) = t(s)X_1(s).$$

Therefore, the arc length of $\sigma$ is

$$L(\sigma) = \int_0^1 t(s)F(\sigma(s),X_1(s))\,ds.$$

Now we select a neighborhood $V_1$ of $x$ in $M$ with compact closure that is contained in $\mathscr{B}_x^+(r)$ and fix a coordinate system $(x_1,x_2,\ldots,x_n)$ in $V_1$. Let $N_1 = \exp^{-1}V_1$. Suppose $\sigma_0 \subset N_1$. Denote by $M(s)$ the matrix of $d(\exp_x)|_{\sigma_0(s)}$ under the basis $\frac{\partial}{\partial x_1},\frac{\partial}{\partial x_2},\ldots,\frac{\partial}{\partial x_n}$. Given any positive number $\delta < \frac{r}{2}$, since $d(\exp_x)|_0 = I_n$ and $\exp$ is $C^1$ smooth, there exists a neighborhood $N_2 \subset N_1$ of the origin of $T_x(M)$ such that for any $C^1$ curve $\sigma_0$ satisfying $\sigma_0(s) \in N_2, \forall s$, we have

$$\|M(s) - I\| < \frac{\delta}{n}, \quad 0 \le s \le 1,$$

where $\|\cdot\|$ denotes the maximum of the absolute values of the entries of a matrix. Write $X(s)$ and $X_1(s)$ as

$$X(s) = \sum_{j=1}^n y_j(s)\frac{\partial}{\partial x_j}\Big|_x;$$

$$X_1(s) = \sum_{j=1}^n y_j'(s)\frac{\partial}{\partial x_j}\Big|_{\sigma(s)}.$$

Then we have

$$|y_j'(s) - y_j(s)| < \delta, \quad 1 \le j \le n.$$

Consider the following subset of $TM_o$:

$$C_0 = \left\{ (w,(d(\exp_x))|_w)y \mid w \in V_1, W = \exp_x^{-1}w, y \in T_W(Tx(M)), F(x,y) = \frac{r}{2} \right\},$$

where we have identified $T_W(T_x(M))$ with $T_x(M)$ in the standard way. Since $\exp$ is $C^1$ smooth, the closure of $C_0$ is compact. Hence the function $F$ is bounded on $C_0$. Suppose $r_1$ is a positive number such that $F < r_1$ on $C_0$. Write the Finsler function $F(w,y)$ as $F(w,y_1,y_2,\ldots,y_n)$ for $y = \sum_{j=1}^n y_j\frac{\partial}{\partial x_j}|_w$. Let $D_1$ be the closure of the set $D_0 = \{(w,y) \in TM \mid w \in V_1, F(x,y) \le \frac{r}{2}+r_1\}$. Since $F$ is continuous and $D_1$ is compact, $F$ is uniformly continuous on $D_1$. Therefore for the given $\varepsilon > 0$, there exist $\delta_1 > 0$ and a neighborhood $V_2 \subset V_1$ of $x$ such that for any $w \in V_2$ and $y,y'$ with $|y_j - y_j'| < \delta_1$, $j = 1,2,\ldots,n$, $F(x,y_1,y_2,\ldots,y_n) < \frac{r}{2}+r_1$ and $F(w,y_1',\ldots,y_n') < \frac{r}{2}+r_1$, we have

$$|F(x,y_1,y_2,\ldots,y_n) - F(w,y_1',y_2',\ldots,y_n')| < \frac{r}{2}\varepsilon.$$

Therefore, if we select the above $\delta$ so small that $\delta < \delta_1$, then for the corresponding $N_2$ and any $C^1$ curve $\sigma_0$, $\sigma_0 \subset N_2 \cap (\exp)^{-1}V_2$, we have

$$\left| \frac{L(\sigma)}{L(\sigma_0)} - 1 \right| = \frac{|\int_0^1 t(s)(F(x,X(s)) - F(\sigma(s),X_1(s)))\,ds|}{|\int_0^1 t(s)F(x,X(s))\,ds|}$$

$$\leq \frac{\int_0^1 t(s)|F(x,X(s)) - F(\sigma(s),X_1(s))|\,ds}{r\int_0^1 t(s)\,ds}$$

$$\leq \frac{\frac{r}{2}\varepsilon \int_0^1 t(s)\,ds}{\frac{r}{2}\int_0^1 t(s)\,ds} = \varepsilon.$$

This completes the proof of the lemma. $\square$

**Theorem 3.1.** *Let $x \in M$ and let $B_x(r)$ be a tangent ball of $T_x(M)$ such that $\exp_x$ is a $C^1$ diffeomorphism from $B_x(r)$ onto $\mathscr{B}_x^+(r)$. For $A, B \in B_x(r)$, $A \neq B$, let $a = \exp_x A$, $b = \exp_x B$. Then we have*

$$\frac{F(x, A - B)}{d(a,b)} \to 1$$

*as $(A,B) \to (0,0)$.*

*Proof.* Set $\mathscr{B}_x^-(r) = \{w \in M \mid d(w,x) < r\}$. Suppose $r$ is so small that each pair of points in $\mathscr{B}_x^+(\frac{r}{2}) \cap \mathscr{B}_x^-(\frac{r}{2})$ can be joined by a unique minimal geodesic contained in $\mathscr{B}_x^+(r)$. Let $\Gamma_0(s)$, $0 \leq s \leq 1$, be the line segment connecting $A$ and $B$, and $\Gamma(s) = \exp_x \Gamma_0(s)$. By Lemma 3.1, we have

$$\frac{L(\Gamma_0)}{L(\Gamma)} = \frac{F(x, A - B)}{L(\Gamma)} \to 1$$

as $(A,B) \to (0,0)$. Now let $a = \exp_x A$, $b = \exp_x B$. Suppose $a, b \in \mathscr{B}_x^+(\frac{r}{2}) \cap \mathscr{B}_x^-(\frac{r}{2})$. Let $\gamma_{ab}(s)$, $0 \leq s \leq 1$, be the unique minimal geodesic of constant speed connecting $a$ and $b$. Let $\gamma_0(s)$, $0 \leq s \leq 1$, be the unique curve in $B_x(r)$ that satisfies $\gamma_{ab}(s) = \exp_x \gamma_0(s)$. Then by Lemma 3.1, we also have

$$\frac{L(\gamma_0)}{L(\gamma_{ab})} \to 1$$

as $(A,B) \to (0,0)$. Since

$$d(a,b) \leq L(\Gamma), \quad L(\gamma_0) \geq F(x, A - B),$$

We have

$$\frac{F(x, A - B)}{L(\Gamma)} \leq \frac{F(x, A - B)}{d(a,b)} \leq \frac{L(\gamma_0)}{L(\gamma_{ab})}.$$

This completes the proof of the theorem. $\square$

Next we prove the differentiability of isometries of a Finsler space. We first consider the special case of a Minkowski space.

**Proposition 3.1.** *Let* $\|\cdot\|_1$, $\|\cdot\|_2$ *be two Minkowski norms on* $\mathbb{R}^n$. *Let* $\phi$ *be a map of* $\mathbb{R}^n$ *into itself such that* $\|\phi(A) - \phi(B)\|_2 = \|A - B\|_1$, $\forall A, B \in \mathbb{R}^n$. *Then* $\phi$ *is an affine transformation of* $\mathbb{R}^n$ *onto itself.*

*Proof.* Consider $\mathbb{R}^n$ endowed with $\|\cdot\|_j$, $j = 1, 2$, as two Finsler spaces, denoted by $(M_1, F_1)$ and $(M_2, F_2)$, respectively. It is obvious that geodesics in $M_j$, $j = 1, 2$, are straight lines, and the distance function of $M_j$ is $d_j(A, B) = \|A - B\|_j$, $j = 1, 2$. Consider $\phi$ as a map from the Finsler space $(M_1, F_1)$ to $(M_2, F_2)$. Then $\phi$ preserves the distance function. Since short geodesics minimize distance between its start and end points, $\phi$ transforms geodesics to geodesics. Therefore $\phi$ transforms straight lines to straight lines. We first treat the case $\phi(0) = 0$. For $A \in \mathbb{R}^n$, $A \neq 0$, the curve $\phi(tA)$, $t \geq 0$, is a ray that coincides with the ray $t\phi(A)$ at $t = 0$ and $t = 1$. Therefore they coincide as point sets. Thus there is a nonnegative function $\mu(t)$ such that $\phi(tA) = \mu(t)\phi(A)$. Since

$$\|\phi(tA) - 0\|_2 = \|tA - 0\|_1 = t\|A\|_1 = \|\mu(t)\phi(A) - 0\|_2 = \mu(t)\|\phi(A)\|_2$$
$$= \mu(t)\|A\|_1, \quad t \geq 0,$$

we have $\mu(t) = t$. Thus $\phi(tA) = t\phi(A)$, for $t \geq 0$. Suppose $A \neq B$. Then a similar argument to that above shows that there exists a nonnegative function $\lambda(t)$ such that $\phi(tA + (1-t)B) = \lambda(t)\phi(A) + (1 - \lambda(t))\phi(B)$, $t \geq 0$. Moreover, we can similarly show that $\lambda(t) = t$. In particular, for $t = \frac{1}{2}$, we have

$$\frac{1}{2}\phi(A + B) = \phi(\frac{1}{2}(A + B)) = \frac{1}{2}\phi(A) + \frac{1}{2}\phi(B).$$

Thus $\phi(A + B) = \phi(A) + \phi(B)$. Taking $A = -B$ in the above equality we have $\phi(-A) = -\phi(A)$. Therefore $\phi$ is a linear transformation. Since $\ker(\phi) = \{0\}$, it is a linear isomorphism. Hence the conclusion holds when $\phi(0) = 0$. Next we suppose $A_1 = \phi(0) \neq 0$. Consider the composition map $\phi_1 = \pi_{A_1} \circ \phi$, where $\phi_{A_1}(A) = A - A_1$ is parallel translation. By the above argument and the facts that $\phi_1(0) = 0$ and $\|\phi_1(A) - \phi_1(B)\|_2 = \|A - B\|_1$, $\phi_1$ is a linear isomorphism. Hence $\phi$ is an affine isomorphism of $\mathbb{R}^n$. $\square$

**Theorem 3.2.** *Let* $(M, F)$ *be a Finsler space and* $\phi$ *a distance-preserving map of* $M$ *onto itself. Then* $\phi$ *is a diffeomorphism.*

*Proof.* Let $p \in M$ and put $q = \phi(p)$. Let $r > 0$, $\varepsilon > 0$ be so small that both $\exp_p$ and $\exp_q$ are $C^1$ diffeomorphisms on the tangent balls $B_p(r + \varepsilon)$, $B_q(r + \varepsilon)$ of $T_p(M)$ and $T_q(M)$, respectively. For any nonzero $X \in T_p(M)$, consider the radial geodesic $\exp_p(tX)$, $0 \leq t \leq \frac{r}{2F(p,X)}$. The image $\gamma(t) = \phi(\exp_p(tX))$ is a geodesic, since $\phi$ is distance-preserving. Let $X'$ denote the tangent vector of $\gamma$ at the point $q$. We have obtained a map $X \to X'$ of $T_p(M)$ into $T_q(M)$. Denoting this map by $\phi'$, we have $\phi'(\lambda X) = \lambda \phi'(X)$, for $X \in T_p(M)$ and $\lambda \geq 0$. Let $A, B \in T_p(M)$, $A \neq B$, and let $t$ be so small that both $tA$ and $tB$ lie in $B_p(r)$. Let $a_t = \exp_p(tA)$, $b_t = \exp_p(tB)$. Then by Theorem 3.1 we have

$$\lim_{t \to 0^+} \frac{F(p, tA - tB)}{d(a_t, b_t)} = 1.$$

On the other hand, by the definition of $\phi'$ we have

$$\exp_q(\phi'(tX)) = \phi(\exp tX),$$

for any $X$ and $t$ small enough. Thus by Theorem 3.1 we also have

$$\lim_{t \to 0^+} \frac{F(q, \phi'(tA) - \phi'(tB))}{d(\phi(a_t), \phi(b_t))} = 1.$$

Since $d(\phi(a_t), \phi(b_t)) = d(a_t, b_t)$, we get

$$\begin{aligned}
1 &= \lim_{t \to 0^+} \frac{F(p, tA - tB)}{F(q, \phi'(tA) - \phi'(tB))} \\
&= \lim_{t \to 0^+} \frac{tF(p, A - B)}{tF(q, \phi'(A) - \phi'(B))} = \frac{F(p, A - B)}{F(q, \phi'(A) - \phi'(B))}.
\end{aligned}$$

Therefore $F(q, \phi'(A) - \phi'(B)) = F(p, A - B)$. By Proposition 3.1, $\phi'$ is a diffeomorphism of $T_p(M)$ onto $T_q(M)$.

Although on $\mathscr{B}_p^+(r) = \exp_p B_r(p)$, we have $\phi = \exp_q \circ \phi' \circ (\exp_p)^{-1}$, we still cannot conclude that $\phi$ is smooth on $\mathscr{B}_p^+(r)$, since in a Finsler space the exponential map is only $C^1$ at the zero section. That is, we can conclude only that $\phi$ is smooth in $\mathscr{B}_p(r) \backslash \{p\}$. To finish the proof, we proceed to take $r$ so small that every pair of points in $\mathscr{B}_p^+(r) \cap \mathscr{B}_p^-(r)$ can be joined by a unique minimizing geodesic. Select $p_1 \in \mathscr{B}_p^+(\frac{r}{2}) \cap \mathscr{B}_p^-(\frac{r}{2})$, $p_1 \neq p$. Consider the tangent ball $B_{p_1}(\frac{r}{2})$ of $T_{p_1}(M)$. The exponential map is a $C^1$ diffeomorphism from $B_{p_1}(\frac{r}{2})$ onto $\mathscr{B}_{p_1}^+(\frac{r}{2})$. The above argument shows that $\phi$ is smooth in $\mathscr{B}_{p_1}^+(\frac{r}{2}) \backslash \{p_1\}$, which is a neighborhood of $p$. This completes the proof of the theorem.                                                      $\square$

Now we study Killing vector fields of Finsler spaces. A smooth vector field $X$ on a Finsler space $(M, F)$ is called a Killing vector field if every local one-parameter transformation group $\varphi_t$ of $M$ generated by $X$ consists of local isometries of $M$.

In the following we will give a geometric description of Killing vector fields, using Chern's orthonormal frame bundle. Let us first introduce the construction of Chern's orthonormal frame bundle of a Finsler space (see [150] for the details). Let $p \in M$. A Chern's orthonormal frame at $p$ is a frame (i.e., a basis of the linear space $T_p(M)$) $\{X_0, X_1, \ldots, X_{n-1}\}$ on $T_p(M)$ such that

(i)  $F(X_0) = 1$.
(ii) The vectors $X_0, X_1, \ldots, X_{n-1}$ form an orthonormal basis of $T_p(M)$ with respect to the inner product $g_{X_0}$, where $g$ is the fundamental form of $F$.

The set of all Chern's orthonormal frames is denoted by $O_F(M)$ and is called Chern's orthonormal frame bundle of $(M,F)$. It is a subbundle of the linear frame bundle $L(M)$ but in general not a principal subbundle of $L(M)$.

The following proposition is a result of [150].

**Proposition 3.2.** *A diffeomorphism* $f : M \to M$ *is an isometry of* $(M,F)$ *if and only if the induced diffeomorphism* $\hat{f}$ *of* $f$ *on* $L(M)$ *preserves Chern's orthonormal frame bundle, i.e.,* $\hat{f}(O_F(M)) \subset O_F(M)$.

Now we prove our next result.

**Proposition 3.3.** *A vector field* $X$ *on a Finsler space* $(M,F)$ *is a Killing vector field if and only if the natural lift* $\hat{X}$ *of* $X$ *to* $L(M)$ *is tangent to Chern's orthonormal frame bundle* $O_F(M)$ *at every point of* $O_F(M)$.

*Proof.* Recall that the natural lift $\hat{X}$ can be obtained in the following way [102, vol. 1, p. 229]: For any point $x \in M$, let $\varphi_t$ be a local one-parameter group of local transformations generated by $X$ in a neighborhood $U$ of $x$. For each $t$, $\varphi_t$ induces a map $\hat{\varphi}_t$ of $\pi^{-1}(U)$ onto $\pi^{-1}(\varphi_t(U))$ in a natural manner, where $\pi : L(M) \to M$ is the natural projection. The local one-parameter transformation groups $\{\hat{\varphi}_t\}$ of $L(M)$ obtained in this way induce a vector field on $L(M)$, which is exactly $\hat{X}$. If $X$ is a Killing vector field, then $\varphi_t$ are all local isometries of $(M,F)$. By Proposition 3.2, $\hat{\varphi}_t$ maps $O_F(M)$ into $O_F(M)$. Thus $\hat{X}$ is tangent to $O_F(M)$ at every point of $O_F(M)$. On the other hand, if $\hat{X}$ is tangent to $O_F(M)$ at every point of $O_F(M)$, then for any $u \in O_F(M)$, the curve $\hat{\varphi}_t(u)$, as an integral curve through $u$ of the vector field $\hat{X}$, must be contained in $O_F(M)$. Thus $\hat{\varphi}_t(O_F(M)) \subset O_F(M)$. By Proposition 3.2, $\varphi_t$ are isometries of $(M,F)$. Hence $X$ is a Killing vector field. □

The proposition implies an important fact, namely that the linear space of all Killing vector fields of $(M,F)$ is closed under the Lie brackets of vector fields. In fact, if $X_1, X_2$ are Killing vector fields, then the corresponding natural lift to $L(M)$, $\hat{X}_1, \hat{X}_2$ are tangent to $O_F(M)$ at every point of $O_F(M)$. Since $O_F(M)$ is a submanifold of $L(M)$, $[\hat{X}_1, \hat{X}_2]$ is tangent to $O_F(M)$ at every point of $O_F(M)$. It is obvious that $[\hat{X}_1, \hat{X}_2]$ is the natural lift of $[X_1, X_2]$. Therefore $[X_1, X_2]$ is also a Killing vector field. This proves our assertion. In the following, we will denote the Lie algebra formed by all Killing vector fields by $\mathfrak{k}(M,F)$ (or simply $\mathfrak{k}(M)$).

The following result will be useful later.

**Proposition 3.4.** *Let* $(M,F)$ *be a Finsler space and let* $\sigma(t)$, $a \leq t \leq b$, *be a geodesic. Let* $X$ *be a Killing vector field. Then the restriction of* $X$ *to* $\sigma$ *is a Jacobi field along* $\sigma$.

*Proof.* For any $t \in [a,b]$, we can find a neighborhood $N_t$ of $\sigma(t)$ and a positive number $\varepsilon_t$ such that $X$ generates a local one-parameter transformation group of $M$ that is defined on $[-\varepsilon_t, \varepsilon_t] \times N_t$. Since the set $C = \{\sigma(t) \mid a \leq t \leq b\}$ is compact, we can find a finite number of such open sets $N_t$ whose union covers $C$. Therefore, we can find a positive number $\varepsilon$ such that the local one-parameter transformation group generated by $X$ is defined on $[-\varepsilon, \varepsilon] \times V$, where $V$ is an open subset of $M$

containing $C$. More precisely, we have a map $\psi_s$ of $[-\varepsilon, \varepsilon] \times V$ into $M$ that satisfies the following conditions:

1. For each $s \in [-\varepsilon, \varepsilon]$, $\psi_t : p \to \psi_t(p)$ is a diffeomorphism of $V$ onto the open set $\psi_s(U)$ of $M$.
2. If $s_1, s_2, s_1 + s_2 \in [-\varepsilon, \varepsilon]$, and if $p, \psi_s(p) \in \psi_s(V)$, then

$$\psi_{s_1+s_2}(p) = \psi_{s_1}(\psi_{s_2}(p)).$$

The induced vector field of $\psi_s$ on $V$ is equal to the restriction of $X$. Therefore $\psi_s$ are local isometries.

Now we define a smooth variation of $\sigma$ by

$$\sigma(t,s) = \psi_s(\sigma(t)), \quad a \le t \le b, \ -\varepsilon < s < \varepsilon.$$

Then all the $t$-curves are geodesics. This can be seen from the following observation. Note that $\sigma$ is locally minimizing, since it is a geodesic. Therefore for any fixed $s \in (-\varepsilon, \varepsilon)$, $\psi_s$ being a local isometry, the curve $\psi_s(t)$, $a \le t \le b$, is locally minimizing. Thus $\psi_s(t)$, $a \le t \le b$, is a geodesic. This proves our assertion. Now it follows from Proposition 1.7 that the variation vector field of this variation, which is just the restriction of $X$ to $\sigma$, is a Jacobi field.                                    $\square$

## 3.2   The Group of Isometries

Theorem 3.2 justifies the following definition of an isometry for a Finsler space.

**Definition 3.1.** Let $(M, F)$ be a Finsler space. A map $\phi$ of $M$ onto itself is called an isometry if $\phi$ is a diffeomorphism and for any $x \in M$ and $X \in T_x(M)$, we have $F(\phi(x), d\phi_x(X)) = F(x, X)$.

In the following we denote the group of isometries of $(M, F)$ by $I(M, F)$, or simply $I(M)$ if the metric is clear.

Let $(N, d)$ be a connected, locally compact metric space and $I(N, d)$ the group of isometries of $(N, d)$. For $x \in N$, denote by $I_x(N, d)$ the isotropy subgroup of $I(N, d)$ at $x$. Van Danzig and van der Waerden [156] proved that $I(N, d)$ is a locally compact topological transformation group on $N$ with respect to the compact-open topology and that the isotropic subgroup $I_x(N, d)$ is compact.

Now on $M$ we have a distance function $d$ defined by the Finsler function $F$. By Theorem 3.2, the group $I(M, F)$ coincides with the group of isometries $I(M, d)$ of $(M, d)$. Although generically $d$ is not a distance, we still have the following result.

**Theorem 3.3.** *Let $(M, F)$ be a connected Finsler space. The compact-open topology turns $I(M)$ into a locally compact transformation group of $M$. Let $x \in M$ and $I_x(M)$ denote the subgroup of $I(M)$ that leaves $x$ fixed. Then $I_x(M)$ is compact.*

*Proof.* A proof of this result for the Riemannian case was given in Helgason [83, pp. 201–204], which is valid in the general cases after some minor changes. Just note that on a Finsler manifold the topology generated by the forward metric balls $\mathscr{B}_p^+(r) = \{x \in M \mid d(p,x) < r\}$, $p \in M$, $r > 0$, is precisely the underlying manifold topology and this is true for the topology generated by the backward metric balls $\mathscr{B}_p^-(r) = \{x \in M \mid d(x,p) < r\}$, $p \in M$, $r > 0$. □

Bochner–Montgomery [29] proved that a locally compact group of differentiable transformations of a manifold is a Lie transformation group. Therefore we have the following theorem.

**Theorem 3.4.** *Let $(M,F)$ be a Finsler space. Then the group of isometries $I(M,F)$ of $M$ is a Lie transformation group of $M$. Let $x \in M$ and let $I_x(M,F)$ be the isotropy subgroup of $I(M,F)$ at $x$. Then $I_x(M,F)$ is compact.*

Now we consider the relationship between the group of isometries $I(M)$ and the Lie algebra of all the Killing fields of $(M,F)$. Recall that a smooth vector field $X$ on $M$ is called complete if the local one-parameter transformation groups generated by $X$ can be extended to global one-parameter transformations. Denote the set of all the complete Killing vector fields of $(M,F)$ by $\mathfrak{i}(M,F)$. We have the following theorem.

**Theorem 3.5.** *Let $(M,F)$ be a connected Finsler space. Then $\mathfrak{i}(M,F)$ is a Lie algebra naturally isomorphic to the Lie algebra of the Lie group $I(M,F)$.*

*Proof.* If $X \in \mathfrak{i}(M,F)$, then the global one-parameter group of transformations generated by $X$ is contained in $I(M,F)$. On the other hand, if $Y$ is an element of the Lie algebra of $I(M,F)$, then the vector field $\widetilde{Y}$ defined by

$$\widetilde{Y}_x = \frac{d}{dt} \exp(tY) \cdot x \big|_{t=0}$$

is a Killing vector field of $(M,F)$. It is obvious that $\widetilde{Y}$ is complete. From this the theorem follows. □

Since on a compact connected manifold every smooth vector field must be complete, we have the following corollary.

**Corollary 3.1.** *Let $(M,F)$ be a compact connected Finsler space. The Lie algebra of all the Killing vector fields of $(M,F)$ is naturally isomorphic to the Lie algebra of the full group of isometries of $(M,F)$.*

## 3.3 A Theorem of H. C. Wang

In the previous sections we have studied the group of isometries of a Finsler space. A natural problem is whether there is a bound on the dimension of the Lie group $I(M,F)$ for an $n$-dimensional Finsler space $(M,F)$. In [163], Wang proved the following important theorem:

**Theorem 3.6 (H.C. Wang).** *If an n-dimensional ($n > 2$, $n \neq 4$) Finsler space $(M, F)$ admits a group $G$ of motions depending on $r > \frac{1}{2}n(n-1) + 1$ essential parameters, then $(M, F)$ is a Riemannian space of constant curvature.*

In view of Theorem 3.4, the above theorem can be restated as:

**Theorem 3.6′** Let $(M, F)$ be an $n$ ($n > 2, n \neq 4$) dimensional Finsler space. If the group of isometries $I(M, F)$ has dimension $> \frac{1}{2}n(n-1) + 1$, then $(M, F)$ is a Riemannian space of constant curvature.

*Proof.* Let $x$ be an arbitrary point in $M$ and let $I_x(M)$ be the subgroup of $I(M)$ that leaves $x$ fixed. Then by Theorem 3.3, $I(M)$ is a Lie transformation group of $M$ with respect to the compact-open topology and $I_x(M)$ is a compact subgroup of $I(M)$. Each $\phi \in I_x(M)$ induces a linear isometry $d\phi_x$ on the Minkowski space $T_x(M)$, and the correspondence $\phi \to d\phi_x$ is a homomorphism from $I_x(M)$ into $\mathrm{GL}(T_x(M))$. It is obvious that this homomorphism is one-to-one. Denote by $I_x^*(M)$ the group consisting of the image of this homomorphism. Let $I \cdot x$ be the orbit of $x$ under the action of $I(M)$. If $\dim I(M) > \frac{1}{2}n(n-1) + 1$, then

$$\dim I_x^*(M) = \dim I_x(M) \geq \dim I(M) - \dim(I \cdot x)$$
$$> \frac{1}{2}n(n-1) + 1 - n = \frac{1}{2}(n-1)(n-2).$$

Now fix a basis of the linear space $T_x(M)$. Then $I_x^*(M)$ is a compact subgroup of $\mathrm{GL}(n, \mathbb{R})$. We assert that the unit component $(I_x^*(M))_e$ of $I_x^*(M)$ is a subgroup of $\mathrm{SL}(n, \mathbb{R})$. In fact, the determinant function is continuous on $\mathrm{GL}(n, \mathbb{R})$, hence must be bounded on the compact subgroup $(I_x^*(M))_e$. Therefore each element in $(I_x^*(M))_e$ has determinant $\leq 1$. On the other hand, if $g \in (I_x^*(M))_e$ has determinant $< 1$ (and of course $> 0$), then $g^{-1} \in (I_x^*(M))_e$ has determinant $> 1$, which is impossible. This proves our assertion. Now, $\mathrm{SL}(n, \mathbb{R})$ is a connected semisimple Lie group and $\mathrm{SO}(n)$ is a maximal subgroup. By the conjugacy of maximal compact subgroups of semisimple Lie groups (see [83, p. 256]), there exists $g \in \mathrm{SL}(n, \mathbb{R})$ such that $g^{-1}(I_x^*(M))_e g \subset \mathrm{SO}(n)$. Note that $\dim(I_x^*(M))_e > \frac{1}{2}(n-1)(n-2)$. According to a lemma of Montgomery and Samelson [121], if $n > 2$ and $n \neq 4$, then $O(n)$ contains no proper subgroup of dimension $> \frac{1}{2}(n-1)(n-2)$ other then $\mathrm{SO}(n)$. Therefore

$$g^{-1}(I_x^*(M))_e g = \mathrm{SO}(n). \tag{3.1}$$

Consider the hypersurface $g \cdot S^n$ of $T_x(M)$ (where $S^n$ is defined by the inner product determined by assuming the above basis to be orthonormal). The group $(I_x^*(M))_e$ acts transitively on it. Hence $F$ is constant on this surface. Therefore $F|_{T_x(M)}$ comes from an inner product of $T_x(M)$. Since $x$ is arbitrary, $F$ is Riemannian. Moreover, it is easily seen from (3.1) that $(I_x^*(M))_e$ acts transitively on the set of planes in $T_x(M)$. Therefore $(M, F)$ is of constant curvature. $\square$

*Remark 3.1.* H.C. Wang's proof of this theorem in [163] is elegant but needs some complicated reasoning. The above proof was given by X. Chen and the author in [39]. This proof is simpler and more direct compared to Wang's original one.

*Remark 3.2.* Recently, V.S. Matveev and M. Troyanov proved that H.C. Wang's theorem holds for all dimensions. They also considered the more generalized class of Finsler metrics that are not necessary smooth. See [116] for the details.

Next we consider the $n$-dimensional ($n > 4$) Finsler spaces whose full groups of isometries have dimension $\frac{1}{2}n(n-1)+1$. Such spaces were studied by Szabó in [151]. We first prove a lemma.

**Lemma 3.2.** *Let* $(M,F)$ *be an n-dimensional connected Finsler space with*

$$\dim I(M,F) = \frac{1}{2}n(n-1)+1,$$

*where* $n > 4$. *Then* $(M,F)$ *must be a homogeneous Finsler space; namely, the group of isometries,* $I(M,F)$, *acts transitively on M.*

*Proof.* Denote by $G$ the unit component of the group $I(M,F)$. Fix a point $x \in M$ and let $H$ be the isotropy subgroup of $G$ at $x$. If the action of $G$ on $M$ is not transitive, then the orbit $G \cdot x$ has dimension less than $n$. Since $G \cdot x = G/H$, we have

$$\dim H > \frac{1}{2}n(n-1)+1-n = \frac{1}{2}(n-1)(n-2).$$

Note that $H$ is isomorphic to a subgroup of the group of linear isometries of the Minkowski space $(T_x(M),F)$. Thus it can also be viewed as a subgroup of $O(n)$. Let $H_0$ be the unit component of $H$. Then $H_0$ is a connected compact subgroup of $O(n)$, and it has dimension larger than $\frac{1}{2}(n-1)(n-2)$. According to Montgomery–Samelson [121], $O(n)$ has no connected compact subgroup of dimension $> \frac{1}{2}(n-1)(n-2)$ other than $SO(n)$. Hence $H_0 = SO(n)$. Thus the action of $H_0$ on $T_x(M)$ is transitive on the unit sphere of $T_x(M)$ with respect to a certain inner product. Hence $F$ must be a Riemannian manifold. Then $(M,F)$ must be a Riemannian globally symmetric space of rank one. In particular, it must be a homogeneous Riemannian manifold. This is a contradiction. Hence the action of $G$ on $M$ must be transitive. □

By Lemma 3.2, $M$ can be written as a coset space $G/H$, where $G$ is the unit component of the full group $I(M,F)$ of isometries and $H$ is the isotropy subgroup of $G$ at a certain $x \in M$. Moreover, $F$ can be viewed as a $G$-invariant Finsler metric on $G/H$. Since $H$ is a compact subgroup of $G$, there exists an inner product in $T_x(M)$ that is invariant under the action of $H$. Then by Proposition 2.8, one can construct a $G$-invariant Riemannian metric $Q$ on $M$. Then $(M,Q)$ is an $n$-dimensional Riemannian manifold whose isometry group has dimension $\geq \frac{1}{2}n(n-1)+1$. By Theorem 3.6, to classify the non-Riemannian Finsler spaces, it suffices to consider the case in which $I(M,Q)$ has dimension $\frac{1}{2}n(n-1)+1$. Riemannian manifolds with the above properties have been studied by Yano [181], Kuiper [109], Obata [125], and Kobayashi [101]. The results imply that $(M,Q)$ and $G$ must be one of the following:

(1) $M = \mathbb{R} \times V$ and $G = \mathbb{R} \times I_0(V)$, or $M = S^1 \times V$ and $G = S^1 \times I_0(V)$, where $V$ is one of the following: $V = \mathbb{R}^{n-1}$, $V = H^{n-1}$ (hyperbolic space), $V = S^{n-1}$, and $V = P^{n-1}(\mathbb{R})$ (projective space); here $I_0(V)$ denotes the unit component of the full group of isometries of the standard metric on $V$.
(2) $(M,Q)$ is the hyperbolic space and $G$ is a subgroup of $I(M,Q)$ that leaves a family of parallel straight lines invariant.
(3) $n = 8$, $M = \mathbb{R}^8$, and $G = \mathbb{R}^8 \cdot \mathrm{Spin}(7)$ (semidirect product), where $\mathbb{R}^8$ denotes the translation group on the Euclidean space $\mathbb{R}^8$ and $\mathrm{Spin}(7)$ is considered as a subgroup of $\mathrm{SO}(8)$.

Therefore the problem of the classification of $n$-dimensional non-Riemannian Finsler spaces whose groups of isometries have order $\frac{1}{2}n(n-1)+1$, $n > 4$, reduces to the problem of finding all the $G$-invariant non-Riemannian Finsler metrics on $M$ in the above cases (1), (2), and (3). Note that in case (3), the subgroup $\mathrm{Spin}(7)$ must be contained in the isotropy subgroup of $G$ at the origin of $\mathbb{R}^8$, and the action of $\mathrm{Spin}(7)$ is transitive on the unit sphere. Therefore, in this case, any $G$-invariant Finsler metric on $M$ must be Riemannian. Consequently, we have the following.

**Theorem 3.7.** *If $(M,F)$ is an n-dimensional $(n > 4)$ non-Riemannian Finsler space with $\dim I(M,F) = \frac{1}{2}n(n-1)+1$, then $M$ and $G = I_0(M,F)$ must be one of* (1) *and* (2) *above.*

It can be shown that in case (1), all the metrics must be Berwaldian and affinely globally symmetric (see Chap. 5 below for the definition), and in case (2), the space must be a $\mathrm{BLF}^n$-space; see [151] for the details.

## 3.4   Two-Point Homogeneous Spaces

In this section we study two-point homogeneous Finsler spaces. We first give the definition.

**Definition 3.2.** Let $(M,F)$ be a connected Finsler manifold. We say that $(M,F)$ is two-point homogeneous if for every two pairs of points $(p_1,q_1)$ and $(p_2,q_2)$ satisfying $d(p_1,q_1) = d(p_2,q_2)$, there exists an isometry $\sigma$ of $(M,F)$ such that $\sigma(p_1) = p_2$ and $\sigma(q_1) = q_2$.

The main result of this section is the following:

**Theorem 3.8.** *A connected two-point homogeneous Finsler space must be Riemannian.*

In the following we will give the proof of this theorem. We first consider the special case of Minkowski spaces. In this case it is closely related to a classical problem in functional analysis, namely the Banach–Mazur rotation problem. This problem can be stated as follows. Let $X$ be a separable Banach space. Suppose the group of linear isometries of $X$ acts transitively on the unit sphere of $X$. Is $X$ necessarily a Hilbert space? The finite-dimensional case was solved affirmatively by Mazur in [117]. Note that the norm in the classical sense must be reversible, but not necessarily smooth on the slit space $V \setminus \{0\}$.

Now we generalize Mazur's result to the nonreversible case, but under the assumption that the norm is Minkowskian.

**Proposition 3.5.** *Let $(\mathbb{R}^n, F)$ be a Minkowski space. If $(\mathbb{R}^n, F)$ is two-point homogeneous when viewed as a Finsler space in the canonical way, then $F$ is the Euclidean norm of an inner product.*

*Proof.* Let $I$ denote the group of isometries of $(\mathbb{R}^n, F)$, and let $I_o$ be the subgroup of $I$ that leaves the origin $o$ fixed. Then by Theorem 3.4, $I$ is a Lie group with respect to the compact-open topology and $I_o$ is a compact subgroup of $I$. Furthermore, from the proof of Proposition 3.1, it is easily seen that $I_o$ consists of linear transformations of $\mathbb{R}^n$. Thus $I_o$ is a compact subgroup of $\mathrm{GL}(n, \mathbb{R})$. Let $I_o^e$ be the unit component of $I_o$. Then a similar argument as in the proof of Theorem 3.6 shows that $I_o^e$ is a subgroup of $\mathrm{SL}(n, \mathbb{R})$. Moreover, there exists $g \in \mathrm{SL}(n, \mathbb{R})$ such that $g^{-1} I_o^e g \subset \mathrm{SO}(n)$. Now consider the indicatrix

$$\mathscr{I}_o = \{ y \in \mathbb{R}^n \mid F(y) = 1 \}.$$

Since $(\mathbb{R}^n, F)$ is two-point homogeneous, for any two points $y_1, y_2 \in \mathscr{I}_0$, there exists $\phi \in I_o$ such that $\phi(y_1) = y_2$. That is, $I_o$ is transitive on $\mathscr{I}_o$. We assert that $I_o^e$ is also transitive on $\mathscr{I}_o$. In fact, for any $y \in \mathscr{I}_o$, the orbit $I_o^e \cdot y$ is an open and closed subset of $I_o \cdot y = \mathscr{I}_0$. Since $\mathscr{I}_o$ is connected, we conclude that $I_o^e \cdot y = \mathscr{I}_o$. Now for any $x_1, x_2 \in S^{n-1}$ (here $S^{n-1}$ is defined with respect to the standard inner product of $\mathbb{R}^n$), we have

$$\frac{g(x_1)}{F(g(x_1))} \in \mathscr{I}_o, \quad \frac{g(x_2)}{F(g(x_2))} \in \mathscr{I}_o.$$

Thus there exists $g_1 \in I_o^e$ such that

$$g_1 \left( \frac{g(x_1)}{F(g(x_1))} \right) = \frac{g(x_2)}{F(g(x_2))}.$$

Then we have

$$g_1 g(x_1) = \frac{F(g(x_1))}{F(g(x_2))} g(x_2).$$

Thus

$$g^{-1} g_1 g(x_1) = \frac{F(g(x_1))}{F(g(x_2))} x_2.$$

By the facts that $g^{-1}g_1g \in SO(n)$, $x_1, x_2 \in S^{n-1}$, and $\frac{F(g(x_1))}{F(g(x_2))} > 0$, one easily deduces that $F(g(x_1))/F(g(x_2)) = 1$. Thus

$$g^{-1}g_1g(x_1) = x_2.$$

This means that $g^{-1}I_o^e g$ is transitive on $S^{n-1}$, or in other words, $I_o^e$ is transitive on the hypersurface $g \cdot S^{n-1}$. Hence $F$ is constant on $g \cdot S^{n-1}$. Suppose $F(g \cdot S^{n-1}) = \lambda > 0$. Then it is easy to check that $F$ coincides with the Euclidean norm of the inner product defined by

$$\langle y_1, y_2 \rangle_1 = \lambda \langle g^{-1} \cdot y_1, g^{-1} \cdot y_2 \rangle,$$

where $\langle \, , \rangle$ is the standard inner product of $\mathbb{R}^n$.                                         □

*Proof of Theorem 3.8.* Let $(M, F)$ be a connected two-point homogeneous Finsler manifold and $p \in M$. We proceed to prove that $(T_p(M), F)$ is a two-point homogeneous Minkowski space. Since for any $y \in T_p(M)$, the parallel translate $\pi_y$: $\pi_y(x) = x - y$ is an isometry with respect to $F$ and $F$ is positively homogeneous of degree one, we need only show that the isotropy subgroup $I_o(T_p(M))$ of $I(T_p(M))$ at the origin acts transitively on the set

$$\mathscr{I}_o(\lambda) = \{x \in T_p(M) | F(x) = \lambda\},$$

for some $\lambda > 0$. Now we select $r > 0$, $\varepsilon > 0$ such that the exponential map $\exp_p$ is a $C^1$-diffeomorphism from the tangent ball

$$B_p(r + \varepsilon) = \{x \in T_p(M) | F(x) < r + \varepsilon\}$$

onto its image. Let $S_p(r) = \{x \in T_p(M) | F(x) = r\}$, $\mathscr{S}_p^+(r) = \{q \in M | d(p, q) = r\}$. Then we have

$$\exp_p[S_p(r)] = \mathscr{S}_p^+(r).$$

Therefore, for any $y_1, y_2 \in \mathscr{I}_o(r)$, we have $\exp_p(y_1), \exp_p(y_2) \in \mathscr{S}_p^+(r)$. Since $(M, F)$ is two-point homogeneous, there exists an isometry $\sigma$ of $(M, F)$ such that $\sigma(p) = p$ and $\sigma(\exp_p(y_1)) = \exp_p(y_2)$. This means that the geodesics $\sigma(\exp_p(ty_1))$, $\exp_p(ty_2)$, $0 \leq t \leq 1$, coincide. Therefore we have

$$d\sigma_p(y_1) = y_2.$$

Since $d\sigma_p$ is a linear isometry of $(T_p(M), F)$, we conclude that $I_o(T_p(M))$ acts transitively on $\mathscr{I}_o(r)$. Thus $(I_p(M), F)$ is a two-point homogeneous Minkowski space. By Proposition 3.5, $F|_{T_p(M)}$ is the Euclidean norm of an inner product. Since $p$ is arbitrary, $F$ is Riemannian.                                         □

Compact two-point homogeneous Riemannian spaces were classified by Wang [164], and noncompact ones were classified by Tits [154]. By Theorem 3.8, their lists are also the classification of two-point homogeneous Finsler spaces. The classifications of Wang and Tits show that any two-point homogeneous Riemannian manifold must be globally symmetric. The first classification-free proof of this result was given by Wolf in 1962; see [177] for the details.

## 3.5 Fixed Points of Isometries

In this section we study the zero points of Killing vector fields in a Finsler space. Since a Killing vector field can be viewed as an infinitesimal isometry, we can also view the zero points as fixed points of isometries.

Let $(M,F)$ be a connected Finsler space. A submanifold $M_1$ of $M$ is called a totally geodesic submanifold if for every $x \in M_1$ and $y \in T_x(M_1)$, the maximal geodesic $\gamma$ with $\gamma(0) = x$ and $\dot{\gamma}(0) = y$ is contained in $M_1$.

**Theorem 3.9.** *Let $(M,F)$ be a connected Finsler space of dimension n, and $\xi$ a Killing vector field of $(M,F)$. Let V be the set of points in M where $\xi$ vanishes and let $V = \cup V_i$, where the $V_i$ are the connected components of V. Assume that V is not empty. Then we have*

1. *Each $V_i$ is a totally geodesic closed submanifold of M and the codimensions of the $V_i$ are even.*
2. *If $(M,F)$ is forwardly complete and $x \in V_i$, $y \in V_j$ with $i \neq j$, then there is a one-parameter family of geodesics connecting x and y. In particular, x and y are conjugate to each other.*
3. *If M is compact, then the Euler characteristic number of M is the sum of the Euler numbers of the $V_i$, i.e.,*

$$\chi(M) = \sum \chi(V_i).$$

*Proof.* (1) Suppose $x \in V$. Then we can take a neighborhood $U$ of $x$ such that $\xi$ generates a local one-parameter transformation group from $U$ into $M$. That is, there is a set of maps $\{\varphi_t \mid |t| < \varepsilon, \varepsilon > 0\}$ such that each $\varphi_t$ is a diffeomorphism from $U$ onto $\varphi_t(U)$ and

$$\varphi_t(\varphi_s(p)) = \varphi_{t+s}(p), \quad \text{for } p \in U, \ \varphi_s(p) \in U, \ |t+s| < \varepsilon.$$

Moreover, for any $p \in U$, $\xi(p)$ is just the initial vector of the curve $\varphi_t(p)$. Since $\xi(x) = 0$, every $\varphi_t$ keeps $x$ fixed. Therefore the differentials $(d\varphi_t)|_x$, $|t| < \varepsilon$, are linear automorphisms of the tangent space $T_x(M)$. Since $\varphi_t$ is a (local) isometry, $(d\varphi_t)|_x$ preserves the length of every vector in $T_x(M)$. Then Proposition 3.1 implies that $(d\varphi_t)|_x$ is a linear isometry of the Minkowski space $(T_x(M), F|_x)$. Let $L$ denote the group of linear isometries of $(T_x(M), F|_x)$. Then $L$ is a compact subgroup of the general linear group $\mathrm{GL}(T_x(M))$. Fix a basis $\{e_1, e_2, \ldots, e_n\}$ of $T_x(M)$. Then $\mathrm{GL}(T_x(M))$ is identified with the group $\mathrm{GL}(n, \mathbb{R})$ of all $n \times n$

real invertible matrices. Since $L$ is a compact subgroup of $\mathrm{GL}(n, \mathbb{R})$, the unit component $L_0$ of $L$ must be contained in the subgroup $\mathrm{SL}(n, \mathbb{R})$. Since $\mathrm{SL}(n, \mathbb{R})$ is a connected semisimple Lie group and $\mathrm{SO}(n)$ is a maximal subgroup of $\mathrm{SL}(n, \mathbb{R})$, there exists $g \in \mathrm{SL}(n, \mathbb{R})$ such that $g^{-1}L_0g \subset \mathrm{SO}(n)$. Now consider the subset $\{(d\varphi_t)|_x\}$ of $L$. It is obvious that it is contained in $L_0$. Thus $g^{-1}\{(d\varphi_t)|_x\}g$ is contained in a one-parameter subgroup of $\mathrm{SO}(n)$. Since every pair of maximal connected commutative subgroups (i.e., the maximal tori) are conjugate and the matrices of the form

$$
\begin{pmatrix}
\cos\theta_1 & \sin\theta_1 & & & & \\
-\sin\theta_1 & \cos\theta_1 & & & & \\
 & & \ddots & & & \\
 & & & \cos\theta_r & \sin\theta_r & \\
 & & & -\sin\theta_r & \cos\theta_r & \\
 & & & & & I_{n-2r}
\end{pmatrix}, \quad r \geq 0,\ \theta_i \in \mathbb{R},
$$

constitute a maximal torus of $\mathrm{SO}(n)$, there are $g_1 \in \mathrm{SO}(n)$, $s \geq 0$, and $\gamma_i \in \mathbb{R}$, $\gamma_i \neq 0$, such that

$$
g_1^{-1}(g^{-1}\{(d\varphi_t)|_x\}g)g_1 =
\begin{pmatrix}
\cos t\gamma_1 & \sin t\gamma_1 & & & & \\
-\sin t\gamma_1 & \cos t\gamma_1 & & & & \\
 & & \ddots & & & \\
 & & & \cos t\gamma_s & \sin t\gamma_s & \\
 & & & -\sin t\gamma_s & \cos t\gamma_s & \\
 & & & & & I_{n-2s}
\end{pmatrix}.
$$

Thus there is a basis $\{w_1, w_2, \ldots, w_n\}$ of $T_x(M)$ such that the matrix of $(d\varphi_t)|_x$ under this basis is like the right-hand side of the above equation.

If $n - 2s = 0$, then $n$ is even and $x$ is an isolated fixed point of $\varphi_t$, hence an isolated zero point of $\xi$. Suppose $n - 2s > 0$. Then for every vector $v \in S$, where $S$ is the span of $w_{2s+1}, w_{2s+2}, \ldots, w_n$, the geodesic emanating from $x$ with the direction $v$ must be kept fixed by the one-parameter group $\varphi_t$. Now we assert that there exists a neighborhood $W$ of $x$ such that the set of such geodesics forms an $(n - 2s)$-dimensional submanifold $W'$ of $W$. Let $c$ be a positive number and

$$
\mathcal{B}_x^+(c) = \{y \in M \mid d(x, y) < c\}, \quad \mathcal{B}_x^-(c) = \{y \in M \mid d(y, x) < c\}.
$$

Then for sufficiently small $c$, the exponential map is a $C^1$ diffeomorphism from $B_x(c) = \{v \in T_x(M) \mid F(v) < c\}$ onto $\mathcal{B}_x^+(c)$. Therefore the exponential map is a $C^1$ diffeomorphism from $S \cap B_x(c)$ onto $\exp(S \cap B_x(c))$. Hence the latter is an $(n - 2s)$-dimensional $C^1$ submanifold of $\mathcal{B}_x^+(c)$. To prove our assertion we need only proceed to take $c$ so small that every pair of points in $\mathcal{B}_x^+(c) \cap \mathcal{B}_x^-(c)$ can be joined by a unique minimizing geodesic. Fix an arbitrary point $z$ in

$\exp(S \cap B_x(c)) \cap (\mathscr{B}_x^+(c) \cap \mathscr{B}_x^-(c))$ and consider the exponential map at $z$. Since each $\varphi_t$ (take $c$ small enough that $\mathscr{B}_x^+(c) \subset U$) keeps $z$ fixed, it is easily seen that there exists an $(n - 2s)$-dimensional subspace $S'$ of $T_z(M)$ such that exp is a (smooth) diffeomorphism from $S' \cap B_z(c) - \{0\}$ onto $\exp(S' \cap B_z(c) - \{0\})$, which is a neighborhood of $x$ in $\exp(S \cap B_x(c))$. This proves our assertion. Now we show that any zero point of $\xi$ in $\mathscr{B}_x^+(c) \cap \mathscr{B}_x^-(c)$ must be in $\exp(S \cap B_x(c))$. Let $y$ be a zero point of $\xi$ in $\mathscr{B}_x^+(c) \cap \mathscr{B}_x^-(c)$. Select a geodesic $\tau$ joining $x$ and $y$. Since $x$ and $y$ are zero points of $\xi$, $\varphi_t$ keeps $x$ and $y$ fixed. Hence $\varphi_t$ must keep the geodesic $\tau$ pointwise fixed. Therefore the initial vector of $\tau$ at $x$ must be in $S$. This proves that $y$ is in $\exp(S \cap B_x(c))$. Hence each $V_i$ is a closed submanifold and the codimension is even. From the above arguments we easily see that each $V_i$ must be a totally geodesic submanifold.

(2) Suppose $x \in V_i$, $y \in V_j$, $i \neq j$, and $(M, F)$ is forwardly complete. By the Hopf–Rinow theorem, there is a geodesic $\tau$ joining $x$ and $y$. By Proposition 3.1.7, the restriction of the Killing vector field $\xi$ to $\tau$ is a Jacobi field along $\tau$. Hence $\xi|_\tau$ is the variation vector field of a one-parameter family of geodesics. Since $x$ and $y$ are zero points of $\xi$, this family of geodesics can be taken to start from $x$ and to end at $y$. However, since $i \neq j$, this family of geodesics cannot be left fixed by the local transformation group generated by $\xi$. Otherwise there would be a curve consisting of zero points of $\xi$ and joining $V_i$ and $V_j$, contradicting the definition of $V_i$.

(3) Since $M$ is compact, every vector field on $M$ is complete. Hence $\xi$ generates a global one-parameter group of transformations $\varphi_t$, $-\infty < t < \infty$. Every $\varphi_t$ is an isometry of the Finsler space $(M, F)$. Let $G$ be the full group of isometries of $(M, F)$. Then $G$ is a Lie transformation group of $M$, and for every $x \in M$, the isotropic subgroup $G_x$ of $G$ at $x$ is a compact subgroup of $G$. Since $M$ is compact, $G$ is a compact Lie group. Let $G_0$ be the unit component of $G$ and $d\mu$ the standard normalized bi-invariant Haar measure of $G_0$. Fix any Riemannian metric $h$ on $M$ and for $x \in M$, $v_1, v_2 \in T_x(M)$ define

$$h_1(x)(v_1, v_2) = \int_{G_0} h(g(x))(dg|_x(v_1), dg|_x(v_2)) \, d\mu(g).$$

Then it is easily seen that $h_1$ is a Riemannian metric on $M$. In fact, the only point we need to check is the smoothness of $h_1$. But this follows easily from the smoothness of the action of $G$ on $M$. By definition, $h_1$ is invariant under the action of $G_0$. Thus every element of $G_0$ is an isometry of $h_1$. It is easily seen that the one-parameter subgroup $\{\varphi_t\}$ must be contained in $G_0$. Thus for any $t$, $\varphi_t$ is an isometry of the Riemannian metric $h_1$. This means that the vector field $\xi$ is also a Killing vector field with respect to the Riemannian metric $h_1$. This reduces the proof of (3) to the Riemannian case.

Now we prove (3) under the assumption that $F$ is a Riemannian metric. The following proof is adapted from [99].

Let $\varepsilon$ be a small positive number. We define $S_x$ to be the set of points $y$ in $M$ such that there is a geodesic from $x$ to $y$ of length not greater than $\varepsilon$ and normal to $V_i$ at $x$.

Thus, to every point $x$ of $V_i$, we attach a solid sphere $S_x$ with center $x$ and radius $\varepsilon$ that is normal to $V_i$ and has dimension $2r$ (equal to the codimension of $V_i$). Let $N_i = \cup_{x \in V_i} S_x$. Taking $\varepsilon$ very small, we may assume that $N_i \cap N_j$ is empty if $i \neq j$ and that every point in $N_i$ lies exactly in one $S_x$. Let $N = \cap N_i$ and denote by $K$ the closure of $M - N$. Then $N \cap K$ is the boundary $dN$ of $N$. We now prove a lemma.

**Lemma 3.3.** $\chi(M) = \chi(N) + \chi(K) - \chi(dN)$.

*Proof.* It is easy to check that if the sequence

$$\to A_k \to B_k - C_k - A_{k-1} \to B_{k-1} \to \cdots$$

of vector spaces is exact, then we have

$$\Sigma = (-1)^k \dim A_k - \sum (-1)^k \dim B_k + \sum (-1)^k \dim C_k = 0.$$

Applying this formula to the exact sequences of homology groups induced by

$$K \to M \to (M, K)$$

and

$$dN \to N \to (N, dN),$$

we obtain

$$\chi(K) - \chi(M) + \chi(M, K) = 0$$

and

$$\chi(dN) - \chi(N) + \chi(N, dN) = 0.$$

By the excision axiom, $(M, K)$ and $(N, dN)$ have the same relative homology. Therefore

$$\chi(M, K) = \chi(N, dN).$$

This completes the proof of the lemma.

The 1-parameter group generated by $x_i$ has no fixed point in $K$ and none in $dN$. By Lefschetz's theorem, $\chi(K) = \chi(dN) = 0$. Hence $\chi(M) = \chi(N)$. Since $N_i$ is a fiber bundle over $V_i$, with solid sphere $S$ as fiber, we have

$$\chi(N_i) = \chi(V_i)\chi(S) = \chi(V_i).$$

Therefore we have

$$\chi(M) = \sum \chi(N_i) = \sum \chi(V_i). \qquad \square$$

*Remark 3.3.* The Riemannian case of Theorem 3.9 was obtained by Kobayashi in [99]. However, Kobayashi's original proof does not apply to the Finslerian case. The above proof was given by the author in [48].

Theorem 3.9 has several interesting corollaries.

**Corollary 3.2.** *Let $\mathfrak{k}$ be an abelian (finite-dimensional) Lie algebra consisting of Killing vector fields, and $V$ the set of points on which every element of $\mathfrak{k}$ vanishes. Then the same conclusions as in the main theorem hold.*

**Corollary 3.3.** *Under the same assumptions as in Corollary 3.2, if $(M, F)$ is a globally symmetric (resp. locally symmetric) Finsler space, then so is each $V_i$.*

*Remark 3.4.* A Finsler space $(M, F)$ is called globally symmetric if each point is the isolated fixed point of an involutive isometry. It is called locally symmetric if for every $x \in M$ there exists a neighborhood $U$ of $x$ such that the local geodesic symmetry is an isometry on $U$ (see Chap. 5 below).

**Corollary 3.4.** *Under the same assumptions as in Corollary 3.2, if $(M, F)$ is forward complete and the flag curvature is nonpositive, then $V$ is either empty or connected.*

**Corollary 3.5.** *Let $(M, F)$ be a compact Finsler space of dimension $2m$. Suppose a torus group of dimension $m$ acts on $M$ differentiably and effectively. Then the Euler number of $M$ is zero or positive according as the fixed point $V$ is empty or not. If $M$ is orientable and $V$ is nonempty, then the Euler number of $M$ is $\geq 2$.*

These corollaries can be proved similarly as in the Riemannian case. We omit the details here; see [99].

# Chapter 4
# Homogeneous Finsler Spaces

In this chapter we study homogeneous Finsler spaces. Let $(M, F)$ be a connected Finsler space. Theorem 3.4 asserts that the group of isometries of $(M, F)$, denoted by $I(M, F)$, is a Lie transformation of $M$. We say that $(M, F)$ is a homogeneous Finsler space if the action of $I(M, F)$ on $M$ is transitive. The purpose of this chapter is to give an exposition of the main results obtained by the author and collaborators on homogeneous Finsler spaces.

In the Riemannian case, there are numerous articles and books on this topic; see, for example, Ambrose–Singer [8], Kobayashi–Nomizu [102], Wolf [172, 174, 175, 177], just to name a few. Moreover, homogeneous Einstein–Riemannian manifolds have been studied rather extensively; see Besse's survey in [28] and the work of Ziller [182], Wang–Ziller [165, 166], and the references therein. One can also find more recent developments of the field in [75, 80, 111], etc.

Our first goal in this chapter is to give an algebraic description of a homogeneous Finsler space. To this end we introduce two new algebraic notions in Sect. 4.1: Minkowski Lie pairs and Minkowski Lie algebras. Then, in Sect. 4.2, we present a sufficient and necessary condition for a coset space to have invariant non-Riemannian Finsler metrics. In Sect. 4.3, we apply our result to study the degree of symmetry of closed manifolds. In particular, we prove that if a closed manifold is not diffeomorphic to a rank-one Riemannian symmetric space, then its degree of symmetry can be realized by a non-Riemannian Finsler metric. In Sect. 4.4, we study homogeneous Finsler spaces of negative curvature and prove that every homogeneous Finsler space with nonpositive flag curvature and negative Ricci scalar must be simply connected. Finally, in Sect. 4.5, we study fourth-root homogeneous Finsler metrics. As an explicit example, we give a classification of all the invariant fourth-root Finsler metrics on the Grassmannian manifolds.

There are many related topics that are not involved in this chapter, but deserve to be studied carefully. Homogeneous Einstein–Finsler spaces are not dealt with here. In fact, up to now, there have appeared only very few results in the literature, mainly on some special types of Finsler metrics; see Sects. 7.3 and 7.4 for some information on homogeneous Einstein–Randers metrics. On the other hand, the formulas for flag curvature and Ricci scalar of homogeneous Finsler spaces have not been obtained

S. Deng, *Homogeneous Finsler Spaces*, Springer Monographs in Mathematics,
DOI 10.1007/978-1-4614-4244-8_4, © Springer Science+Business Media New York 2012

up to now. A deep study of this problem would lead to many interesting and important results. For example, it would be very nice to give a classification of homogeneous Finsler spaces of constant flag curvature. Note that by the theorem of Akbar–Zadeh (see Sect. 1.8), a homogeneous Finsler space of negative constant flag curvature must be Riemannian, since the Cartan tensor and Landsberg tensor of a homogeneous Finsler space must be bounded. Similarly, a homogeneous Finsler space of vanishing flag curvature must be locally Minkowski. Therefore one need consider only the homogeneous Finsler spaces of positive constant flag curvature.

The standard references of this chapter are [9, 30, 32, 50, 53, 56, 61, 63, 72, 114]; see [52, 68, 69, 90, 107, 108, 110, 134] for further study on related topics.

## 4.1  Algebraic Description

In this section, we will give an algebraic description of the invariant Finsler metrics on homogeneous manifolds (not necessarily reductive). Similar to the proof of Proposition 2.8, we can easily prove the following result.

**Proposition 4.1.** *Let $G$ be a Lie group and $H$ a closed subgroup of $G$, $\operatorname{Lie} G = \mathfrak{g}$, $\operatorname{Lie} H = \mathfrak{h}$. Then there is a one-to-one correspondence between the $G$-invariant Finsler metric on $G/H$ and the Minkowski norm on the quotient space $\mathfrak{g}/\mathfrak{h}$ satisfying*

$$F(\operatorname{Ad}_{\mathfrak{g}/\mathfrak{h}}(h)(w)) = F(w), \quad \forall h \in H, w \in \mathfrak{g}/\mathfrak{h},$$

*where $\operatorname{Ad}_{\mathfrak{g}/\mathfrak{h}}$ is the representation of $H$ on $\mathfrak{g}/\mathfrak{h}$ induced by the adjoint representation of $H$ on $\mathfrak{g}$.*

**Corollary 4.1.** *Let $G/H$ be a reductive homogeneous manifold with a reductive decomposition $\mathfrak{g} = \mathfrak{h} + \mathfrak{m}$. Then there is a one-to-one correspondence between the $G$-invariant Finsler metric on $G/H$ and the Minkowski norm on $\mathfrak{m}$ satisfying*

$$F(\operatorname{Ad}(h)w) = F(w), \quad \forall h \in H, w \in \mathfrak{m}.$$

To state the next result, we first give a definition.

**Definition 4.1.** *Let $\mathfrak{g}$ be a real Lie algebra and $\mathfrak{h}$ a subalgebra of $\mathfrak{g}$. If $F$ is a Minkowski norm on the quotient space $\mathfrak{g}/\mathfrak{h}$ such that*

$$g_y(\operatorname{ad}_{\mathfrak{g}/\mathfrak{h}}(w)u, v) + g_y(u, \operatorname{ad}_{\mathfrak{g}/\mathfrak{h}}(w)v) + 2C_y(\operatorname{ad}_{\mathfrak{g}/\mathfrak{h}}(w)(y), u, v) = 0,$$

*where $y, u, v \in \mathfrak{g}/\mathfrak{h}$, $y \neq 0$, $w \in \mathfrak{h}$, $g_y$ is the fundamental tensor of $F$, and $C_y$ is the Cartan tensor of $F$, then $\{\mathfrak{g}, \mathfrak{h}, F\}$ (or simply $\{\mathfrak{g}, \mathfrak{h}\}$) is called a Minkowski Lie pair. In particular, if $F$ is a Euclidean norm, then $\{\mathfrak{g}, \mathfrak{h}\}$ is called a Euclidean Lie pair.*

**Theorem 4.1.** *Let $G$ be a Lie group and $H$ a closed subgroup of $G$, $\operatorname{Lie} G = \mathfrak{g}$, $\operatorname{Lie} H = \mathfrak{h}$. Suppose there exists an invariant Finsler metric on the homogeneous* *manifold $G/H$. Then there exists a Minkowski norm $F$ on the quotient space $\mathfrak{g}/\mathfrak{h}$ such that $\{\mathfrak{g}, \mathfrak{h}, F\}$ is a Minkowski Lie pair. On the other hand, if $H$ is connected and there exists a Minkowski norm on $\mathfrak{g}/\mathfrak{h}$ such that $\{\mathfrak{g}, \mathfrak{h}, F\}$ is a Minkowski Lie pair, then there exists an invariant Finsler metric on $G/H$.*

*Proof.* Let $F$ be an invariant Finsler metric on $G/H$. Identifying $\mathfrak{g}/\mathfrak{h}$ with $T_o(G/H)$, where $o$ is the origin of $G/H$, we get a Minkowski norm on $\mathfrak{g}/\mathfrak{h}$ (still denoted by $F$). Since $F$ is $G$-invariant, we have

$$F(\mathrm{Ad}_{\mathfrak{g}/\mathfrak{h}}(h)(u)) = F(u), \quad \forall h \in H, u \in \mathfrak{g}/\mathfrak{h}.$$

By the definition of $g_y$, we have

$$g_y(u,v) = g_{\mathrm{Ad}_{\mathfrak{g}/\mathfrak{h}}(h)(y)}(\mathrm{Ad}_{\mathfrak{g}/\mathfrak{h}}(h)(u), \mathrm{Ad}_{\mathfrak{g}/\mathfrak{h}}(h)(v)), \quad \forall y, u, v \in \mathfrak{g}/\mathfrak{h},\ y \neq 0,\ h \in H.$$

For $w \in \mathfrak{h}$, considering the one-parameter subgroup $\exp tw$ of $H$, we get

$$g_y(u,v) = g_{\mathrm{Ad}_{\mathfrak{g}/\mathfrak{h}}(\exp tw)(y)}(\mathrm{Ad}_{\mathfrak{g}/\mathfrak{h}}(\exp tw)(u), \mathrm{Ad}_{\mathfrak{g}/\mathfrak{h}}(\exp tx)(v)), \quad \forall t \in \mathbb{R}.$$

Taking the derivative with respect to $t$, we obtain

$$g_y(\mathrm{ad}_{\mathfrak{g}/\mathfrak{h}}(w)u, v) + g_y(u, \mathrm{ad}_{\mathfrak{g}/\mathfrak{h}}(w)v) + 2C_y(\mathrm{ad}_{\mathfrak{g}/\mathfrak{h}}(w)(y), u, v) = 0.$$

Therefore $\{\mathfrak{g}, \mathfrak{h}, F\}$ is a Minkowski Lie pair. On the other hand, suppose $F$ is a Minkowski norm on $\mathfrak{g}/\mathfrak{h}$ such that $\{\mathfrak{g}, \mathfrak{h}, F\}$ is a Minkowski Lie pair. Given $w \in \mathfrak{h}$, $y, u, v \in \mathfrak{g}/\mathfrak{h}$, $y \neq 0$, we define a function

$$\psi(t) = g_{\mathrm{Ad}_{\mathfrak{g}/\mathfrak{h}}(\exp tw)(y)}(\mathrm{Ad}_{\mathfrak{g}/\mathfrak{h}}(\exp tw)(u), \mathrm{Ad}_{\mathfrak{g}/\mathfrak{h}}(\exp tw)(v)).$$

Then for any $t_0 \in \mathbb{R}$ we have

$$\begin{aligned}
\psi'(t_0) &= g_{\mathrm{Ad}_{\mathfrak{g}/\mathfrak{h}}(\exp t_0 w)(y)}(\mathrm{ad}_{\mathfrak{g}/\mathfrak{h}}(t_0 w)(\mathrm{Ad}_{\mathfrak{g}/\mathfrak{h}}(\exp t_0 w)(u)), \mathrm{Ad}_{\mathfrak{g}/\mathfrak{h}}(\exp t_0 w)(v)) \\
&\quad + g_{\mathrm{Ad}_{\mathfrak{g}/\mathfrak{h}}(\exp t_0 w)(y)}(\mathrm{Ad}_{\mathfrak{g}/\mathfrak{h}}(\exp t_0 w)(u), \mathrm{ad}_{\mathfrak{g}/\mathfrak{h}}(w)(\mathrm{Ad}_{\mathfrak{g}/\mathfrak{h}}(\exp t_0 w)(v))) \\
&\quad + 2C_{\mathrm{Ad}_{\mathfrak{g}/\mathfrak{h}}(\exp t_0 w)(y)}(\mathrm{ad}_{\mathfrak{g}/\mathfrak{h}}(w)(\mathrm{Ad}_{\mathfrak{g}/\mathfrak{h}}(\exp t_0 w)(y)), u, v) = 0.
\end{aligned}$$

Therefore the function $\psi$ is a constant. Hence

$$g_y(u,v) = g_{\mathrm{Ad}_{\mathfrak{g}/\mathfrak{h}}(\exp tw)(y)}(\mathrm{Ad}_{\mathfrak{g}/\mathfrak{h}}(\exp tw)(u), \mathrm{Ad}_{\mathfrak{g}/\mathfrak{h}}(\exp tw)(v)), \forall t \in \mathbb{R}.$$

As a connected subgroup, $H$ is generated by elements of the form $\exp tw$, $w \in \mathfrak{h}$, $t \in \mathbb{R}$. Therefore for any $h \in H$, we have

$$g_y(u,v) = g_{\mathrm{Ad}_{\mathfrak{g}/\mathfrak{h}}(h)(y)}(\mathrm{Ad}_{\mathfrak{g}/\mathfrak{h}}(h)(u), \mathrm{Ad}_{\mathfrak{g}/\mathfrak{h}}(h)(v)).$$

To complete the proof, we need some computation. Let $\alpha_1, \alpha_2, \ldots, \alpha_n$ be a basis of the linear space $\mathfrak{g}/\mathfrak{h}$. For $u \in \mathfrak{g}/\mathfrak{h} - \{0\}$, define $g_{ij}(u) = g_u(\alpha_i, \alpha_j)$. Then we have

$$F^2(u) = \sum_{i,j=1}^{n} g_{ij}(u)u^i u^j,$$

where $u = \sum u^i \alpha_i$. Now for any $h \in H$, we have

$$F^2(\mathrm{Ad}_{\mathfrak{g}/\mathfrak{h}}(h)(u)) = \sum_{i,j=1}^{n} g_{ij}(\mathrm{Ad}_{\mathfrak{g}/\mathfrak{h}}(h)(u))\bar{u}^i \bar{u}^j,$$

where $\bar{u}^i$, $i = 1, 2, \ldots, n$, are defined by $\mathrm{Ad}_{\mathfrak{g}/\mathfrak{h}}(h)(u) = \sum \bar{u}^i \alpha_i$. Let $(m_{ij})_{n \times n}$ be the matrix of $\mathrm{Ad}_{\mathfrak{g}/\mathfrak{h}}(h)$ under the basis $\alpha_1, \ldots, \alpha_n$, and $(m^{ij})_{n \times n}$ the inverse of $(m_{ij})$. Then

$$\bar{u}^i = \sum_{k=1}^{n} m_{ik} u^k.$$

Now

$$\begin{aligned}
g_{ij}(\mathrm{Ad}_{\mathfrak{g}/\mathfrak{h}}(h)(u)) &= g_{\mathrm{Ad}_{\mathfrak{g}/\mathfrak{h}}(h)(u)}(\alpha_i, \alpha_j) \\
&= g_u((\mathrm{Ad}_{\mathfrak{g}/\mathfrak{h}}(h))^{-1}\alpha_i, (\mathrm{Ad}_{\mathfrak{g}/\mathfrak{h}}(h))^{-1}\alpha_j) \\
&= g_u\left(\sum_{k=1}^{n} m^{ik}\alpha_k, \sum_{l=1}^{n} m^{jl}\alpha_l\right).
\end{aligned}$$

Therefore, taking into account the fact that $g_u$ is bilinear, we have

$$\begin{aligned}
F^2(\mathrm{Ad}_{\mathfrak{g}/\mathfrak{h}}(h)(u)) &= \sum_{i,j=1}^{n} g_u\left(\sum_{k=1}^{n} m^{ik}\alpha_k, \sum_{l=1}^{n} m^{jl}\alpha_l\right)\left(\sum_{s=1}^{n} m_{is}u^s\right)\left(\sum_{t=1}^{n} m_{jt}u^t\right) \\
&= \sum_{i,j=1}^{n} g_u(\alpha_i, \alpha_j)u^i u^j.
\end{aligned}$$

Thus

$$F^2(\mathrm{Ad}_{\mathfrak{g}/\mathfrak{h}}(h)(u)) = F^2(u).$$

Since $F \geq 0$, we have

$$F(\mathrm{Ad}_{\mathfrak{g}/\mathfrak{h}}(h)(u)) = F(u).$$

Therefore, by the correspondence of Proposition 4.1, there exists an invariant Finsler metric on $G/H$.                                                                                    □

To find a sufficient and necessary condition for $G/H$ to have invariant Finsler metrics, we first make an observation.

**Theorem 4.2.** *Let $G$ be a Lie group, $H$ a closed subgroup of $G$. Suppose there exists an invariant Finsler metric on $G/H$. Then there exists an invariant Riemannian metric on $G/H$.*

*Proof.* Let $o = H$ be the origin of $G/H$ and $T_o(G/H)$ the tangent space of $G/H$ at $o$. Then $F$ defines a Minkowski norm on $T_o(G/H)$ (still denoted by $F$). The linear isotropic group $\text{Ad}(H) = \{\text{Ad}_{\mathfrak{g}/\mathfrak{h}}(h) \mid h \in H\}$ keeps the indicatrix

$$\mathscr{I}_o = \{x \in T_o(G/H) \mid F(x) = 1\}$$

invariant. Let $G_1$ be the subgroup of the general linear group $\text{GL}(T_o(G/H))$ consisting of the elements keeping $I_o$ invariant. Then $G_1$ is a compact Lie group and $\text{Ad}(H)$ is a subgroup of $G_1$. Therefore, by Weyl's unitary trick, there exists a $G_1$-invariant inner product $\langle\,,\,\rangle$ on $T_o(G/H)$. Then $\langle\,,\,\rangle$ is $\text{Ad}\,(H)$-invariant. Therefore, by Proposition 2.8, there exists a $G$-invariant Riemannian metric on $G/H$. $\qquad\square$

By Proposition 2.9, we have the following corollary.

**Corollary 4.2.** *Let $G$ be a Lie group and $H$ a closed subgroup of $G$ such that $G$ acts almost effectively on $G/H$. Define $\text{Lie}\,G = \mathfrak{g}$, $\text{Lie}\,H = \mathfrak{h}$. Suppose there exists a $G$-invariant Finsler metric on $G/H$ and $\mathfrak{h}$ contains no nonzero ideal of $\mathfrak{g}$. Then*

1. *The Killing form $B_{\mathfrak{h}}$ of $\mathfrak{h}$ is negative semidefinite.*
2. *The restriction of the Killing form of $\mathfrak{g}$ to $\mathfrak{h}$ is negative definite.*

Next we consider invariant Finsler metrics on Lie groups. Let $G$ be a Lie group. Then we can write $G = G/H$ with $H = \{e\}$ and the action is the left translation of $G$. Therefore, by Proposition 4.1, every Minkowski norm on $\mathfrak{g}$ can produce a left-invariant Finsler metric on $G$. It is easy to see that there exist non-Riemannian ones if $\dim G \geq 2$. One can also consider the right action. However, we are interested in bi-invariant Finsler metrics. For this purpose, we consider the product group $G \times G$ and the subgroup

$$G^* = \{(g,g) \in G \times G \mid g \in G\}.$$

Then $G \times G/G^*$ is isomorphic to $G$ under the map

$$(g_1, g_2)G^* \mapsto g_1 g_2^{-1}.$$

Under this isomorphism, a Finsler metric on $G$ is bi-invariant if and only if the corresponding Finsler metric on $G \times G/G^*$ is $G \times G$-invariant. Now $G \times G/G^*$ can be viewed as a reductive homogeneous manifold with the reductive decomposition

$$(u,v) = \left(\frac{1}{2}(u+v), \frac{1}{2}(u+v)\right) + \left(\frac{1}{2}(u-v), -\frac{1}{2}(u-v)\right), \quad u, v \in \mathfrak{g}.$$

Let $\mathfrak{h} = \{(u,u) \in \mathfrak{g} + \mathfrak{g} \mid u \in \mathfrak{g}\}$, $\mathfrak{m} = \{(u,-u) \mid u \in \mathfrak{g}\}$. Then by Theorem 4.1 we have the following result.

**Proposition 4.2.** *Let $G$ be a connected Lie group. Then there is a one-to-one correspondence between the bi-invariant Finsler metric on $G$ and the Minkowski norm $F$ on $\mathfrak{m}$ such that $\{\mathfrak{g} + \mathfrak{g}, \mathfrak{h}, F\}$ is a Minkowski Lie pair.*

Since $\mathfrak{h}$ is isomorphic to $\mathfrak{g}$ as a Lie algebra under the map $\sigma : (u,u) \mapsto u$ and $\mathfrak{m}$ is linear isomorphic to $\mathfrak{g}$ as a vector space under the map $\tau : (u,-u) \mapsto u$, we can restate Proposition 4.2 without introducing $\mathfrak{h}$ and $\mathfrak{m}$. We first give a definition.

**Definition 4.2.** Let $\mathfrak{g}$ be a real Lie algebra and $F$ a Minkowski norm on $\mathfrak{g}$. Then $\{\mathfrak{g}, F\}$ (or simply $\mathfrak{g}$) is called a Minkowski Lie algebra if the following condition is satisfied:

$$g_y([w,u],v) + g_y(u,[w,v]) + 2C_y([w,y],u,v) = 0,$$

where $y \in \mathfrak{g} - \{0\}, w,u,v \in \mathfrak{g}$.

*Remark 4.1.* In the theory of Lie algebras, we have the notion of quadratic Lie algebras. Let $\mathfrak{g}$ be a Lie algebra over a field $F$ and let $B$ be a nondegenerate symmetric bilinear form on $\mathfrak{g}$. Then $(\mathfrak{g}, B)$ is called a quadratic Lie algebra if the following condition is satisfied:

$$B([z,x],y) + B(x,[z,y]) = 0, \quad \forall x,y,z \in \mathfrak{g}.$$

It is easy to see that the notion of Minkowski Lie algebras is the natural generalization of quadratic Lie algebras. Quadratic Lie algebras have important applications in physics.

*Remark 4.2.* In [118], Mestag studied invariant Lagrangian systems on a Lie group. It is proved that bi-invariant Lagrangian systems on a Lie group can also be described by the notion of Minkowski Lie algebras. This is an interesting application of Finsler geometry to mechanics and physics.

**Proposition 4.3.** *Let $\mathfrak{g}$, $\mathfrak{h}$, $\mathfrak{m}$ be as above. Then a Minkowski norm $F$ on $\mathfrak{m}$ makes $\{\mathfrak{g} + \mathfrak{g}, \mathfrak{h}, F\}$ a Minkowski Lie pair if and only if the induced (by $\tau$) Minkowski norm on $\mathfrak{g}$ makes $\{\mathfrak{g}, F\}$ a Minkowski Lie algebra.*

The proof is easy and we omit it.

In summarizing, we have proved the following:

**Theorem 4.3.** *Let $G$ be a connected Lie group. Then there exists a bi-invariant Finsler metric on $G$ if and only if there exists a Minkowski norm $F$ on $\mathfrak{g}$ such that $\{\mathfrak{g}, F\}$ is a Minkowski Lie algebra.*

Finally, we point out that the new notions of Minkowski Lie pairs and Minkowski Lie algebras may be generalized to the nondegenerate case; see [63]. It will be interesting to generalize this notion to Lie algebras over an arbitrary field and study such structures from the purely algebraic point of view.

## 4.2 Existence Theorem

In this section we consider the existence problem of invariant Finsler metrics on a coset space of a Lie group. In particular, we will obtain a sufficient and necessary condition for a homogeneous manifold $G/H$, where $G$ is a connected Lie group and $H$ is a compact subgroup of $G$, to admit $G$-invariant non-Riemannian Finsler metrics.

We first define the notion of a Minkowski representation of a Lie group. This will be useful for describing invariant Finsler metrics on homogeneous manifolds.

**Definition 4.3.** Let $G$ be a compact Lie group. A triple $(V, \rho, F)$ is said to be a Minkowski representation of $G$, where $V$ is a finite-dimensional real vector space, $\rho$ is a continuous homomorphism from $G$ to $GL(V)$, and $F$ is a Minkowski norm on $V$ such that $F(\rho(g)(v)) = F(v)$, $\forall g \in G$ and $v \in V$. Two Minkowski representations $(V_1, \rho_1, F_1)$ and $(V_2, \rho_2, F_2)$ of $G$ are called isomorphic if there exists a linear isomorphism $\varphi$ from $V_1$ onto $V_2$ such that $\varphi \circ \rho_1(g) = \rho_2(g) \circ \varphi$ for every $g \in G$, and $F_1(v) = F_2(\varphi(v))$ for every $v \in V_1$.

If the Minkowski norm $F$ is a Euclidean norm, namely, $F(v) = \sqrt{\langle v, v \rangle}$, $v \in V$, for some inner product $\langle , \rangle$, then $(V, \rho)$ is an orthogonal representation of $G$ on the inner product space $(V, \langle , \rangle)$. Therefore, Minkowski representation is a natural generalization of orthogonal representation.

Corollary 4.1 can be restated as follows.

**Theorem 4.4.** *Let $G/H$ be a coset space of a Lie group $G$ with $H$ compact. Suppose*

$$\mathfrak{g} = \mathfrak{h} + \mathfrak{m}$$

*is a reductive decomposition of the Lie algebra. Then there exists a one-to-one correspondence between the invariant Finsler metric on $G/H$ and the Minkowski norm $F$ on $\mathfrak{m}$ such that $(\mathfrak{m}, \mathrm{Ad}, F)$ is a Minkowski representation of $H$.*

In the next chapter we will define the infinitesimal version of Minkowski representations. We now give some examples of Minkowski representations.

*Example 4.1.* Let $G$ be a Lie group and $H$ a closed connected subgroup of $G$, Lie $G = \mathfrak{g}$, Lie $H = \mathfrak{h}$. Suppose $F$ is a Minkowski norm on the quotient space $\mathfrak{g}/\mathfrak{h}$ such that $(\mathfrak{g}, \mathfrak{h}, F)$ is a Minkowski Lie pair. That is,

$$g_y(\mathrm{ad}_{\mathfrak{g}/\mathfrak{h}}(w)(u), v) + g_y(u, ad_{\mathfrak{g}/\mathfrak{h}}(w)v) + 2C_y(ad_{\mathfrak{g}/\mathfrak{h}}(w)y, u, v) = 0,$$

for every $y (\neq 0)$, $u, v \in \mathfrak{g}/\mathfrak{h}$, $w \in \mathfrak{h}$, where $\mathrm{ad}_{\mathfrak{g}/\mathfrak{h}}$ is the representation of $\mathfrak{h}$ on $\mathfrak{g}/\mathfrak{h}$ induced by the adjoint representation. It is proved in Theorem 4.1 that in this case $(\mathfrak{g}/\mathfrak{h}, \mathrm{Ad}_{\mathfrak{g}/\mathfrak{h}}, F)$ is a Minkowski representation of $H$.

*Example 4.2.* Let us give an explicit example of Minkowski representation of the classical simple group $G = \mathrm{SU}(n)$ with $n \geq 3$. On the Lie algebra $\mathfrak{su}(n)$, for each positive real number $\lambda$, we define

$$F_\lambda(A) = \sqrt{\sqrt{\sum_{i=1}^{n} |\mu_i(A)|^4} + \lambda \sum_{i=1}^{n} |\mu_i(A)|^2}, \quad A \in \mathfrak{su}(n),$$

where $\mu_i(A)$, $i = 1, 2, \ldots, n$, are all the eigenvalues of $A$ and $|\cdot|$ is the length function. It is easy to check that $(\mathfrak{su}(n), \mathrm{Ad}, F_\lambda)$ is a series of Minkowski representations of $G$ that are not isomorphic to each other (see also[16, p. 21]).

The following result is useful for studying Minkowski representations.

**Theorem 4.5.** *Let $G$ be a compact Lie group and $(V, \rho, F)$ a Minkowski representation of $G$. Then there exists an inner product $\langle , \rangle$ on $V$ such that $(V, \rho, \langle , \rangle)$ is an orthogonal representation of $G$.*

*Proof.* The proof of this theorem is similar to that of Theorem 4.2. Just note that $\rho(G)$ is contained in the group $K$ of linear isometries of $(V, F)$, which is a compact subgroup of $\mathrm{GL}(V)$.                                                                                  $\square$

We now consider the condition for the existence of Minkowski representations. By Theorem 4.5, we need only treat the case that the Lie group $G$ has an orthogonal representation on a Euclidean space $V$ and find the condition for the existence of a non-Euclidean Minkowski norm on $V$ that makes the orthogonal representation a Minkowski representation. We first prove the following result.

**Theorem 4.6.** *Let $G$ be a compact Lie group and $(V, \rho, \langle , \rangle)$ an orthogonal representation of $G$. If the representation $V$ is not irreducible, then there exists a non-Euclidean Minkowski norm $F$ on $V$ such that $(V, \rho, F)$ is a Minkowski representation of $G$.*

*Proof.* By the assumption, $V$ has a decomposition

$$V = V_0 + V_1 + \cdots + V_n,$$

where $V_0$ is the subspace of fixed points of $G$ and $V_i$, $i = 1, 2, \ldots, n$,s are irreducible invariant subspaces. We have the following cases:

1. $n \geq 1$. In this case we construct a Minkowski norm $F$ on $V$ as follows:

$$F(w) = \sqrt{|w|^2 + \sqrt[s]{|w_0|^{2s} + |w_1|^{2s} + \cdots + |w_n|^{2s}}},$$

where $|\cdot|$ denotes the length with respect to $\langle , \rangle$, $w = w_0 + \cdots + w_n$ is the decomposition of $w$ corresponding to the above decomposition of $V$, and $s$ is an integer greater than 2. Then it is easy to check that $F$ satisfies the condition.
2. $n = 0$. Then $\dim V_0 \geq 2$. In this case any non-Euclidean Minkowski norm $F_0$ on $V_0$ is invariant under $G$.

This completes the proof of the theorem.                                            $\square$

Next we consider the irreducible case. We first prove the following theorem.

**Theorem 4.7.** *Let $K$ be a compact Lie group and $\rho : K \to O(V)$ an orthogonal representation of $K$ on a Euclidean space $V$. Suppose the representation $(V, \rho)$ is irreducible. Then there exists a non-Euclidean Minkowski norm $F$ on $V$ such that $(V, \rho, F)$ is a Minkowski representation of $K$ if and only if the restriction of the action of $K$ on the unit sphere of $V$ is nontransitive.*

*Proof.* First consider the case in which $K$ is connected. Suppose the action of $K$ on the unit sphere $S$ is nontransitive. We assert that there exists a $K$-invariant nonnegative smooth function $f$ on $S$ that is not a constant. In fact, let $x$ and $y$ be two points in $S$ that do not lie in the same orbit of the action of $K$ on $S$. Then the orbits $K \cdot x$ and $K \cdot y$ are two closed subsets of $S$ that do not intersect. By the theorem of partitions of unity, there exists a nonnegative smooth function $f_1$ on $S$ that is equal to 1 on $K \cdot x$ and is equal to 0 on $K \cdot y$. Let $dv$ be the standard normalized bi-invariant Haar measure of $K$ and define

$$f(z) = \int_K f_1(k(z)) \, dv(k), \quad z \in S.$$

Then $f$ is a smooth function on $S$ that is invariant under the action of $K$. It is obvious that $f$ is also equal to 1 on $K \cdot x$ and equal to 0 on $K \cdot y$; hence it is not a constant function. Now we consider the function $F$ on $V \setminus \{0\}$ defined by

$$F(u) = |u| f\left(\frac{u}{|u|}\right), \quad u \in V \setminus \{0\}.$$

Then $F$ is a smooth function on $V \setminus \{0\}$ that is positively homogeneous of degree 1. Fix an orthonormal basis $v_1, v_2, \ldots, v_n$ of $V$ and consider the Hessian matrix

$$M(w) = \left. \left[ \frac{1}{2} [F^2]_{v^i v^j} \right] \right|_w.$$

The entries of $M$ are smooth functions on $V \setminus \{0\}$ that are positively homogeneous of degree 0, that is, $M(\lambda w) = M(w), \forall \lambda > 0$. Now let $\lambda_1(w), \lambda_2(w), \ldots, \lambda_n(w)$ be all the eigenvalues of $M(w)$. Then we have

$$\sum_{i=1}^n |\lambda_i(w)| = \sqrt{\sum_{i=1}^n |\lambda_i(w)|^2 + 2 \sum_{i<j} |\lambda_i(w)||\lambda_j(w)|}$$

$$\leq \sqrt{n} \sqrt{\sum_{i=1}^n (\lambda_i(w))^2}$$

$$= \sqrt{n} \sqrt{\mathrm{tr}(M(w)^2)}.$$

Since the function $\mathrm{tr}((M(w)^2)$ is smooth on the compact set $S$, it is bounded on $S$. Hence there exists a positive number $c_0$ such that

$$\sum_{i=1}^{n} |\lambda_i(w)| \le c_0, \quad \forall w \in S.$$

Now given any positive number $c$, we define a function $F_c$ by

$$F_c(u) = \sqrt{F^2(u) + (c+c_0)\langle u,u \rangle}, \quad u \in V \setminus \{0\}.$$

Then the Hessian of $F_c$ is

$$M_1 = M + (c+c_0)I_n,$$

where $I_n$ is the $n \times n$ identity matrix. By the above arguments, $M_1$ has positive eigenvalues everywhere. Therefore, $F_c$ is a Minkowski norm on $V$. It is obvious that $F_c$ is invariant under the action of $K$. Hence $(V,\rho,F_c)$ is a Minkowski representation of $K$.

Now we prove that $F_c$ is non-Euclidean for every $c > 0$. Suppose, to the contrary, that this is not true. Then there exists an inner product $\langle,\rangle'$ such that $F_c(u) = \sqrt{\langle u,u \rangle'}$, $u \in V \setminus \{0\}$. Since $F_c$ is invariant under $K$, the inner product $\langle,\rangle'$ is invariant under $K$. Since the representation is irreducible, by Schur's lemma there must be a positive number $b$ such that $\langle,\rangle' = b\langle,\rangle$. Then we have

$$F^2(u) + (c+c_0)\langle u,u \rangle = b\langle u,u \rangle, \quad u \in V \setminus \{0\}.$$

In particular, on the unit sphere $S$ we have

$$F^2(w) = b - c - c_0, \quad w \in S.$$

This means that $F$ is a constant on $S$, contradicting the construction of $f$.

Conversely, if the action of $K$ on $S$ is transitive, then every $K$-invariant function on $S$ must be constant. In particular, every $K$-invariant Minkowski norm on $V$ must be a positive multiple of the Euclidean norm $\sqrt{\langle,\rangle}$. This completes the proof of the theorem in the connected case.

Next we consider the case in which $K$ is not connected. The proof for this case is similar to that of the connected case but needs some modifications. Suppose the action of $K$ on $S$ is not transitive. Select $x,y$ in $S$ that do not lie in the same orbit of $K$ and define a smooth function $f_1$ on $S$ which is equal to 1 on $K \cdot x$ and equal to 0 on $K \cdot y$. Let $K_0$ be the unity connected component of $K$ and $d\mu$ the standard normalized bi-invariant Haar measure of $K_0$. Define

$$f_2(z) = \int_{K_0} f_1(k(z)) \, d\mu(k), \quad z \in S.$$

Moreover, let $k_0 K_0, k_1 K_0, k_2 K_0, \ldots, k_m K_0$ $(k_0 = e)$ be all the distinct elements of the quotient group $K/K_0$. Define

$$f(z) = \frac{1}{m+1} \sum_{i=0}^{m} f_2(k_i(z)), \quad z \in S.$$

Then it is easily seen that $f$ is a smooth function on $S$ that is also equal to 1 on $K \cdot x$ and equal to 0 on $K \cdot y$. Further, it is easy to check that $f$ is invariant under the action of $K$. Using the function $f$, we can prove the theorem in the nonconnected case similarly as in the connected case.                                                    □

As an application of Theorem 4.7, we give a sufficient and necessary condition for a coset space to admit invariant non-Riemannian Finsler metrics. Suppose $G/H$ is a coset space of a connected Lie group $G$ with $H$ compact. Then the Lie algebra $\mathfrak{g}$ of $G$ has a reductive decomposition

$$\mathfrak{g} = \mathfrak{h} + \mathfrak{m} \quad \text{(direct sum of subspaces)},$$

where $\mathfrak{h} =\mathrm{Lie}\, H$, and $\mathrm{Ad}(h)(\mathfrak{m}) \subset \mathfrak{m}$, $\forall h \in H$. Since $H$ is compact, there exists an inner product $\langle\, ,\rangle$ on $\mathfrak{m}$ that makes the linear isotropic representation into an orthogonal representation. Let $S$ denote the unit sphere of $\mathfrak{m}$ with respect to the inner product $\langle\, ,\rangle$.

**Theorem 4.8.** *Let $G$, $H$, $\mathfrak{m}$, $\langle\, ,\rangle$, and $S$ be as above. Then there exists a $G$-invariant non-Riemannian Finsler metric on the coset space $G/H$ if and only if the restriction of the linear isotropic representation of $H$ on $\mathfrak{m}$ to the sphere $S$ is not transitive.*

*Proof.* By Theorem 4.4, there exists a $G$-invariant non-Riemannian Finsler metric on $G/H$ if and only if there exists a non-Euclidean Minkowski norm $F$ on $\mathfrak{m}$ such that $(\mathfrak{m}, \mathrm{Ad}, F)$ is a Minkowski representation of $H$. Hence the theorem follows from Theorem 4.7.                                                    □

## 4.3   The Degree of Symmetry of Closed Manifolds

In this section we will use the results in the previous sections to study the relationship between Finsler metrics and the degree of symmetry of a closed manifold.

Let $M$ be a connected compact smooth manifold. The degree of symmetry of $M$ is defined to be the maximum of the dimensions of the groups of isometries of all possible Riemannian metrics on $M$. This notion was introduced by Hsiang. By the Meyers–Steenrod theorem, the degree of symmetry of $M$ is equal to the maximum of the dimensions of Lie groups that act smoothly and effectively on it. The degree of symmetry is an important geometric invariant of closed manifolds. Since the introduction of this invariance, many interesting results have been established; see,

for example, [84, 85, 112, 178]. One of the most important problems in this direction
is to determine the degree of symmetry of an explicit manifold. Let $M$ be an
$n$-dimensional connected compact manifold and denote its degree of symmetry by
$N(M)$. Then by classical results, one easily deduces that $N(M) \leq \frac{1}{2}n(n+1)$, with
equality holding if and only if $M$ admits a Riemannian metric of positive constant
sectional curvature. As another explicit example, Hsiang proved that the degree
of symmetry of the homogeneous manifold $SO(n)/SO(n-2)$ (the second Stiefel
manifold), with $n$ odd and $\geq 20$, is equal to $\frac{1}{2}n(n-1)+1$, and its degree of symmetry
can be realized only by its natural metric [84]. However, up to now only very few
results have been obtained on explicit manifolds.

The purpose of this section is to study the relationship between Finsler metrics
and the degree of symmetry of a closed (i.e., connected compact without boundary)
manifold. Let $F$ be a Finsler metric on a closed manifold $M$. By a simple
observation, we first prove the following.

**Theorem 4.9.** *The degree of symmetry of a compact connetced manifold $M$ is equal
to the maximum of the dimensions of the groups of isometries of all the possible
Finsler metrics on $M$.*

Let $F$ be a Minkowski norm on a real vector space $V$. Then by Proposition 3.1 and
Theorem 3.3, the group of linear isometries of $F$, denoted by $L(V, F)$, is a compact
subgroup of $GL(V)$. Hence the unity component of $L(V, F)$, $L_0(V, F)$, is a connected
compact subgroup of $SL(V)$. Fix an inner product on $V$. Then by the conjugacy of
compact subgroups of semisimple Lie groups, $L_0(V, F)$ is conjugate to a subgroup
of $SO(V)$. We assert that $\dim L_0(V, F) = \dim SO(V)$ if and only if $F$ is a Euclidean
norm. In fact, if $\dim L_0(V, F) = \frac{1}{2}n(n-1)$, where $n = \dim V$, then by Lemma 3.2,
the action of $L_0(V, F)$ on the indicatrix

$$\mathscr{I} = \{w \in V \mid F(w) = 1\}$$

is transitive (see [163]). On the other hand, as a compact Lie group, $L_0(V, F)$
also leaves an inner product $\langle , \rangle$ invariant. Then it is easy to see that $F$ is equal
to a positive multiple of the Euclidean norm $\sqrt{\langle , \rangle}$. This proves the assertion.
This fact means that Euclidean norms possess more symmetry than non-Euclidean
Minkowski norms. Intuitively, this fact may imply that on a closed manifold
the most symmetric Finsler metrics should be Riemannian metrics. However, the
following theorem says that this intuition is not true in almost all cases.

**Theorem 4.10.** *Let $M$ be a connected compact manifold and $N(M)$ the degree
of symmetry of $M$. If $M$ is not diffeomorphic to a compact rank-one Riemannian
symmetric space, then there exists a non-Riemannian Finsler metric on $M$ such that
the group of isometries of $F$, $I(M, F)$, has dimension $N(M)$.*

The rank-one Riemannian symmetric spaces include the spheres $S^n$, the real
projective spaces $\mathbb{R}P^n$, the complex projective spaces $\mathbb{C}P^n$, the quaternion projective
spaces $\mathbb{H}P^n$, and the Cayley projective plane $\text{Cay}\,P^2$, all endowed with their natural
metrics ([164]; see also [83]). These are the most symmetric compact Riemannian

manifolds. Thus the merit of Theorem 4.10 is to assert that one can find non-Riemannian Finsler metrics that are most symmetric among all the Finsler metrics (including Riemannian metrics) on a manifold with only very few exceptions.

The remainder of this section is devoted to the proof of Theorems 4.9 and 4.10. We first prove a result on the nontransitive actions of a Lie group on a manifold.

**Theorem 4.11.** *Let G be a connected compact Lie group acting smoothly and effectively on a compact connected manifold M. If the action of G on M is nontransitive, then there exist infinitely many G-invariant non-Riemannian Finsler metrics on M that are mutually nonisometric to one another.*

*Proof.* We first assert that there exists a $G$-invariant Riemannian metric on $M$. In fact, suppose $Q_0$ is a fixed Riemannian metric on $M$. Let $d\mu$ be the standard normalized bi-invariant Haar measure of $G$. For $x \in M$, $X, Y \in T_x(M)$, define

$$Q_1(X,Y) = \int_G Q_0(\mathrm{d}k(X), \mathrm{d}k(Y)) \, \mathrm{d}\mu(k).$$

Then it is easily seen that $Q_1$ is a $G$-invariant Riemannian metric on $M$. This proves our assertion. Now suppose $Q$ is a $G$-invariant Riemannian metric on $M$. Let $x \in M$ be such that the orbit $G \cdot x$ is a principal orbit. By Proposition 2.4, there exists a sufficiently small $\varepsilon > 0$ such that for every $y \in B(x, \varepsilon)$, the orbit $G \cdot y$ is also a principal orbit, where

$$B(x, \varepsilon) = \{z \in M \mid d(x, z) < \varepsilon\},$$

and $d$ is the distance function with respect to the Riemannian metric $Q$. Let $U$ be the union of all the orbits $G \cdot y$, $y \in B(x, \varepsilon)$. Then from the fact that $Q$ is $G$-invariant, it follows that $U$ is an open subset of $M$ and it is invariant under the action of $G$. Now we define a series of non-Riemannian Finsler metrics on the submanifold $U$. Given $y \in U$, we have $\dim G \cdot y < \dim M$. Therefore the tangent space $T_y(M)$ has an orthogonal decomposition

$$T_y(M) = T_y(G \cdot y) + (T_y(G \cdot y))^{\perp}. \tag{4.1}$$

Define a series of Minkowski norms on $T_y(M)$ by

$$L_m(Y) = \sqrt{Q(Y,Y) + \sqrt[m]{Q(Y_1,Y_1)^m + Q(Y_2,Y_2)^m}}, \quad Y \in T_y(U),$$

where $m \geq 2$ is an integer and $Y_1, Y_2$ are determined by the decomposition $Y = Y_1 + Y_2$ with respect to (4.1). It is easy to check that $L_m$ is actually a non-Euclidean Minkowski norm. Moreover, since the orbit of any point in $U$ is principal, and the action of $G$ on $M$ is smooth, for any fixed $m \geq 2$, the collection of the $L_m$ on the tangent spaces of $U$ defines a (smooth) non-Riemannian Finsler metric on $U$. We still denote this Finsler metric by $L_m$. It follows easily from the definition that $L_m$ is invariant under the action of $G$.

By the theorem of partitions of unity of smooth manifolds, there exists a nonnegative smooth function $f_1$ on $M$ that is equal to 1 on the closure of the subset $G \cdot B(x, \frac{1}{2}\varepsilon)$ and equal to 0 outside the subset $G \cdot B(x, \frac{3}{4}\varepsilon)$. Now we define a new function $f$ by

$$f(y) = \int_G f_1(k(y)) \, d\mu(k), \quad y \in M.$$

Then $f$ is a $G$-invariant nonnegative smooth function on $M$ that is still equal to 1 on $G \cdot B(x, \frac{1}{2}\varepsilon)$ and equal to 0 outside $G \cdot B(x, \frac{3}{4}\varepsilon)$. Hence $f(y)L_m$ are globally defined smooth functions on the slit tangent bundle $T(M) \setminus \{0\}$. Now we define a series of smooth functions on $T(M) \setminus \{0\}$ as the following:

$$F_{m,\lambda}(Y) = \sqrt{f(y)L_m^2(Y) + \lambda Q(Y,Y)}, \quad Y \in T_y(M), \ \lambda > 0.$$

Then for any $m \geq 2$, $F_{m,\lambda}$ is a non-Riemannian Finsler metric on $M$. In fact, the positive definiteness of the Hessian can be easily checked either in the open subset $G \cdot B(x, \varepsilon)$, where $f \geq 0$ and $L_m$ is a Finsler metric, or outside the open subset $G \cdot B(x, \varepsilon)$, where $f = 0$. On the other hand, it is easily seen that for any $x \in M$, the Cartan tensor of the Minkowski norm of $F_{m,\lambda}$ at $x$ is equal to that of $L_m$, which is nonzero. Hence $F_{m,\lambda}$ is non-Riemannian. Moreover, from the invariance of $f$, $L_m$, and $Q$, it follows that $F_{m,\lambda}$ is $G$-invariant.

Now we assert that for a fixed $m \geq 2$, and $\lambda_1 \neq \lambda_2$, we cannot have $F_{m,\lambda_1} = \mu F_{m,\lambda_2}$ for any $\mu > 0$. In fact, otherwise we have

$$(1 - \mu^2)f(y)L_m^2(Y) = (\lambda_2\mu^2 - \lambda_1)Q(Y,Y), \quad Y \in T_y(M), \ y \in M.$$

In particular, for $y \in B(x, \frac{1}{2}\varepsilon)$, we have

$$(1 - \mu^2)L_m^2(Y) = (\lambda_2\mu^2 - \lambda_1)Q(Y,Y), \quad Y \in T_y(M).$$

This implies that $L_m$ is a Euclidean norm on $T_y(M)$, which is a contradiction to the definition.

Finally, we prove that for a fixed $m$, there are infinitely many Finsler metrics $F_{m,\lambda}$ that are not isometric to one another. In fact, since

$$F_{m,\lambda}(X) \geq \sqrt{\lambda Q(X,X)}, \quad \forall X \in TM \setminus \{0\},$$

we have $D(M, F_{m,\lambda}) \geq \sqrt{\lambda}D(M, Q)$, where $D$ denotes the diameter with respect to the Finsler metrics. Thus there exist infinitely many numbers $\lambda_1 < \lambda_2 < \cdots < \lambda_k < \cdots$ such that

$$D(M, F_{m,\lambda_1}) < D(M, F_{m,\lambda_2}) < \cdots < D(M, F_{m,\lambda_k}) < \cdots.$$

Since isometric Finsler spaces must have the same diameter, the assertion follows. This completes the proof of the theorem.                                              □

Now we can prove the main results of this section.

*Proof of Theorem 4.9.* Let $M$ be a closed manifold and $F$ a Finsler metric on $M$. By Theorem 3.4, the group of isometries of $F$, $I(M,F)$, is a Lie transformation group of $M$ and for $x \in M$, the isotropy subgroup of $I(M,F)$ at $x$, $I_x(M,F)$, is a compact subgroup. Since $M$ is compact, the action of $I(M,F)$ on $M$ is proper. Therefore, the orbit $I(M,F) \cdot x$ of $x$ is a closed submanifold of $M$. In particular, $I(M,F) \cdot x$ is compact. Since $I(M,F)/I_x(M,F) = I(M,F) \cdot x$, $I(M,F)$ is a compact Lie group. Denote by $G$ the unity component of $I(M,F)$. Then $G$ is a connected compact Lie transformation group of $M$. By the proof of Theorem 4.11, there exists a Riemannian metric $g$ on $M$ that is invariant under $G$. In particular, $G \subset I(M,g)$. Thus

$$\dim I(M,F) = \dim G \leq \dim I(M,g).$$

This completes the proof of Theorem 4.9. □

*Proof of Theorem 4.10.* Let $M$ be a connected compact manifold with degree of symmetry $N(M)$. Select a Riemannian metric $Q$ on $M$ with full isometry group $G$ such that $\dim G = N(M)$. If the action of $G$ on $M$ is not transitive, then the action of the unity component $G_0$ of $G$, is not transitive on $M$ either. By Theorem 4.11, there exists a non-Riemannian Finsler metric $F$ on $M$ that is invariant under $G_0$. Then we have $\dim I(M,F) \geq \dim G_0$. Since $\dim G_0 = \dim G = N(M)$, we have $\dim I(M,F) = N(M)$. On the other hand, if the action of $G$ on $M$ is transitive, then the action of $G_0$ on $M$ is also transitive. Hence $M$ is diffeomorphic to a coset space $G_0/H_0$ of $G_0$, where $H_0$ is the isotropic subgroup of $G_0$ at certain point of $M$. Fix a $G_0$-invariant Riemannian metric $Q$ on $M$ and let $S$ be the unit sphere of $T_o(M)$ with respect to $Q$, where $o = eH_0$ is the origin of $G_0/H_0$. If the action of $H_0$ at $T_o(M)$ is not transitive on $S$, then Theorem 4.11 asserts that there exists a non-Riemannian Finsler metric $F$ on $M$ that is $G_0$-invariant. Hence $\dim I(M,F) = N(M)$. If $H_0$ acts transitively on $S$, then $M = G_0/H_0$ endowed with the Riemannian metric $Q$ is a homogeneous Riemannian manifold. We assert further that $(M,Q)$ is a two-point homogeneous Riemannian manifold, i.e., for every two pairs of points $(p_1,q_1)$ and $(p_2,q_2)$ with $d(p_1,q_1) = d(p_2,q_2)$, one can find an isometry $\sigma$ such that $\sigma(p_1) = p_2$ and $\sigma(q_1) = q_2$. In fact, by homogeneity we can first select an isometry $\sigma_1$ such that $\sigma_1(p_1) = p_2$. Then

$$d(p_2,\sigma_1(q_1)) = d(\sigma_1(p_1),\sigma_1(q_1)) = d(p_1,q_1) = d(p_2,q_2).$$

Let $H_{p_2}$ be the isotropy subgroup of the full group of isometries of $(M,g)$ at $p_2$. Then $H_{p_2}$ acts transitively on the unit sphere of $T_{p_2}(M)$. By the completeness of $(M,Q)$, we can select unit vectors $X_1, X_2 \in T_{p_2}(M)$ such that

$$\sigma_1(q_1) = \exp(t_0 X_1), \quad q_2 = \exp(t_0 X_2),$$

where $t_0 = d(p_2,q_2)$. Now select $\sigma_2 \in H_{p_2}$ such that $d\sigma_2(X_1) = X_2$. Then it is easy to check that $\sigma_2(\sigma_1(q_1)) = q_2$. Let $\sigma = \sigma_2 \circ \sigma_1$. Then we have $\sigma(p_1) = p_2$ and $\sigma(q_1) = q_2$. This proves our assertion. By the classification result of H.C. Wang on compact two-point homogeneous Riemannian manifolds [164], $M$ is diffeomorphic to a rank-one Riemannian symmetric space (see also [83, p. 535]).                    □

## 4.4    Homogeneous Finsler Spaces of Negative Curvature

The Cartan–Hadamard theorem (Theorem 1.5) asserts that the universal covering space of a connected complete Finsler space of nonpositive flag curvature is diffeomorphic to Euclidean space. In particular, a connected simply connected complete Finsler space of nonpositive flag curvature must be diffeomorphic to a Euclidean space. It is therefore interesting to find out under what conditions a Finsler space of nonpositive flag curvature is necessarily simply connected. The purpose of this section is to prove the following theorem.

**Theorem 4.12.** *Let $(M, F)$ be a connected homogeneous Finsler space of nonpositive flag curvature. If the Ricci scalar is everywhere strictly negative, then $M$ is simply connected.*

The Riemannian case of this result is due to Kobayashi [100]. To prove the theorem in the general case, we need a result about Killing vector fields on homogeneous Finsler spaces. Let $M$ and $N$ be Finsler spaces and $p : N \to M$ a locally isometric covering projection (see [16] for the fundamental properties of covering projections between Finsler spaces). Let $G$ be a connected Lie group of isometries on $M$ that acts transitively on $M$, $\mathfrak{g} = \mathrm{Lie}\,G$. Each $w \in \mathfrak{g}$ generates a one-parameter group of isometries of $M$, hence can be viewed as a Killing vector field on $M$, denoted by $\widetilde{w}$. Let $w^*$ be the lift of $\widetilde{w}$ to $N$ (by the projection $p$). The set of all such $w^*$ forms a Lie algebra, denoted by $\mathfrak{g}^*$. Since $p$ is a locally isometric covering projection, each $w^*$ is a complete Killing vector field on $N$. Thus $\mathfrak{g}^*$ is a Lie subalgebra of $\mathfrak{i}(N)$. Let $G^*$ be the (unique) connected Lie subgroup of $I(N)$ corresponding to $\mathfrak{g}^*$. Then we have the following.

**Lemma 4.1.** *$G^*$ acts transitively on $N$.*

*Proof.* The Riemannian case of this lemma is obvious, since it is easy to check that the $G^*$-orbit of any point in $N$ is both a closed and an open subset of $N$. For the general case, recall (Theorem 4.2) that if a coset space $G_1/H_1$ of a Lie group $G_1$ admits a $G_1$-invariant Finsler metric, then it also admits a $G_1$-invariant Riemannian metric. Now the Lie group $G$ acts transitively and isometrically on $M$, so $M$ can be written as $G/H$, where $H$ is the isotropic subgroup of $G$ at some point. Moreover, the metric on $M$ is invariant under $G$. Therefore we can find a Riemannian metric $g$ on $M$ which is $G$-invariant. This means that each $w \in \mathfrak{g}$ can also generate a Killing vector field on $M$ with respect to the Riemannian metric $g$. Let $g^* = p^*g$. Then $g^*$ is a Riemannian metric on $N$ and $p : (N, g^*) \to (M, g)$ is a locally isometric covering projection. This reduces the general case to the Riemannian case, and the lemma is proved.                    □

Now we turn to the proof of the main theorem. It is just a careful and technical modification of Kobayashi's argument in [100]. We first need to prove two lemmas. Similarly to the Riemannian case, we call an isometry $\sigma$ of a Finsler space a Clifford translation if for every two points $a, b$ we have $d(a, \sigma(a)) = d(b, \sigma(b))$. The Riemannian case of the next lemma was obtained by Wolf in [171].

**Lemma 4.2.** *Let $N$ and $M$ be Finsler spaces and $p : N \to M$ a locally isometric covering projection. If $M$ is homogeneous, then every homeomorphism $\phi$ of $N$ onto itself satisfying $p \circ \phi = p$ is a Clifford translation of $N$.*

*Proof.* By the generalized Myers–Steenrod theorem, a distance-preserving map of $M$ onto itself is necessarily an isometry. Thus we need only prove that for every two points $a, a' \in N$,

$$d(a', \phi(a')) = d(a, \phi(a)),$$

where $d$ is the distance function on $N$. Let $G$, $G^*$ be as in Lemma 4.1. Then $G^*$ is transitive on $N$. Hence there exists a $\psi \in G^*$ such that $a' = \psi(a)$. Since $p \circ \phi = p$, $\phi$ induces the identity map on $N$. Hence $\phi_* w^* = w^*$, $\forall w^* \in \mathfrak{g}^*$. Therefore, in the Lie group $G^*$, we have

$$\exp(\phi_*(tw^*)) = \exp tw^*, \quad \forall t \in \mathbb{R}.$$

Hence

$$\phi^{-1} \exp(tw^*) \phi = \exp tw^*, \quad \forall t \in \mathbb{R}.$$

From this we conclude that $\phi$ commutes with every element of the form $\exp tw^*$, $t \in \mathbb{R}$, $w^* \in \mathfrak{g}^*$. Since $G^*$ is connected, it is generated by $\exp U$, where $U$ is a neighborhood of the origin in $\mathfrak{g}^*$. Therefore $\phi$ commutes with each element of $G^*$. In particular, we have $\phi \circ \psi = \psi \circ \phi$. Now

$$d(a', \phi(a')) = d(\psi(a), \phi(\psi(a))) = d(\psi(a), \psi(\phi(a))) = d(a, \phi(a)).$$

Thus $\phi$ is a Clifford translation. $\qquad\square$

**Lemma 4.3.** *Let $M$, $N$, and $\phi$ be as in Lemma 4.2. Given $a_0 \in N$, let $a_1 = \phi(a_0)$ and $\gamma^*(t)$, $0 \leq t \leq L$, be a unit-speed-minimizing geodesic from $a_0$ to $a_1$. Set $\gamma(t) = p(\gamma^*(t))$. Then $\gamma(t)$ is a unit-speed smooth closed geodesic.*

*Proof.* Since $p$ is a local isometry and $\gamma^*$ is minimizing, $\gamma$ is locally minimizing. Therefore $\gamma$ is a geodesic and it is obviously with unit speed. Hence we need only prove that $\gamma(t)$ is smooth at $\gamma(0) = \gamma(L)$. Suppose conversely that this is not true. Let $\varepsilon > 0$ be so small that the forward metric ball $\mathscr{B}_{a_i}^+(\varepsilon)$, $i = 0, 1$, in $N$ is diffeomorphic (by the map $p$) to the forward metric ball $\mathscr{B}_{\gamma(0)}^+(\varepsilon)$ in $M$. Let $\delta > 0$ be so small that both $\gamma(L - \delta)$ and $\gamma(\delta)$ are contained in $\mathscr{B}_{\gamma(0)}^+(\varepsilon)$. Then there exists a curve $\sigma$ in $\mathscr{B}_{\gamma(0)}^+(\varepsilon)$ from $\gamma(L - \delta)$ to $\gamma(\delta)$ with length strictly less than the length of $\gamma$ from $\gamma(L - \delta)$ to $\gamma(\delta)$. Thus (note that $p$ is a local isometry)

$$l(\sigma) < l[\gamma(t)|_{[L-\delta, \delta]}] \leq l[\gamma(t)|_{[L-\delta, L]}] + l[\gamma(t)|_{[0, \delta]}] = \delta + \delta = 2\delta.$$

Let $\sigma*$ be the curve in $\mathscr{B}_{a_1}^+(\varepsilon)$ such that $p(\sigma^*) = \sigma$ and let $a^*$ be the end point of $\sigma^*$. Then $a^* = \phi(\gamma^*(\delta))$. Now by the triangle inequality of the distance function in Finsler spaces, we have

$$d(\gamma^*(\delta), a^*) \leq d(\gamma^*(\delta), \gamma^*(L-\delta)) + d(\gamma^*(L-\delta), a^*).$$

Since $\gamma^*$ is minimizing, we have

$$d(\gamma^*(\delta), \gamma^*(L-\delta)) = L - 2\delta.$$

On the other hand,

$$d(\gamma^*(L-\delta), a^*) \leq l(\sigma^*) = l(\sigma) < 2\delta.$$

Hence

$$d(\gamma^*(\delta), \phi(\gamma^*(\delta))) = d(\gamma^*(\delta), a^*) < 2L = d(a_0, a_1) = d(a_0, \phi(a_0)).$$

This is a contradiction to Lemma 4.2.                                                         □

**Proof of Theorem 4.12.** Suppose conversely that $M$ is not simply connected. Let $N$ be the universal covering manifold of $M$ and $p : N \to M$ the covering projection. Endow $N$ with the Finsler metric $F^*$ defined by

$$F^*(y) = F(dp(y)), \quad y \in TN.$$

Then $p : N \to M$ is a locally isometric covering projection. Since $M$ is not simply connected, $p$ is not a diffeomorphism. This means that there exists a nontrivial (i.e., not equal to the identity transformation) homeomorphism $\phi$ of $N$ such that $p \circ \phi = p$. By Lemmas 4.2 and 4.3, we can find a closed smooth unit-speed geodesic on $M$, say $\gamma(t)$, $0 \leq t \leq L$, where $L > 0$. Let $T$ be the tangent vector field of $\gamma$ and let $V$ be any Killing vector field on $M$. Set $T_t = T|_{\gamma(t)}$. Define a nonnegative function $f(t)$, $-\infty < t < \infty$, as follows:

$$f(t) = g_{T_t}(V,V), \quad \text{for } 0 \leq t \leq L,$$

and extend $f(t)$ to a periodic function of period $L$. Note that $T_t$ is everywhere nonzero, since $\gamma(t)$ is unit speed. This means that $f(t)$ is well defined and is smooth on $(0,L)$. By Lemma 4.3, $f(t)$ is smooth for all $t$.

Let

$$V' = D_T V, \quad V'' = D_T V',$$

where the covariant derivatives are taken with reference vector $T$. By Proposition 3.4, $V$ is a Jacobi field along $\gamma$, i.e., $V'' = -R((V,T)T$. Therefore we have (see Sect. 1.5)

$$f'(t) = 2g_{T_t}(V,V'),$$
$$f''(t) = 2g_{T_t}(V',V') - 2g_{T_t}(R(V,T)T,V).$$

Since $(M, F)$ has nonpositive flag curvature, we have $g_{T_t}(R(V,T)T,V) \leq 0, \forall t$. Thus $f''(t) \geq 0$. Since $f$ is a periodic smooth function, this implies that $f(t)$ is a constant function. Thus $f''(t) = 0$. Therefore we have $g_{T_t}(V',V') = 0$ and

$$g_{T_t}(R(V,T)T,V) = 0, \quad \forall t.$$

Now, $M$ is a homogeneous Finsler space. Therefore we can write $M = G/H$, where $G$ is a connected Lie group of isometries that is transitive on $M$ and $H$ is the isotropic subgroup of $G$ at $\gamma(0)$. By Theorem 4.2, there exists a $G$-invariant Riemannian metric on $G/H$. In particular, $G/H$ is a reductive homogeneous manifold. Hence there exists a subspace $\mathfrak{m}$ of $\mathfrak{g}$ with $\mathrm{Ad}(H)(\mathfrak{m}) \subset \mathfrak{m}$ such that

$$\mathfrak{g} = \mathfrak{h} + \mathfrak{m} \quad \text{(direct sum of subspaces)}.$$

There is a canonical way to identify $\mathfrak{m}$ with the tangent space $T_{eH}(G/H) = T_{\gamma(0)}(M)$. For every $w \in \mathfrak{m}$, the one-parameter subgroup $\exp tw$ induces a Killing vector field on $M = G/H$ that is equal to $w$ at $\gamma(0) = eH$. Hence

$$g_{T_0}(R(w,T)T,w) = 0, \quad \forall w \in \mathfrak{m}.$$

But this is in contradiction to the assumption that $\mathrm{Ric}(T_0) > 0$. The contradiction comes from the assumption that $M$ is not simply connected. □

## 4.5 Examples: Invariant Fourth-Root Metrics

In this section we study a special type of invariant Finsler metric on homogeneous manifolds. This gives numerous explicit examples of homogeneous Finsler spaces. The metrics we will consider are currently called fourth-root metrics. Such metrics are a special case of the $m$th-root metrics, which are defined by the fundamental function

$$F = \sqrt[m]{a_{i_1 i_2 \cdots i_m}(x) y^{i_1} y^{i_2} \cdots y^{i_m}}. \tag{4.2}$$

This class of Finsler metrics was first studied by Shimada in [148]. Recently, $m$th-root metrics, in particular fourth-root metrics, have been taken as a model of space-time in physics (see, e.g. [129]). See also Li and Shen's paper [113] on projectively flat fourth-root Finsler metrics.

Let $G/H$ be a reductive homogeneous space with the reductive decomposition $\mathfrak{g} = \mathfrak{h} + \mathfrak{m}$, i.e., $\mathrm{Ad}(h)\mathfrak{m} \subset \mathfrak{m}, \forall h \in H$, where $\mathfrak{g}$ and $\mathfrak{h}$ are the Lie algebras of $G$ and $H$, respectively. By Proposition 4.1, the $G$-invariant Finsler metric is in one-to-one correspondence with the Minkowski norm on $\mathfrak{m}$ satisfying

$$F(\mathrm{Ad}(h)y) = F(y), \quad \forall h \in H, \ y \in \mathfrak{m}. \tag{4.3}$$

If $F$ is an invariant $m$th-root Finsler metric on $G/H$, then $f = F^m$ is an $\mathrm{Ad}(H)$-invariant positive polynomial function of degree $m$ on $\mathfrak{m}$. Therefore $m$ must be even. Hence an invariant $m$th-root Finsler metric must be reversible. For $m = 2$, the invariant square-root Finsler metrics are invariant Riemannian metrics. It is well known that there exist invariant Riemannian metrics on homogeneous spaces $G/H$ when $H$ is compact. Let $\alpha = \sqrt{a_{ij}y^i y^j}$ be an invariant Riemannian metric. Then $F = \sqrt[2n]{\alpha^{2n}}$ is an invariant $2n$th-root metric and is Riemannian. So we are more concerned with the non-Riemannian metrics of the form (4.2). By (4.3), in order to construct an invariant $m$th-root Finsler metric on $G/H$, we need to accomplish the following two steps:

1. Find all $H$-invariant polynomials of degree $m$ on $\mathfrak{m}$.
2. Determine which of them can define a Minkowski norm on $\mathfrak{m}$.

In this section, we investigate invariant fourth-root Finsler metrics on the real Grassmannian manifolds $SO(p+q)/SO(p) \times SO(q)$.

**Definition 4.4.** A Finsler metric $F$ is called a $2m$th-root metric if in a local coordinate system, $F$ can be expressed in the following form:

$$F = \sqrt[2m]{a_{i_1 i_2 \cdots i_{2m}}(x)y^{i_1} y^{i_2} \cdots y^{i_{2m}}}.$$

Let $T$ denote the $2m$th power of a $2m$th-root metric $F$. Then a direct computation shows that

$$F = T^{\frac{1}{2m}},$$

$$F_{y^i} = \frac{1}{2m}T^{\frac{1}{2m}-1}T_{y^i},$$

$$F_{y^i y^j} = \frac{1}{4m^2}T^{\frac{1}{2m}-2}((1-2m)T_{y^i}T_{y^j} + 2mTT_{y^i y^j}),$$

$$g_{ij} = F_{y^i}F_{y^j} + FF_{y^i y^j} = \frac{1}{2m^2}T^{\frac{1}{m}-2}((1-m)T_{y^i}T_{y^j} + mTT_{y^i y^j}).$$

From the last equation we easily get the following.

**Proposition 4.4.** *The Hessian matrix* $(g_{ij})$ *is positive definite if and only if* $((1-m)T_{y^i}T_{y^j} + mTT_{y^i y^j})$ *is positive definite.*

We now consider invariant fourth-root Finsler metrics on the Grassmannian manifolds. Recall that they are the symmetric spaces $G/H = SO(p+q)/SO(p) \times SO(q)$, and are usually denoted by $G_p^+(\mathbb{R}^{p+q})$. Without loss of generality, we suppose $p \geq q$ and $p+q \geq 3$. Let $\mathfrak{g} = \mathfrak{h} + \mathfrak{m}$ be the standard Cartan decomposition, namely,

$$\mathfrak{g} = \mathfrak{so}(p+q) = \left\{ \begin{pmatrix} A & C \\ -C^T & B \end{pmatrix} \middle| A+A^T = 0, B+B^T = 0, A \in R^{p\times p}, B \in R^{q\times q} \right\},$$

$$\mathfrak{h} = \mathfrak{so}(p) + \mathfrak{so}(q) = \left\{ \begin{pmatrix} A & O \\ O & B \end{pmatrix} \middle| A+A^T = 0, B+B^T = 0, A \in R^{p\times p}, B \in R^{q\times q} \right\},$$

$$\mathfrak{m} = \left\{ \begin{pmatrix} O & C \\ -C^T & O \end{pmatrix} \middle| C \in R^{p \times q} \right\}.$$

Let $k[x_{ij}]$ be a polynomial ring over a field $k$ in $pq$ variables $x_{ij}$, $1 \le i \le p$, $1 \le j \le q$. The matrix $X = (x_{ij})$ is called a generic $p \times q$ matrix over $k$. Then the vector space $\mathfrak{m}$ can be identified with the set of generic matrices.

**Lemma 4.4.** *Let $f$ be a polynomial on $\mathfrak{m}$. Then $f$ is $\mathrm{Ad}(H)$-invariant if and only if*

$$f(X) = f(AX) = f(XB^T), \quad \forall A \in \mathfrak{so}(p), \ B \in \mathfrak{so}(q). \tag{4.4}$$

*Proof.* Recall that the adjoint representation is given by

$$\mathrm{Ad}(h) \begin{pmatrix} O & C \\ -C^T & O \end{pmatrix} = \begin{pmatrix} A_1 & O \\ O & A_2 \end{pmatrix} \begin{pmatrix} O & C \\ -C^T & O \end{pmatrix} \begin{pmatrix} A_1^T & O \\ O & A_2^T \end{pmatrix} = \begin{pmatrix} O & A_1 C A_2^T \\ -A_2 C^T A_1^T & O \end{pmatrix},$$

where

$$h = \begin{pmatrix} A_1 & O \\ O & A_2 \end{pmatrix} \in \mathrm{SO}(p) \times \mathrm{SO}(q), \quad \begin{pmatrix} O & C \\ -C^T & O \end{pmatrix} \in \mathfrak{m}.$$

It follows that $f$ is $\mathrm{Ad}(H)$-invariant if and only if

$$f(X) = f(AXB^T), \quad \forall A \in \mathfrak{so}(p), B \in \mathfrak{so}(q).$$

This is equivalent to (4.4).        □

Next we introduce some notation from the invariant theory of Lie groups. Let $W$ be a finite-dimensional real vector space and let $R[W]$ be the $R$-algebra of the ring of polynomial functions on $W$. A function $f \in R[W]$ is called $G$-invariant if $f(gw) = f(w)$ for all $g \in G$ and $w \in W$. The set of $G$-invariants forms a subalgebra of $R[W]$, denoted by $R[W]^G$. We need the following famous first fundamental theorem for $\mathrm{SO}(p)$ (see [130]).

**Theorem 4.13.** *Let $V_p$ be a $p$-dimensional real linear space. Then the invariant algebra $R[V_p^q]^{\mathrm{SO}(p)}$ is generated by the invariants $c(i,j)$, $1 \le i \le j \le q$, together with the determinants $\det[i_1, i_2, \ldots, i_p]$, $1 \le i_1 < i_2 < \cdots < i_p \le q$, where $c(i,j) = \sum_{a=1}^{p} x_{ai} x_{aj}$ and $\det[i_1, i_2, \ldots, i_p]$ means the determinant of the matrix with columns $i_1, i_2, \ldots, i_p$.*

By Lemma 4.4 and the above theorem, we have the following result.

**Proposition 4.5.** *A polynomial $f$ on $\mathfrak{m}$ is $\mathrm{Ad}(H)$-invariant if and only if*

$$f \in R[V_p^q]^{\mathrm{SO}(p)} \bigcap R[V_p^q]^{\mathrm{SO}(q)_r}. \tag{4.5}$$

Here $R[V_p^q]^{SO(q)_r}$ stands for the algebra of the invariant polynomials of the right action of $SO(q)$ on $V_p$. It is easily seen that

$$R[V_p^q]^{SO(q)_r} = \langle r(i,j), \det[i_1,\ldots,i_q] : 1 \le i \le j \le p, 1 \le i_1 < \cdots < i_q \le p \rangle, \quad (4.6)$$

where $r(i,j) = \sum_{a=1}^q x_{ia} x_{ja}$ and the determinants are composed of some row vectors.
In particular, if $q = 1$, then (4.4) becomes

$$f \in R[V_p]^{SO(p)} = \left\langle \sum_{a=1}^p x_{a1}^2 \right\rangle.$$

In this case, the corresponding invariant $2m$th-root metric is Riemannian.

Next we consider the special case of $p = q = 2$. It is obvious that $SO(4)/SO(2) \times SO(2)$ is isometric to the Riemannian product $SO(3)/SO(2) \times SO(3)/SO(2)$. This implies that the isotropy representation is not irreducible.

In general, let $G/H$ be a homogeneous manifold with $H$ compact. Suppose the adjoint representation of $H$ on $\mathfrak{g}/\mathfrak{h}$ is not irreducible. Then $\mathfrak{g}/\mathfrak{h}$ has a decomposition

$$\mathfrak{g}/\mathfrak{h} = \mathfrak{m}_1 + \mathfrak{m}_2,$$

where $\mathfrak{m}_i$ are nontrivial invariant subspaces. Since $H$ is compact, there exists an $\mathrm{Ad}(H)$-invariant inner product $\alpha_i$ on $\mathfrak{m}_i$, $i = 1,2$. Then we can construct a fourth-root Finsler metric using the following result (see [44]).

**Proposition 4.6.** *Let $M_1, M_2$ be connected manifolds endowed with a Riemannian metric $\alpha_1, \alpha_2$, respectively. Let $f : [0,\infty) \times [0,\infty) \to [0,\infty)$ be a $C^\infty$ function satisfying*

$$f(\lambda s, \lambda t) = \lambda f(s,t), \quad \forall \lambda > 0,$$

*and*

$$f > 0, \quad f_s > 0, \quad f_t > 0, \quad f_s + 2s f_{ss} > 0, \quad f_t + 2t f_{tt} > 0, \quad f_s f_t - 2 f f_{st} > 0,$$

*for all $(s,t) \ne (0,0)$. Then*

$$F(x,y) := \sqrt{f([\alpha_1(x_1,y_1)]^2, [\alpha_2(x_2,y_2)]^2)}$$

*defines a reversible Finsler metric on $M = M_1 \times M_2$, where $x = (x_1, x_2) \in M$, $y = y_1 \oplus y_2 \in T_x M$.*

In order to produce a polynomial of degree four, $f(s,t)$ must have the form

$$f(s,t) = c\sqrt{s^2 + 2\lambda st + \mu t^2} \quad (c > 0). \quad (4.7)$$

It is easy to see that the metric $F$ constructed as above is an invariant Finsler metric if and only if

$$0 < \lambda < 3\sqrt{\mu}. \tag{4.8}$$

When $\lambda^2 \neq \mu$, the metric is non-Riemannian and Berwaldian.

A coset space $G/H$ is called isotropy irreducible if the isotropic representation of $H$ on $T_o(G/H)$, where $o$ is the origin of $G/H$, is irreducible. The above arguments lead to the following result.

**Theorem 4.14.** *If the homogeneous $G/H$ is not isotropy irreducible, then there exist infinitely many different invariant fourth-root non-Riemannian Finsler metrics on $G/H$. In particular, On $SO(4)/SO(2) \times SO(2)$ there exist infinitely many invariant fourth-root non-Riemannian Finsler metrics.*

Next we show how to construct invariant fourth-root metrics on $SO(4)/SO(2) \times SO(2)$. This actually gives a classification of all such metrics on it.

**Proposition 4.7.** *Let $F = f^{1/4}$ be an invariant fourth-root Finsler metric on $G_2^+(\mathbb{R}^4)$ and $\mathfrak{m} = \left( \begin{smallmatrix} x_1 & x_2 \\ x_3 & x_4 \end{smallmatrix} \right)$. Then*

$$f = c\left(A^2 + 2tAB + 4sB^2\right), \tag{4.9}$$

*where $c > 0$, $A = x_1^2 + x_2^2 + x_3^2 + x_4^2$, $B = x_1 x_4 - x_2 x_3$, and $s$, $t$ satisfy*

$$0 < 1 - s < 3\sqrt{(1+s)^2 - t^2}.$$

*Proof.* Let

$$\mathfrak{m} = \mathfrak{m}_1 + \mathfrak{m}_2 = \frac{1}{2}\begin{pmatrix} x_1 + x_4 & x_2 - x_3 \\ x_3 - x_2 & x_1 + x_4 \end{pmatrix} + \frac{1}{2}\begin{pmatrix} x_1 - x_4 & x_2 + x_3 \\ x_2 + x_3 & x_4 - x_1 \end{pmatrix}$$

be the decomposition of $\mathfrak{m}$ into irreducible invariant subspaces with respect to the adjoint representation. Then the invariant Riemannian metrics on $\mathfrak{m}_1$ and $\mathfrak{m}_2$ are $(x_1 + x_4)^2 + (x_2 - x_3)^2$ and $(x_1 - x_4)^2 + (x_2 + x_3)^2$, respectively. By (4.7), (4.8), and a simple computation we get (4.9). This completes the proof. $\qquad\square$

Now we consider the other cases. Without loss of generality, we can assume that $p > 2, q \geq 2$.

**Lemma 4.5.** *Let $f$ be an $\mathrm{Ad}(SO(p) \times SO(q))$-invariant polynomial of degree four on $\mathfrak{m}$. Then*

$$f = \lambda \left( \sum_{\substack{1 \leq u \leq p \\ 1 \leq i \leq q}} x_{ui}^2 \right)^2 + \mu \sum_{1 \leq i,j \leq q} \left( \sum_{1 \leq u \leq p} x_{ui}x_{uj} \right)^2 + \nu \delta_{4p}\delta_{4q}\det(x_{ij})_{4 \times 4}, \tag{*}$$

*where* $\lambda, \mu, \nu$ *are real numbers. In other words, the real vector space* $V$ *of the* $\mathrm{Ad}(SO(p) \times SO(q))$-*invariant polynomials of degree four is spanned by the following three polynomials:*

$$\alpha_1 = \left( \sum_{\substack{1 \le u \le p \\ 1 \le i \le q}} x_{ui}^2 \right)^2, \quad \alpha_2 = \sum_{1 \le i, j \le q} \left( \sum_{1 \le u \le p} x_{ui} x_{uj} \right)^2, \quad \alpha_3 = \delta_{4p} \delta_{4q} \det(x_{ij})_{4 \times 4}.$$

*Proof.* It is obvious that the above-defined $f$ satisfies (4.5). Hence we need only prove that if $f$ satisfies (4.5), then it must be of the form as stated in the lemma. Since $f \in R[V_p^q]^{SO(p)}$ by the left action, by Theorem 4.13, $f$ must be a real linear span of $c(i, j)c(k, l)$, $1 \le i \le j \le q$, $1 \le k \le l \le q$, $i \le k$ and the determinant $\alpha_3$. Next we consider the right action. By (4.6), $f$ must be a real span of the following polynomials:

$$r(u, v)r(s, t), \quad 1 \le u \le v \le p, \quad 1 \le s \le t \le p, \quad u \le s;$$

$$\alpha_3 = \delta_{4p} \delta_{4q} \det(x_{ij})_{4 \times 4};$$

$$\delta_{2q} r(s, t) \begin{vmatrix} x_{u1} & x_{u2} \\ x_{v1} & x_{v2} \end{vmatrix}, \quad 1 \le s \le t \le p, \quad 1 \le u < v \le p;$$

$$\delta_{2q} \begin{vmatrix} x_{u1} & x_{u2} \\ x_{v1} & x_{v2} \end{vmatrix} \begin{vmatrix} x_{s1} & x_{s2} \\ x_{t1} & x_{t2} \end{vmatrix}, \quad 1 \le u < v \le p, \quad 1 \le s < t \le p, u \le s.$$

Since $\alpha_3$ appears in both cases, we need treat only the following cases:

(1) There is a quadruple $(u, v, s, t)$, with $u \ne v$, such that the coefficient of $r(u, v)r(s, t)$ in the expansion of $f$ is nonzero. Since

$$r(u, v)r(s, t) = \sum_{m, n} x_{um} x_{vm} x_{sn} x_{tn},$$

by the invariance of $f$ under the left action, for any pair $(m, n)$, there must be a quadruple $(i, j, k, l)$ such that $x_{um} x_{vm} x_{sn} x_{tn}$ appears in the expansion of $c(i, j)c(k, l)$. Then we have $u = s$, $v = t$ and $i = k = m$, $j = l = n$. Now

$$r(u, v)^2 = \left( \sum_m x_{um} x_{vm} \right)^2 = \sum_{m, n} x_{um} x_{vm} x_{un} x_{vn}.$$

Note that among the polynomials $c(i, j)c(k, l)$ and $\alpha_3$, the monomial $x_{um} x_{vm} x_{un} x_{vn}$ appears only in the expansion of $c(m, n)^2$. Thus all the monomials $c(m, n)^2$ will appear in the expansion in $f$ with the same coefficients when we write $f$ as a linear combination of $c(i, j)c(k, l)$ and $\alpha_3$. Therefore the polynomial

$$\sum_{m, n} c(m, n)^2 = \sum_{m, n} \left( \sum_a x_{am} x_{an} \right)^2 = \alpha_2$$

must appear in the expansion of $f$ with nonzero coefficient. On the other hand, it is easily seen that

$$\alpha_2 = \sum_{m,n,a,b} x_{ma}x_{mb}x_{na}x_{nb} = \sum_{m,n} r(m,n)^2.$$

A similar argument as above shows that the monomial $x_{ma}x_{mb}x_{na}x_{nb}$ can appear only in the expansion of $r(m,n)^2$ when we write $f$ as a linear combination of the right invariants above. This argument shows that if the coefficient of $r(u,v)r(s,t)$, $u \neq v$, in the expansion of $f$ is nonzero, then $u = s$, $v = t$, and all the $r(u,v)^2$ appear in $f$ with the same coefficients.

(2) There is a pair $(u,s)$ such that the coefficient of $r(u,u)r(s,s)$ in the expansion of $f$ is nonzero. As in case (1), for any pair $(m,n)$, there must be a quadruple $(i,j,k,l)$ such that the monomial $x_{um}^2 x_{sn}^2$ appears in the expansion of $c(i,j)c(k,l)$. From this it follows that $i = j = m$, $k = l = n$. Similar to case (1), we can prove that all the $c(i,i)c(j,j)$ appear in the expression of $f$ when $f$ is written as a linear combination of the left invariants. Hence in this case, the coefficient of

$$\sum_{i,j} c(i,i)c(j,j)$$

in the expression of $f$ as a linear combination of the invariants of the left action must be nonzero. Moreover, this shows that all the $r(a,a)r(b,b)$ also appear in the expression of $f$ as a linear combination of right invariants, with the same nonzero coefficient. Note that

$$\sum_{a,b} r(a,a)r(b,b) = \alpha_1.$$

(3) $q = 2$ and the coefficient of

$$\begin{vmatrix} x_{u1} & x_{u2} \\ x_{v1} & x_{v2} \end{vmatrix} \begin{vmatrix} x_{s1} & x_{s2} \\ x_{t1} & x_{t2} \end{vmatrix}$$

in the expansion of $f$ is nonzero. Since in the expansion of $c(i,j)c(k,l)$, each monomial has at most two row indices, we have $u = s$ and $v = t$. Then

$$\begin{vmatrix} x_{u1} & x_{u2} \\ x_{v1} & x_{v2} \end{vmatrix} \begin{vmatrix} x_{s1} & x_{s2} \\ x_{t1} & x_{t2} \end{vmatrix} = \begin{vmatrix} x_{u1} & x_{u2} \\ x_{v1} & x_{v2} \end{vmatrix}^2 = \begin{vmatrix} r(u,u) & r(u,v) \\ r(u,v) & r(v,v) \end{vmatrix}.$$

This reduces the proof to cases (1) and (2).

(4) $q = 2$ and the coefficient of

$$r(s,t) \begin{vmatrix} x_{u1} & x_{u2} \\ x_{v1} & x_{v2} \end{vmatrix}$$

in the expression of $f$ is nonzero. Similarly to case (3), we have $u = s$, $v = t$. However,

$$r(u,v) \begin{vmatrix} x_{u1} & x_{u2} \\ x_{v1} & x_{v2} \end{vmatrix} = (x_{u1}x_{v1} + x_{u2}x_{v2})(x_{u1}x_{v2} - x_{u2}x_{v1}),$$

and it is easily seen that for any given quadruple $(i,j,k,l)$, the monomials in the expansion of the above polynomial cannot appear in the expansion of $c(i,j)c(k,l)$. This is a contradiction. Therefore this case cannot happen.

Combining cases (1)–(4), we see that $f$ can be expressed as a linear combination of $\alpha_1$, $\alpha_2$, $\alpha_3$. This completes the proof of the lemma.                    $\square$

*Remark 4.3.* It is easy to check that a polynomial $f$ as in $(*)$ of the above lemma can be rewritten as

$$f = \lambda [\operatorname{tr}(X^T X)]^2 + \mu \operatorname{tr}(X^T X)^2 \pm \nu \delta_{4p} \delta_{4q} \sqrt{\det(X^T X)}. \qquad (4.10)$$

*Remark 4.4.* The first section in the expression of $f$ in (4.10) corresponds to the standard Riemannian metric.

**Theorem 4.15.** *When $\varepsilon$ is small enough, the function*

$$f = \begin{cases} \{\operatorname{tr}(X^T X)\}^m + \varepsilon \\ \{\operatorname{tr}((X^T X)^2)\}^{\frac{m}{2}}, & \text{for } m \text{ even,} \\ \{\operatorname{tr}(X^T X)\}^m + \varepsilon\{\operatorname{tr}((X^T X)^2)\}^{\frac{m-1}{2}} \operatorname{tr}(X^T X), & \text{for } m \text{ odd,} \end{cases}$$

*defines an invariant $2m$th-root Finsler metric on $\mathrm{SO}(p+q)/\mathrm{SO}(p) \times \mathrm{SO}(q)$.*

*Proof.* It is sufficient to notice the fact that $\operatorname{tr}(X^T X)$ and $\operatorname{tr}(X^T X)^2$ are invariant polynomials from the proof of the above lemma.                    $\square$

Combining Theorems 4.14 and 4.15, we get the following result.

**Theorem 4.16.** *For $G/H = \mathrm{SO}(p+q)/\mathrm{SO}(p) \times \mathrm{SO}(q)$ with $p \geq q$ and $p+q \geq 3$, we have*

1. *If $q = 1$, then there do not exist any invariant non-Riemannian fourth-root Finsler metrics on $G/H$.*
2. *If $p = q = 2$, then the isotropic representation is reducible and there exist infinitely many different invariant non-Riemannian fourth-root Finsler metrics on $G/H$.*
3. *If $p > q \geq 2$, then the isotropic representation is irreducible and there exist infinitely many distinct invariant non-Riemannian fourth-root Finsler metrics on $G/H$.*

# Chapter 5
# Symmetric Finsler Spaces

In this chapter, we study the symmetry of Finsler spaces. By Theorem 1.9, if a connected Finsler space $(M, F)$ is a Berwald space, then the Chern connection of $(M, F)$ is the Levi-Civita connection of a Riemannian metric. We call a Berwald space $(M, F)$ locally (resp. globally) affine symmetric if its linear connection is locally (resp. globally) affine symmetric. On the other hand, a connected Finsler space $(M, F)$ is called locally symmetric if for each point $x$ of the manifold, there exists a neighborhood $U$ of $x$ such that the geodesic symmetry is a local isometry of $U$. It is called globally symmetric if each point is the isolated point of an involutive isometry of $(M, F)$.

Section 5.1 is devoted to studying the geometric properties of locally affine symmetric and globally affine symmetric Berwald spaces. Here an important result of Cartan, Theorem 2.28, is generalized to Berwald spaces; namely, we prove that a Berwald space is locally affine symmetric if and only if the flag curvature is invariant under all parallel displacements. In Sect. 5.2, we prove that every globally symmetric Finsler space must be a Berwald space. In Sect. 5.3, we introduce the notion of a Minkowski symmetric Lie algebra to give an algebraic description of affine symmetric Berwald spaces as well as globally symmetric Finsler spaces. Another important result of Cartan, asserting that a globally symmetric Riemannian manifold has nonnegative (resp. nonpositive) sectional curvature if it is of compact type (resp. noncompact type), is generalized to the Finslerian case in this section. In Sect. 5.4, we prove some important rigidity results on symmetric Finsler spaces. In particular, we prove that a connected complete locally symmetric Finsler space whose flag curvature is everywhere nonzero must be Riemannian. Finally, in Sect. 5.1, we study complex structures on symmetric Finsler spaces and obtain a complete classification of the complex symmetric Finsler spaces.

Regarding further development of the topics in this chapter, we should mention that the general geometric properties of symmetric Finsler spaces (or more generally, affine symmetric Berwald spaces) deserve to be studied thoroughly. For example, there has been a great deal of work on the properties of geodesics and totally geodesic submanifolds of Riemannian symmetric spaces (see [40, 41, 82, 83]), but for the Finslerian case this problem has not been considered. On the other

S. Deng, *Homogeneous Finsler Spaces*, Springer Monographs in Mathematics, 105
DOI 10.1007/978-1-4614-4244-8_5, © Springer Science+Business Media New York 2012

hand, a series of papers as appeared on the structure and classification of pseudo-Riemannian symmetric spaces (see, for example, Cahen–Parker [36], and Kath–Olbrich [96,97]). The Finslerian case of this problem is very interesting.

The standard references of this chapter are [1,57,58,62]; see [50,54,65,76,149, 155,157,167] for further information.

## 5.1  Affine Symmetric Berwald Spaces

We first study some properties of isometries of a Berwald space. This will be useful in our study of symmetric Finsler spaces. The following results are well known for Riemannian manifolds. However, for a Berwald space, they are far from obvious.

**Theorem 5.1.** *Let $(M,F)$ be a connected Berwald space and $\psi$ an isometry of $(M,F)$ onto itself. Then $\psi$ is an affine transformation with respect to the connection of $F$.*

*Proof.* We first deduce a formula for the connection $D$ of $F$. For any vector fields $T,V,W$ on $M$, we have (see [16, p. 260])

$$T g_W(V,W) = g_W(D_T V, W) + g_W(V, D_T W). \tag{5.1}$$

Similarly,

$$V g_W(T,W) = g_W(D_V T, W) + g_W(T, D_V W), \tag{5.2}$$

$$W g_W(V,W) = g_W(D_W V, W) + g_W(V, D_W W). \tag{5.3}$$

Subtracting (5.2) from the sum of (5.1) and (5.3), we get

$$g_W(V, D_{W+T}W) + g_W(W - T, D_V W)$$
$$= T g_W(V,W) - V g_W(T,W) + W g_W(V,W) - g_W([T,V], W) - g_W([W,V], W),$$

where we have used the symmetry of the connection, i.e., $D_V W - D_W V = [V,W]$. Setting $T = W - V$ in the above equation, we obtain

$$2 g_W(V, D_W W) = 2W g_W(V,W) - V g_W(W,W) - 2 g_W([W,V], W). \tag{5.4}$$

Since $\psi$ is an isometry, $d\psi$ is a linear isometry between the spaces $T_p(M)$ and $T_{\psi(p)}(M)$, $\forall p \in M$. Therefore for any vector fields $X,Y,Z$ on $M$, we have

$$g_{d\psi(X)}(d\psi(Y), d\psi(Z)) = g_X(Y,Z).$$

By (5.4),

$$g_{d\psi(W)}(d\psi(V), D_{d\psi(W)} d\psi(W)) = g_W(V, D_W W).$$

Consequently,

$$g_{d\psi(W)}(d\psi(V), D_{d\psi(W)}d\psi(W)) = g_{d\psi(W)}(d\psi(V), d\psi(D_W W)).$$

Since $V$ is arbitrary and $g_{d\psi(W)}(\cdot, \cdot)$ is an inner product, we have

$$D_{d\psi(W)}d\psi(W) = d\psi(D_W W).$$

Now using the identity

$$D_V W = \frac{1}{2}(D_{V+W}(V + W) - D_V V - D_W W - [W, V]),$$

we get

$$D_{d\psi(V)}d\psi(W) = d\psi(D_V W).$$

Therefore, $\psi$ is an affine transformation with respect to $D$. $\qquad\square$

The following result is a generalization of a well-known result in Riemannian geometry.

**Theorem 5.2.** *Let $(M_i, F_i)$, $i = 1, 2$, be two Berwald spaces, and $\psi$ an affine diffeomorphism from $(M_1, F_1)$ onto $(M_2, F_2)$ with respect to the connections of $F_1$ and $F_2$. If there exists $p \in M_1$ such that $d\psi$ is a linear isometry from $T_p(M_1)$ onto $T_{\psi(p)}(M_2)$, then $\psi$ is an isometry.*

*Proof.* Let $q \in M_1$. We prove that $F_1(y) = F_2(d\psi(y))$, $\forall y \in T_q(M_1)$. Join $q$ to $p$ by a curve $\gamma$. Let $\tau$ denote the parallel transformation from $q$ to $p$ along $\gamma$. Then for every $u, v \in T_q(M_1)$, we have

$$g_y(u, v) = g_{\tau(y)}(\tau(u), \tau(v)) = g_{d\psi(\tau(y))}(d\psi(\tau(u)), d\psi(\tau(v))).$$

Now $\tau(y), \tau(u), \tau(v)$ is the result of the parallel displacement (along $\gamma$) of $y, u, v$, respectively. Since $\psi$, being an affine diffeomorphism, transforms vectors that are parallel along $\gamma$ into vectors that are parallel along $\psi(\gamma)$, the vectors $d\psi(\tau(y))$, $d\psi(\tau(u))$, $d\psi(\tau(v))$ must be the result of the parallel displacement (along $\psi(\gamma)$) of $d\psi(y), d\psi(u), d\psi(v)$, respectively. Thus

$$g_{d\psi(\tau(y))}(d\psi(\tau(u)), d\psi(\tau(v))) = g_{d\psi(y)}(d\psi(u), d\psi(v)).$$

Therefore

$$g_y(u, v) = g_{d\psi(y)}(d\psi(u), d\psi(v)).$$

Now the theorem follows from the facts that $F_1(y) = \sqrt{g_y(y, y)}$ and that

$$F_2(d\psi(y)) = \sqrt{g_{d\psi(y)}(d\psi(y), d\psi(y))}. \qquad\square$$

Next we generalize Cartan's results on locally symmetric Riemannian manifolds to Berwald spaces. The following result is a generalization of Theorem 2.28 on Riemannian manifolds. However, the proof is more complicated. Let us first explain the notion of parallel displacements of flags. Let $(M, F)$ be a Berwald space and $(P, y)$ a flag in a tangent space $T_p(M)$, $p \in M$. Let $q \in M$ and let $c$ be a piecewise smooth curve in $M$ connecting $p$ and $q$. Suppose $\tau$ is the parallel displacement along $c(t)$. Then $\tau(P)$ is a plane in $T_q(M)$ containing $\tau(y)$ and $\tau(y) \neq 0$ (since $\tau$ is a linear isomorphism). Therefore $(\tau(P), \tau(y))$ is a flag in $T_q(M)$. We say that the flag curvature is invariant under the parallel displacement $\tau$ if $K(P, y) = K(\tau(P), \tau(y))$, for every flag $(P, y)$ in $T_p(M)$.

**Theorem 5.3.** *Let $(M, F)$ be a Berwald space. Then $M$ is locally affine symmetric if and only if the flag curvature is invariant under all parallel displacements.*

*Proof.* Let $(M, F)$ be a locally affine symmetric Berwald space. Then its connection is locally affine symmetric. So the curvature tensor $R$ is invariant under all parallel displacements. On the other hand, suppose $p, q \in M$ and $\gamma$ is a curve connecting $p$ and $q$. Let $\tau$ be the parallel displacement along $\gamma$. Then for any $y\, (\neq 0), u, v \in T_p(M)$, we have $g_{\tau(y)}(\tau(u), \tau(v)) = g_y(u, v)$. Therefore, by the definition of flag curvature, we have

$$K(P, y) = K(\tau(P), \tau(y)),$$

where $P$ is any plane in $T_p(M)$ containing $y$. That is, the flag curvature is invariant under all parallel displacements.

Conversely, suppose the flag curvature is invariant under all parallel displacements. Let $p$, $q$, $\gamma$, $\tau$, $y$, $u$, $v$ be as above and suppose $u$ is linearly independent of $y$. Let $l = y/F$. Consider the quantity

$$K(l, u, v) = \frac{g_l(R(l, u)l, v)}{g_l(u, v) - g_l(l, u)g_l(l, v)}. \tag{5.5}$$

Then $K(P, y) = K(l, u, u)$. Since $K(P, y)$ is invariant under parallel displacements, we have

$$K(l, u, u) = K(\tau(l), \tau(u), \tau(u)).$$

By Theorem 1.11, any parallel displacement of a Berwald space is a linear isometry between the tangent spaces. Thus the denominator of $K(l, u, v)$ in (5.5) is invariant under $\tau$. Therefore we have

$$g_l(R_p(l, u)l, u) = g_{\tau(l)}(R_q(\tau(l), \tau(u))\tau l, \tau(u)).$$

Set $R(l, u, v) = g_l(R(l, u)l, v)$. Then it is easy to check that the following polarization identity holds (see [16, p. 70] and the erratum by Bao):

$$R(l, u, v) = \frac{1}{4}(R(l, u + v, u + v) - R(l, u - v, u - v)).$$

Then it follows that

$$g_{\tau(l)}(\tau(R_p(l,u)l), \tau(v)) = g_{\tau(l)}(R_q(\tau(l), \tau(u))\tau(l), \tau(v)).$$

This implies that

$$\tau(R_p(l,u)l) = R_q(\tau(l), \tau(u))\tau(l). \tag{5.6}$$

It is obvious that the above equality still holds if $u$ is linearly dependent of $y$. Now suppose $Q$ is a Riemannian metric on $M$ with the same connection as $F$. Consider the quadrilinear form $B$ defined by

$$B(u,v,z,t) = Q(R_q(\tau(u), \tau(v))\tau(z), \tau(t)) - Q(\tau(R_p(u,v)z), \tau(t)),$$

where $u,v,z,t \in T_p(M)$. Then we have

(a)  $B(u,v,z,t) = -B(v,u,z,t)$;
(b)  $B(u,v,z,t) = -B(u,v,t,z)$;
(c)  $B(u,v,z,t) + B(v,z,u,t) + B(z,u,v,t) = 0$;
(d)  $B(u,v,u,v) = 0$.

In fact, (a) is the well-known property of the curvature tensor of a Riemannian manifold. Moreover, it is easily seen that (b) and (c) follow from the fact

$$Q(\tau(R_p(u,v)z), \tau(t)) = Q(R_p(u,v)z, t),$$

since $\tau$ is also the parallel displacement with respect to $Q$, and that (d) follows from (5.6). By a well-known result in Riemannian geometry, we have $B \equiv 0$. Thus

$$\tau(R_p(u,v)z) = R_q(\tau(u), \tau(v))\tau(z),$$

i.e., $\tau R_p = R_q$. Therefore $D_X R = 0$ for each vector field $X$. Thus $D$ is locally affine symmetric.                                                                                    □

**Theorem 5.4.** *Let $(M,F)$ be a locally affine symmetric Berwald space. Then for any $p \in M$ there exist a globally affine symmetric Berwald space $(\widetilde{M}, \widetilde{F})$, a neighborhood $N_p$ of $p$ in $M$, and an isometry $\varphi$ of $N_p$ onto an open neighborhood of $\varphi(p)$ in $\widetilde{M}$.*

*Proof.* Let $D$ denote the connection of $F$. Suppose $Q$ is a Riemannian metric on $M$ such that $D$ is the Levi-Civita connection of $Q$. Then $(M,Q)$ is a Riemannian locally symmetric space. Thus by [83, Theorem 5.1, Chap. IV], there exist a Riemannian globally symmetric space $(\widetilde{M}, \widetilde{Q})$, a neighborhood $N_p$ of $p$ in $M$, and an isometry (with respect to $Q$ and $\widetilde{Q}$) $\varphi$ from $N_p$ onto a neighborhood of $\varphi(p)$ in $\widetilde{M}$. Let $\widetilde{D}$ be the Levi-Civita connection of $\widetilde{Q}$ and denote by $H$, resp. $\widetilde{H}$, the holonomy group of $D$ (at $p$), resp. $\widetilde{D}$ (at $\widetilde{p} = \varphi(p)$). Then $d\varphi_p$ induces an isomorphism between the holonomy algebra $\mathfrak{h}$ of $H$ and $\widetilde{\mathfrak{h}}$ of $\widetilde{H}$. Hence there exist a neighborhood $U_e$ of the unit element $e$ of $H$ and a neighborhood $\widetilde{U}_{\widetilde{e}}$ of the unit element $\widetilde{e}$ of $\widetilde{H}$ such that $\mathrm{d}\varphi_p$ induces a local isomorphism between $U_e$ and $\widetilde{U}_{\widetilde{e}}$. Without loss of generality, we can assume that $\widetilde{M}$ is simply connected. Then $\widetilde{H}$ is connected and is generated by the elements of $\widetilde{U}_{\widetilde{e}}$.

Now we define a Berwald metric on $\widetilde{M}$. Identifying $T_{\widetilde{p}}(\widetilde{M})$ with $T_p(M)$ through $\mathrm{d}\varphi_p$, we get a Minkowski norm $\widetilde{F}$ on $T_{\widetilde{p}}(\widetilde{M})$ by

$$\widetilde{F}(\widetilde{x}) = F((\mathrm{d}\varphi_p)^{-1}(\widetilde{x})), \quad \widetilde{x} \in T_{\widetilde{p}}(\widetilde{M}).$$

Since $F$ is invariant under $H$, $\widetilde{F}$ is invariant under $\widetilde{U}_{\widetilde{e}}$. Therefore, using the fact that $\widetilde{H}$ is generated by the elements of $\widetilde{U}_{\widetilde{e}}$, we see that $\widetilde{F}$ is invariant under $\widetilde{H}$. Now for any $\widetilde{q} \in \widetilde{M}$, join $\widetilde{q}$ to $\widetilde{o}$ by a curve $\widetilde{\gamma}$. Let $\widetilde{\tau}_{\widetilde{\gamma}}$ be the parallel displacement of $\widetilde{D}$ along $\widetilde{\gamma}$. Define a Finsler metric (still denoted by $\widetilde{F}$) on $\widetilde{M}$ by

$$\widetilde{F}(\widetilde{u}) = \widetilde{F}(\widetilde{\tau}_{\widetilde{\gamma}}(\widetilde{u})), \quad \widetilde{u} \in T_{\widetilde{q}}(\widetilde{M}).$$

It is easily seen that $\widetilde{F}$ is well-defined and $(\widetilde{M}, \widetilde{F})$ is a Berwald space with connection $\widetilde{D}$. Therefore it is a globally affine symmetric Berwald space.

It remains to prove that $\varphi$ is an isometry. But this follows easily from the fact that $\varphi$ is an affine diffeomorphism and $\mathrm{d}\varphi_p$ is a linear isometry between the Minkowski spaces $(T_pM, F)$ and $(T_{\widetilde{p}}\widetilde{M}, \widetilde{F})$ (see Theorem 5.2).                                          $\square$

**Corollary 5.1.** *Let $(M, F)$ be a locally affine symmetric Berwald space and $(\widetilde{M}, \pi)$ the universal covering manifold of $M$. Then $\widetilde{M}$ with the metric $\pi^*(F)$ is a globally affine symmetric Berwald space.*

## 5.2  Globally Symmetric Finsler Spaces

Intuitively, the definition of locally and globally affine symmetric Berwald spaces has no direct relation with geometry and it is limited to Berwald spaces. It is well known that a locally affine symmetric Riemannian space is actually locally symmetric. However, this is no longer true for Berwald spaces, since there exist many examples of nonreversible locally affine symmetric Berwald spaces, and a locally symmetric Finsler space must be reversible. For example, any Minkowski space, when viewed as a Finsler space in the canonical way, must be globally affine symmetric, and a Minkowski space need not be reversible. In this section, we will consider the symmetry of Finsler spaces from the geometric point of view.

**Definition 5.1.** Let $(M, F)$ be a connected Finsler space. If for any $x \in M$ there exists a neighborhood $U$ of $x$ such that the geodesic symmetry is a local isometry of $U$, then $(M, F)$ is called a locally symmetric Finsler space. If for any $x \in M$ there exists an involutive isometry $\sigma_x$ of $(M, F)$ such that $x$ is an isolated fixed point of $\sigma_x$, then $(M, F)$ is called a globally symmetric Finsler space.

Note that the exponential map of a Finsler space is generally not smooth. In fact, it is only $C^1$ at the zero section, and it is smooth if and only if the Finsler space is of the Berwald type; see [16]. However, using Theorem 1.3 and an argument similar to that used in the proof of Theorem 3.2, one can easily prove that the local geodesic isometry is well defined and is a smooth map. In fact, the notion of locally symmetric Finsler spaces appeared already in Foulon's paper [71].

We first give an effective method to construct globally symmetric Finsler spaces.

**Theorem 5.5.** *Let $G/K$ be a symmetric coset space. Then any $G$-invariant reversible Finsler metric (if it exists) $F$ on $G/K$ makes $(G/K, F)$ a globally symmetric Finsler space.*

*Proof.* We first define some diffeomorphisms of $G/K$ onto itself. Let $o = eK$ be the origin of $G/K$ and $\sigma$ the involutive automorphism of $G$ such that $(G_\sigma)_0 \subset K \subset G_\sigma$. Define a map $\sigma_o$ of $G/K$ onto itself by

$$\sigma_o(aK) = \sigma(a)K, \quad a \in G.$$

For any $x \in G/K$, select $a \in G$ such that $x = \pi(a)$, where $\pi$ is the natural projection of $G$ onto $G/K$. Define a map $\sigma_x$ of $G/K$ onto itself by

$$\sigma_x = \tau_a \sigma_o \tau_a^{-1},$$

where $\tau_a$ is defined by $\tau_a : gK \to agK$, $g \in G$. By the definition of symmetric coset spaces, it is easily seen that $\sigma_x$ is independent of the choice of $a$. Obviously $\sigma_x$ is an involutive diffeomorphism of $G/K$ with $x$ as an isolated fixed point. Next we prove that it is an isometry. Since $F$ is $G$-invariant, $\tau_a$ keeps $F$ invariant. Therefore we need treat only the case of $\sigma_o$. Let $\mathfrak{g}$, $\mathfrak{k}$ be the Lie algebras of $G$, $K$, respectively. Then $\sigma$ induces an involutive automorphism (still denoted by $\sigma$) of $\mathfrak{g}$. By definition, $\mathfrak{k}$ coincides with the set of fixed points of $\sigma$. Let $\mathfrak{m}$ be the eigenspace of $\sigma$ with eigenvalue $-1$. Then we have

$$\mathfrak{g} = \mathfrak{k} + \mathfrak{m} \quad \text{(direct sum of subspaces)}.$$

Therefore we can identify the tangent space $T_o(G/K)$ with $\mathfrak{m}$. Under this identification, $F$ corresponds to a $K$-invariant norm on $\mathfrak{m}$. By the assumption, $F(y) = F(-y)$, $\forall y \in \mathfrak{m}$. Thus $F(\sigma(y)) = F(y)$, $\forall y \in \mathfrak{m}$. By the invariance of the metric we see that $\sigma$ preserves the length of every tangent vector. Thus $\sigma$ is an isometry of $(G/K, F)$. This completes the proof of the theorem. □

Using the above theorem, we can construct a large number of globally symmetric Finsler metrics that are non-Riemannian. Let us give an explicit example.

*Example 5.1.* Let $G_1/K_1$, $G_2/K_2$ be two symmetric coset spaces with $K_1, K_2$ compact. Suppose $g_1, g_2$ are invariant Riemannian metrics on $G_1/K_1$, $G_2/K_2$, respectively. Let $M = (G_1/K_1) \times (G_2/K_2)$ and denote by $o_1, o_2$ the origins of $G_1/K_1$,

$G_2/K_2$, respectively. Let $o = (o_1, o_2)$ (the origin of $M$). Now for $y = y_1 + y_2 \in T_o(M) = T_{o_1}(G_1/K_1) \dotplus T_{o_2}(G_2/K_2)$, we define

$$F(y) = \sqrt{g_1(y_1, y_1) + g_2(y_2, y_2) + \sqrt[s]{g_1(y_1, y_1)^s + g_2(y_2, y_2)^s}},$$

where $s \geq 2$ is an integer. Then $F$ is a $K_1 \times K_2$-invariant Minkowski norm on $T_o(M)$ (see Example 1.3 of Sect. 1.1). Hence it defines a $G$-invariant Finsler metric on $M$. By Theorem 5.5, $(M, F)$ is a globally symmetric Finsler space.

Next we consider the converse of Theorem 5.5. For this purpose, we first need to study the geometric properties of globally symmetric Finsler spaces.

**Theorem 5.6.** *Let $(M, F)$ be a globally symmetric Finsler space. For $x \in M$, denote the involutive isometry of $(M, F)$ at $x$ by $\sigma_x$. Then we have*

(a) *For every $x \in M$, $(d\sigma_x)_x = -\mathrm{id}$. In particular, $F$ must be reversible.*
(b) *$(M, F)$ is (forward and backward) complete.*
(c) *$(M, F)$ is homogeneous. That is, the group $I(M, F)$ of isometries of $(M, F)$ acts transitively on $M$.*
(d) *Let $\widetilde{M}$ be the universal covering manifold of $M$ and $\pi$ the projection map. Then $(\widetilde{M}, \pi^*(F))$ is a globally symmetric Finsler space.*

*Proof.* (a) There exists a neighborhood $U$ of the origin of $T_x(M)$ such that the exponential map $\exp_x$ is a $C^1$ diffeomorphism from $U$ onto its image, and for any $u \in U$, $\exp_x(tu)$, $t < |\varepsilon|$, is a geodesic through $x$. Since $\sigma_x$ is an isometry, it maps geodesics into geodesics. Now, $\sigma_x(\exp_x(tu))$ and $\exp_x((d\sigma_x)_x tu)$ are two geodesics through $x$ with the same initial vector $(d\sigma_x)_x u$. Therefore they coincide. In particular, we have $\sigma_x(\exp_x(u)) = \exp_x((d\sigma_x)_x u)$. Since $\sigma_x^2 = \mathrm{id}$, we have $(d\sigma_x)_x^2 = \mathrm{id}$. To prove $(d\sigma_x)_x = -\mathrm{id}$, we need only prove that the number 1 is not an eigenvalue of $(d\sigma_x)_x$. Suppose conversely that there exists $u \neq 0$ such that $(d\sigma_x)_x(u) = u$. Then $(d\sigma_x)_x(tu) = tu$, $t \in \mathbb{R}$. Therefore for every $t$ we have

$$\sigma_x(\exp_x(tu)) = \exp_x((d\sigma_x)_x)(tu) = \exp_x(tu).$$

But this contradicts the assumption that $x$ is an isolated fixed point of $\sigma_x$. Therefore $(d\sigma_x)_x = -\mathrm{id}$.

(b) Let $\gamma(t)$, $0 \leq t \leq l$, be a geodesic parametrized to have constant Finslerian speed. We construct a curve $\widetilde{\gamma}$ by $\widetilde{\gamma}(t) = \gamma(t)$, for $0 \leq t \leq l$; $\widetilde{\gamma}(t) = \sigma_{\gamma(l)}(\gamma(2l - t))$, for $l \leq t \leq 2l$. Similarly as in (a), $\sigma_{\gamma(l)}(\gamma(2l - t))$ is a geodesic. By (a), the incoming vector of the geodesic $\gamma$ at $\gamma(l)$ coincides with the initial vector of the geodesic $\widetilde{\gamma}$ at $\gamma(l)$. Therefore, by the uniqueness of the geodesics, $\widetilde{\gamma}(t)$ is smooth and $\widetilde{\gamma}(t)$, $0 \leq t \leq 2l$, is a geodesic. It is obvious that this geodesic still has constant Finslerian speed (since $\sigma_{\gamma(l)}$ is an isometry). Therefore $(M, F)$ is forward geodesically complete. By the Hopf–Rinow theorem, $(M, F)$ is forward complete. Since $F$ is reversible, $(M, F)$ is complete.

(c) Since $(M,F)$ is complete, for any $x,y \in M$, there exists a unit-speed minimal geodesic $\gamma(t)$, $0 \leq t \leq T$, that realizes the distance of each pair of points in $\gamma$. Let $x_0 = \gamma(\frac{T}{2})$. Since $F$ is reversible, we have $d(x_0,x) = d(x,x_0) = d(x_0,y)$. By the proof of (a), we know that $\sigma_{x_0}(x) = y$. Therefore $(M,F)$ is homogeneous.

(d) Let $(M,F)$ be a globally symmetric Finsler space. Then it is easily seen that $(\widetilde{M},\pi^*(F))$ is a Finsler space. It is well known that any diffeomorphism $\sigma$ of $M$ can be lifted to a diffeomorphism $\widetilde{\sigma}$ of $\widetilde{M}$ such that $\sigma\pi = \pi\widetilde{\sigma}$. Furthermore, if $\sigma(x) = x$, then for any $\widetilde{x} \in \widetilde{\pi}^{-1}(x)$, we can take $\widetilde{\sigma}$ such that $\widetilde{\sigma}(\widetilde{x}) = \widetilde{x}$. Now for $\widetilde{y} \in \widetilde{M}$, set $y = \pi(\widetilde{y})$. Then there exists a diffeomorphism $\widetilde{\sigma}_{\widetilde{y}}$ of $\widetilde{M}$ such that $\sigma_y\pi = \pi\widetilde{\sigma}_{\widetilde{y}}$ and $\widetilde{\sigma}_{\widetilde{y}}(\widetilde{y}) = \widetilde{y}$. Since $\sigma_x^2 = \mathrm{id}_M$, we have $\widetilde{\sigma}_{\widetilde{y}}^2\pi^{-1}(y) = \pi^{-1}(y)$, $\forall y \in M$. Since $\widetilde{\sigma}_{\widetilde{y}}$ is a diffeomorphism keeping $\widetilde{y}$ fixed, we see that $\widetilde{\sigma}_{\widetilde{y}}^2 = \mathrm{id}_{\widetilde{M}}$. It is obvious that $\widetilde{y}$ is an isolated fixed point of $\widetilde{\sigma}_{\widetilde{y}}$. By the definition of $\pi^*(F)$, $\widetilde{\sigma}_{\widetilde{y}}$ is an isometry. Therefore $(\widetilde{M},\pi^*(F))$ is a globally symmetric Finsler space. $\square$

**Corollary 5.2.** *Let $(M,F)$ be a globally symmetric Finsler space. Then for any $x \in M$, $\sigma_x$ is a local geodesic symmetry at $x$. In particular, $(M,F)$ is locally symmetric and for every $x \in M$, the symmetry $\sigma_x$ is unique.*

*Proof.* By (a) of Theorem 5.6, we know that for every $x \in M$, $(d\sigma_x)_x = -\mathrm{id}$. Therefore, for every $u \in T_x(M)$, we have

$$\sigma_x(\exp_x(tu)) = \exp_x((d\sigma_x)_x(tu)) = \exp_x(-tu). \tag{5.7}$$

This means that $\sigma_x$ is the local geodesic symmetry at $x$. Hence $(M,F)$ is locally symmetric. Since $(M,F)$ is complete (Theorem 5.6 (b)), $\exp_x$ is defined on the whole space $T_x(M)$ and it is surjective. Therefore, by (5.7), $\sigma_x$ is unique. $\square$

**Theorem 5.7.** *Let $(M,F)$ be a globally symmetric Finsler space. Then there exists a Riemannian symmetric pair $(G,K)$ such that $M$ is diffeomorphic to $G/K$ and $F$ is invariant under $G$.*

*Proof.* By (c) of Theorem 5.6, the group $I(M,F)$ of isometries of $(M,F)$ acts transitively on $M$. Since $M$ is connected, the identity component $G = I_0(M,F)$ of $I(M,F)$ is also transitive on $M$, and the isotropy subgroup $K$ of $G$ at $x$ fixed is compact. Furthermore, $M$ is diffeomorphic to $G/K$ under the map $gH \to g \cdot x, g \in G$.

Similarly to the Riemannian case, we define a map $\sigma$ of $G$ into itself by $\sigma(g) = \sigma_x g \sigma_x$, where $\sigma_x$ denotes the (unique) involutive isometry of $(M,F)$ with $x$ as an isolated fixed point. Then it is easily seen that $\sigma$ is an involutive automorphism of $G$ and the group $K$ lies between the closed subgroup $K_\sigma$ of fixed points of $\sigma$ and the identity component of $K_\sigma$. Furthermore, the group $K$ contains no normal subgroup of $G$ other than $\{e\}$. That is, $(G,K)$ is a symmetric pair. Since $K$ is compact, $(G,K)$ is a Riemannian symmetric pair. $\square$

Now we can prove the main result of this section.

**Theorem 5.8.** *Let $(M,F)$ be a globally symmetric Finsler space. Then $(M,F)$ is a Berwald space. Furthermore, the connection of $F$ coincides with the Levi-Civita* *connection of a Riemannian metric $Q$ such that $(M,Q)$ is a Riemannian globally symmetric space.*

*Proof.* We first prove that $F$ is a Berwald metric. By Theorem 5.7, there exists a Riemannian symmetric pair $(G,K)$ such that $M$ is diffeomorphic to $G/K$ and $F$ is invariant under $G$. Fix a $G$-invariant Riemannian metric $Q$ on $G/K$. Without loss of generality, we can assume that $(G,K)$ is effective. Since being a Berwald space is a local property, we can assume further that $G/K$ is simply connected. Then we have the decomposition

$$G/K = E \times G_1/K_1 \times G_2/K_2 \times \cdots \times G_n/K_n,$$

where $E$ is a Euclidean space, and $G_i/K_i$ s, $i = 1,2,\ldots,n$, are simply connected irreducible Riemannian globally symmetric spaces. Now we determine the holonomy group of $Q$ at the origin of $G/K$. According to the de Rham decomposition theorem, it is equal to the product of the holonomy group of $E$ and that of the $G_i/K_i$'s at the origin. Obviously, $E$ has trivial holonomy group. Next we determine the holonomy group of the factors $G_i/K_i$. By the holonomy theorem of Ambrose–Singer [7], the Lie algebra $\mathfrak{h}_i$ of the holonomy group $H_i$ is linearly spanned by the linear maps of the form $\{\tilde{\tau}^{-1}R_o(u,v)\tilde{\tau}\}$, where $\tau$ denotes any piecewise smooth curve starting from $o$, $\tilde{\tau}$ denotes the parallel displacements (with respect to the restricted Riemannian metric) along $\tau$, $R_o$ is the curvature tensor of $G_i/K_i$ of the restricted Riemannian metric, and $u,v \in T_o(G_i/K_i)$. Since $G_i/K_i$ is a globally Riemannian symmetric space, the curvature tensor is invariant under parallel displacements. Therefore

$$\mathfrak{h}_i = \text{span}\{R_o(u,v) \mid u,v \in T_o(G_i/K_i)\}.$$

On the other hand, Since $G_i$ is a semisimple group, the Lie algebra of $K_i^* = \text{Ad}(K_i) \simeq K$ is also equal to the span of $R_o(u,v)$ (see [83, p. 207]). Since $G_i/K_i$ is simply connected, the groups $H_i$ and $K_i^*$ are connected [102]. Therefore $H_i = K_i^*$. Consequently the holonomy group $H_o$ of $G/K$ at the origin is

$$K_1^* \times K_2^* \times \cdots \times K_n^*.$$

Now $F$ defines a Minkwoski norm $F_o$ on $T_o(G/K)$ that is invariant under $H_o$. By Theorem 1.7, we can construct a Finsler metric $\bar{F}$ on $G/K$ by parallel translations of $Q$. Then Theorem 1.8 implies that $\bar{F}$ is a Berwald metric. Now, for any point $p_0 = aK \in G/K$, there exists a geodesic of the Riemannian manifold $(G/K,Q)$, say $\gamma(t)$, such that $\gamma(0) = o$ and $\gamma(1) = p_0$. Suppose the initial vector of $\gamma$ is $u_0$ and take $u \in \mathfrak{p}$ such that $d\pi(u) = u_0$. Then $\gamma(t) = \exp tu \cdot p_0$ and $d\tau(\exp tu)$ is the parallel translate of $(G/K,Q)$ along $\gamma$. Since $F$ is $G$-invariant, it is invariant under this parallel displacement. This means that the two Finsler metrics $F$ and $\bar{F}$ coincide in the tangent space $T_{p_0}(G/K)$. Consequently, they coincide everywhere. Thus $F$ is a Berwald metric.

Now we prove the next assertion. Since $(M,F)$ is a Berwald space, there exists a Riemannian metric $Q_1$ on $M$ with the same connection as $F$. Since the above $G$ acts isometrically on $M$, by Theorem 5.1, $G$ acts as a group of affine transformations on the connection. Since $(G,K)$ is a (Riemannian) symmetric pair, $M$ endowed with this connection must be globally affine symmetric. Therefore $(M, Q_1)$ is a Riemannian globally symmetric space. □

From the proof of Theorem 5.8, we obtain the following corollary.

**Corollary 5.3.** *Let $(G/K, F)$ be a globally symmetric Finsler space and $\mathfrak{g} = \mathfrak{k} + \mathfrak{p}$ the corresponding decomposition of the Lie algebra. Let $\pi$ be the natural map of $G$ onto $G/K$. Then $(d\pi)_e$ maps $\mathfrak{p}$ isomorphically onto the tangent space of $G/K$ at $p_0 = eK$. If $u \in \mathfrak{p}$, then the geodesic emanating from $p_0$ with initial tangent vector $(d\pi)_e u$ is given by*

$$\gamma_{d\pi \cdot u}(t) = \exp tu \cdot p_0.$$

*Furthermore, if $v \in T_{p_0}(G/K)$, then $(d\exp tu)_{p_0}(v)$ is the parallel of $v$ along the geodesic.*

## 5.3 Minkowski Symmetric Lie Algebras

In this section, we introduce the notion of Minkowski symmetric Lie algebras to give an algebraic description of symmetric Finsler spaces. By Theorem 5.8, each globally symmetric Finsler space must be a Berwald space. Therefore, in the following we will consider locally and globally affine symmetric Berwald spaces. The results are all valid for globally symmetric Finsler spaces. We first give a new definition.

**Definition 5.2.** Let $(\mathfrak{g}, \sigma)$ be a symmetric Lie algebra and $\mathfrak{g} = \mathfrak{k} + \mathfrak{m}$ the canonical decomposition of $\mathfrak{g}$ with respect to the involution $\sigma$. If $F$ is a Minkowski norm on $\mathfrak{m}$ such that for every $y \neq 0$, $y, u, v \in \mathfrak{m}$, and $w \in \mathfrak{k}$, we have

$$g_y([w,u],v) + g_y(u,[w,v]) + 2C_y([w,y],u,v) = 0, \tag{5.8}$$

where $g_y$ (resp. $C_y$) is the fundamental tensor (resp. Cartan tensor), then $(\mathfrak{g}, \sigma, F)$ is called a Minkowski symmetric Lie algebra.

Using the notion of a Minkowski symmetric Lie algebra, we can give an algebraic description of globally affine symmetric Finsler spaces. In the following, for a symmetric pair $(G,K)$, we use $\sigma$ to denote the involutive automorphism of $G$ as well as that of the Lie algebra $\mathfrak{g}$ of $G$. Let $\mathfrak{g} = \mathfrak{k} + \mathfrak{m}$ be the canonical decomposition of $\mathfrak{g}$ with respect to $\sigma$. As usual, we identify the tangent space $T_o(G/K)$ with $\mathfrak{m}$.

**Theorem 5.9.** *Let $(G/K, F)$ be a globally affine symmetric Berwald space. Then $(\mathfrak{g}, \sigma, F)$ is a Minkowski symmetric Lie algebra. Conversely, let $(\mathfrak{g}, \sigma, F)$ be a Minkowski symmetric Lie algebra and suppose $(G,K)$ is a pair associated with $(\mathfrak{g}, \sigma)$ such that $K$ is closed and connected. Then there exists a Finsler metric (still denoted by $F$) on $G/K$ such that $(G/K, F)$ is a locally affine symmetric Berwald* *space. Furthermore, if $(G,K)$ is a symmetric pair, then $G/K$ with this metric is globally affine symmetric.*

*Proof.* For the first assertion, we need only verify that $F$ satisfies (5.8). Since $F$ is $G$-invariant, we have

$$F(\mathrm{Ad}(h)u) = F(u), \quad \forall h \in H, \, u \in \mathfrak{m}.$$

Therefore, for any $y \, (\neq 0), u, v \in \mathfrak{m}$, and $w \in \mathfrak{k}, \, t, r, s \in \mathbb{R}$, we have

$$F^2(\mathrm{Ad}(\exp(tw))(y + ru + sv)) = F^2(y + ru + sv).$$

By definition,

$$g_y(u,v) = \frac{1}{2} \frac{\partial^2}{\partial r \partial s} F^2(y + ru + sv)|_{r=s=0}.$$

Thus

$$g_y(u,v) = \frac{1}{2} \frac{\partial^2}{\partial r \partial s} F^2(\mathrm{Ad}(\exp(tw))(y + ru + sv))|_{r=s=0}. \qquad (5.9)$$

Now for $u \in \mathfrak{m}$, we have

$$\mathrm{Ad}(\exp tw)u = u + t[w,u] + O(t^2).$$

Therefore

$$g_y(u,v) = \frac{1}{2} \frac{\partial^2}{\partial r \partial s} F^2(y + ru + sv + t[w, y + ru + sv] + O(t^2))|_{r=s=0}.$$

Taking the derivative with respect to $t$ at $t = 0$, we get

$$0 = g_y([w,u],v) + g_y(u,[w,v]) + 2C_y([w,y],u,v).$$

Therefore, $(\mathfrak{g}, \sigma, F)$ is a Minkowski symmetric Lie algebra.

Conversely, let $(\mathfrak{g}, \sigma, F)$ be a Minkowski symmetric Lie algebra and suppose $(G,H)$ is a pair associated with $(\mathfrak{g}, \sigma)$, with $H$ closed and connected. We assert that there exists a $G$-invariant Finsler metric on $G/H$. In fact, given $y \, (\neq 0), u, v \in \mathfrak{m}$, and $w \in \mathfrak{h}$, consider the function

$$\psi(t) = g_{\mathrm{Ad}(\exp(tw))y}(\mathrm{Ad}(\exp(tw))u, \mathrm{Ad}(\exp(tw))v).$$

Taking the derivative with respect to $t$, we easily see that

$$\psi'(t) = 0.$$

Therefore $\psi(t) = \psi(0)$, $\forall t \in \mathbb{R}$. Since $H$ is connected, it is generated by elements of the form $\exp(tw)$, $w \in \mathfrak{h}$, $t \in \mathbb{R}$. Thus

$$g_y(u,v) = g_{\mathrm{Ad}(h)y}(\mathrm{Ad}(h)u, \mathrm{Ad}(h)v), \quad \forall h \in H.$$

Using the formula $F(y) = \sqrt{g_y(y,y)}$, we have

$$F(\mathrm{Ad}(h)u) = F(u), \quad \forall h \in H, u \in \mathfrak{m}. \tag{5.10}$$

Now for any $gH \in G/H$, the map $\tau_g : g_1 H \to g g_1 H$ is a diffeomorphism of $G/H$ sending $o$ to $gH$. Define a Finsler metric $F$ on $G/H$ by

$$F(y) = F(d\tau_{g^{-1}}(y)), \quad y \in T_{gH}(G/H).$$

By (5.10), $F$ is well defined. It is obvious that $F$ is a $G$-invariant Finsler metric on $G/H$.

To prove the last two assertions, we only need to show that if $(G,H)$ is a pair associated with $(\mathfrak{g}, \sigma)$ and $G$ is simply connected, then $(G/H, F)$ is a globally symmetric affine Berwald space. This can be verified similarly to what was done in Theorem 5.5. We omit the details.  □

By Theorem 5.9, if $(\mathfrak{g}, \sigma, F)$ is a Minkowski symmetric Lie algebra, then $(\mathfrak{g}, \sigma)$ is an orthogonal symmetric Lie algebra. We say that a Minkowski symmetric Lie algebra $(\mathfrak{g}, \sigma, F)$ is of Euclidean type, compact type or noncompact type according to the type of the corresponding orthogonal symmetric Lie algebra. Moreover, $(\mathfrak{g}, \sigma, F)$ is called irreducible if $(\mathfrak{g}, \sigma)$ is irreducible as an orthogonal symmetric Lie algebra.

As in the Riemannian case, there exists a remarkable duality among locally or globally symmetric Finsler spaces. By Theorem 5.9, we need only treat Minkowski symmetric Lie algebras.

**Proposition 5.1.** *Let* $(\mathfrak{g}, \sigma, F)$ *be a Minkowski symmetric Lie algebra and* $\mathfrak{g} = \mathfrak{h} + \mathfrak{m}$ *the canonical decomposition. Let* $\mathfrak{g}^* = \mathfrak{h} + \sqrt{-1}\mathfrak{m}$ *be the real subalgebra of* $(\mathfrak{g}^{\mathbb{C}})_R$. *Define a Minkowski norm on* $\sqrt{-1}\mathfrak{m}$ *by*

$$F^*(\sqrt{-1}u) = F(u), \quad u \in \mathfrak{m}.$$

*Let* $\sigma^*$ *be the involutive automorphism of* $\mathfrak{g}^*$ *induced by* $\sigma$. *Then* $(\mathfrak{g}^*, \sigma^*, F^*)$ *is a Minkowski symmetric Lie algebra. Moreover, if* $(\mathfrak{g}, \sigma, F)$ *is of compact type (resp. noncompact type), then* $(\mathfrak{u}, \sigma^*, F^*)$ *is of noncompact type (resp. compact type), and conversely.*

*Proof.* The only point we need to prove is that $F^*$ satisfies (5.8). By the definitions of fundamental form and Cartan tensor, we have

$$g_{\sqrt{-1}y}(\sqrt{-1}u_1, \sqrt{-1}u_2) = g_y(u_1, u_2),$$
$$C_{\sqrt{-1}y}(\sqrt{-1}u_1, \sqrt{-1}u_2, \sqrt{-1}u_3) = C_y(u_1, u_2, u_3),$$

where $y(\neq 0)$, $u_i \in \mathfrak{m}$. From this it is easily seen that $F^*$ satisfies (5.8).  □

**Definition 5.3.** The triple $(\mathfrak{g}^*, \sigma^*, F^*)$ defined in Proposition 5.1 is called the dual of the Minkowski symmetric Lie algebra $(\mathfrak{g}, \sigma, F)$.

It is clear that if $(\mathfrak{g}_1, \sigma_1, F_1)$ is the dual of $(\mathfrak{g}_2, \sigma_2, F_2)$, then $(\mathfrak{g}_2, \sigma_2, F_2)$ is the dual of $(\mathfrak{g}_1, \sigma_1, F_1)$.

**Definition 5.4.** Two Minkowski symmetric Lie algebras $(\mathfrak{g}_1, \sigma_1, F_1)$ and $(\mathfrak{g}_2, \sigma_2, F_2)$ are called isomorphic if there exists a Lie algebra isomorphism $\varphi$ of $\mathfrak{g}_1$ onto $\mathfrak{g}_2$ such that $\varphi \circ \sigma_1 = \sigma_2 \circ \varphi$ and $F_1(u) = F_2(\varphi(u)), \forall u \in \mathfrak{g}_1$.

**Proposition 5.2.** Let $(\mathfrak{g}_i, \sigma_i, F_i)$ be two Minkowski symmetric Lie algebras. Then $(\mathfrak{g}_1, \sigma_1, F_1)$ is isomorphic to $(\mathfrak{g}_2, \sigma_2, F_2)$ if and only if $(\mathfrak{g}_1^*, \sigma_1^*, F_1^*)$ is isomorphic to $(\mathfrak{g}_2^*, \sigma_2^*, F_2^*)$.

The proof is similar to the Riemannian case, so we omit it.

In the Riemannian case, every (simply connected) Riemannian globally symmetric space can be uniquely decomposed into the product of a Euclidean space, a Riemannian symmetric space of compact type, and a Riemannian space of noncompact type. Furthermore, each Riemannian space of compact type or noncompact type can be uniquely decomposed into the product of irreducible ones. It is natural to consider the same problem for symmetric Finsler spaces. However, we must be careful because in this case, we do not have orthogonality. Therefore, the product of two symmetric Finsler spaces may not be unique. To give a valid definition, we need the following result:

**Theorem 5.10.** Let $(M, F)$ be a globally affine symmetric Berwald space and $(G, K)$ an effective Riemannian symmetric pair such that $M \simeq G/K$. Then the connection of $F$ (as a Berwald metric) coincides with the canonical connection of $G/K$. In particular, for any effective Riemannian symmetric pair $(G, K)$, the connection is the same for all $G$-invariant Finsler metrics on $G/K$.

*Proof.* By Theorem 5.8, the connection of $F$ coincides with the Levi-Civita connection of a $G$-invariant Riemannian metric $Q$ on $G/K$. But the Levi-Civita connection on $G/K$ is the same for all $G$-invariant Riemannian metrics on $G/K$. Therefore the theorem follows.                                                              $\square$

By Theorem 5.10, to consider the decomposition of symmetric Finsler spaces, we do not need to consider the connection. Therefore we have the following definition:

**Definition 5.5.** Let $(M, F)$, $(M_1, F_1)$, $(M_2, F_2)$ be globally affine symmetric Berwald spaces. A Finsler space $(M, F)$ is called the product of $(M_1, F_1)$ and $(M_2, F_2)$ if $M \simeq M_1 \times M_2$ and $F_i = F|_{M_i}$, $i = 1, 2$. A globally affine symmetric Berwald space of compact or noncompact type is called irreducible if it cannot be written as the product of two globally affine symmetric Berwald spaces

Similarly, we can define the notion of product of finitely many globally symmetric Finsler spaces. Now we have the following theorem.

**Theorem 5.11.** *Let $(M,F)$ be a simply connected globally affine symmetric Berwald space. Then $(M,F)$ can be decomposed into the product of a Minkwoski space, a globally affine symmetric Berwald space of compact type and a globally affine symmetric Berwald space of noncompact type. Moreover, every simply connected globally affine symmetric Finsler space of compact or noncompact type can be decomposed into the product of irreducible globally affine symmetric Berwald spaces. The decomposition is unique as manifolds (in general not unique as Finsler spaces).*

*Proof.* By Theorem 5.8, the only point we need to check is that the coset space $\mathbb{R}^n \simeq \mathbb{R}^n/\{0\}$ endowed with a Finsler metric invariant under the parallel translation is a Minkowski space. But this is obvious. □

*Remark 5.1.* Although we have the decomposition theorem, the classification of symmetric Finsler spaces is not reduced to the case of irreducible ones, because the product of two or more symmetric Finsler spaces is not unique.

Next we give a formula to compute the flag curvature and Ricci scalar of locally affine symmetric Berwald spaces.

**Theorem 5.12.** *Let $(\mathfrak{g},\sigma,F_0)$ be a Minkowski symmetric Lie algebra and $(G,H)$ a pair associated with $(\mathfrak{g},\sigma)$. Suppose there exists an invariant Finsler metric $F$ on $G/H$ such that the restriction of $F$ to $\mathfrak{m}$ is $F_0$. Then the curvature tensor of $F$ is given by*

$$R_o(u,v)w = -[[u,v],w], \quad \forall u,v,w \in \mathfrak{m},$$

*and the flag curvature of the flag $(P,y)$, $y \neq 0$, $y \in P$, is given by*

$$K(P,y) = g_l([[l,v],l],v),$$

*where $l = y/F$ (the distinguished section) and $l,u$ is an orthonormal basis of the plane $P$ with respect to $g_l$.*

*Proof.* Since flag curvature is a local invariance, we can assume, without loss of generality, that $G$ is simply connected. Then by Theorem 5.9, $G/H$ with $F$ is a globally affine symmetric Berwald space. Let $D$ be the (linear) connection of $(G/H,F)$. Then $D$ is the canonical connection of the globally affine symmetric space $G/H$. Therefore,

$$R_o(u,v)w = -[[u,v],w], \quad u,v,w \in \mathfrak{m}.$$

The last conclusion is the direct consequence of the definition of flag curvature. □

**Corollary 5.4.** *Let $(\mathfrak{g},\sigma,F_0)$ be a Minkowski symmetric Lie algebra and $(\mathfrak{g}^*,\sigma^*,F_0^*)$ its dual. Let $(G,H)$ and $(G^*,H^*)$ be two pairs associated with $(\mathfrak{g},\sigma,F_0)$ and $(\mathfrak{g}^*,\sigma^*,F_0^*)$, respectively. Suppose there exist invariant Finsler metrics $F$ on*

$G/H$ and $F^*$ on $G^*/H^*$ such that the restrictions of $F$ and $F^*$ on $\mathfrak{m}$ and $\sqrt{-1}\mathfrak{m}$ are equal to $F_0$ and $F_0^*$, respectively. Then the flag curvatures of $(G/H, F)$ and $(G^*/H^*, F^*)$ satisfy

$$K(P, y) = -K(\sqrt{-1}P, \sqrt{-1}y), \quad 0 \neq y \in \mathfrak{m},$$

where $P$ is a plane in $\mathfrak{m}$ containing $y$ and $\sqrt{-1}P$ is the plane in $\sqrt{-1}\mathfrak{m}$ spanned by $\sqrt{-1}u$, $u \in P$.

It is a result of Cartan (see [83, Theorem 3.1, Chap. V]) that the sectional curvature of a Riemannian symmetric space is everywhere 0, everywhere $\geq 0$, or everywhere $\leq 0$, according as it is of Euclidean type, compact type, or noncompact type. It is easy to see that the flag curvature of a (locally or globally) affine symmetric Berwald space of Euclidean type is everywhere 0. However, in the cases of compact type or noncompact type, the situation is somehow complicated. The reason for this complexity is that the proof of the Riemannian case depends on the invariance of the inner product in the formula of the sectional curvature. But in the formula of flag curvature the inner product $g_y$ (or $g_l$) is not invariant under the action of $H$. Therefore the original proof is not applicable. However, using Weyl's unitary trick, we can generalize this elegant result to the Finslerian case.

**Theorem 5.13.** Let $(G/H, F)$ be a locally affine symmetric Berwald space with $G$ semisimple and $(\mathfrak{g}, \sigma, F)$ the associated Minkowski symmetric Lie algebra with the canonical decomposition $\mathfrak{g} = \mathfrak{h} + \mathfrak{m}$.

1. If $(\mathfrak{g}, \sigma)$ is of compact type, then the flag curvature of $(G/H, F)$ is everywhere $\geq 0$.
2. If $(\mathfrak{g}, \sigma)$ is of noncompact type, then the flag curvature of $(G/H, F)$ is everywhere $\leq 0$.

*Proof.* By the duality, we need treat only the noncompact case. Then $H$ is a connected compact Lie group (see Sect. 2.5). Given $y \neq 0$ and a vector $u$ that is linearly independent from $y$, let $P$ be the plane spanned by $y$, $u$. Then the flag curvature is given by

$$K(P, y) = \frac{g_l([[l, u], l], u)}{g_l(u, u) - g_l(l, u)^2}.$$

Accordingly, we need only prove that $g_l([[l, u], l], u) \leq 0$. Consider the pullback bundle $\pi^*(TM)$ over $TM \setminus \{0\}$. It is easily seen that in the notation of [16, pp. 73–74], we have $g_l([[l, u], l], u) = (l, u)$. Moreover, it is also true that $(l, u) = (u, l)$. Thus

$$g_l([[l, u], l], u) = g_l([[u, l], u], l).$$

Set $v = [[u, l], u]$. Then by the identity

$$C_l(z_1, z_2, l) = 0, \quad \forall z_1, z_2 \in \mathfrak{m},$$

we have

$$g_l([w,v],l) + g_l(v,[w,l]) = -2C_l([w,l],v,l) = 0, \ \forall w \in \mathfrak{h}.$$

This means that for the given $w$, the derivative of the function

$$g_l(\text{Ad}(\exp(tw))v, \text{Ad}(\exp(tw))l)$$

is equal to 0 everywhere. Therefore

$$g_l(\text{Ad}(\exp(tw))v, \text{Ad}(\exp(tw))l) = g_l(v,l), \quad \forall w \in \mathfrak{h}, t \in \mathbb{R}.$$

Since $H$ is connected and compact, the exponential map is surjective. Thus

$$g_l(\text{Ad}(h)v, \text{Ad}(h)l) = g_l(v,l), \quad \forall h \in H.$$

Now we define a new inner product $Q_0$ on $\mathfrak{m}$ by

$$Q_0(u_1,v_1) = \int_H g_l(\text{Ad}(h)u_1, \text{Ad}(h)v_1) \, \mathrm{d}h, \quad u_1, v_1 \in \mathfrak{m},$$

where $\mathrm{d}h$ is the normalized standard bi-invariant Haar measure of $H$. By the definition, $Q_0$ is $\text{Ad}(H)$-invariant and

$$Q_0(v,l) = g_l(v,l).$$

Using $Q_0$ we can define a $G$-invariant Riemannian metric $Q$ on the coset space $G/H$ such that the restriction of $Q$ to $\mathfrak{m}$ coincides with $Q_0$. Then $(G/H, Q)$ is a Riemannian globally symmetric space of noncompact type. Therefore the sectional curvature of $(G/H, Q)$ is everywhere $\leq 0$. Thus

$$Q_0(v,l) = g_l([[u,l],u],l) \leq 0.$$

Consequently

$$g_l([[u,l],u],l) = g_l(v,l) = Q_0(v,l) \leq 0.$$

Therefore

$$K(P,y) \leq 0. \qquad \square$$

*Remark 5.2.* In the above theorem, the equality is strict if and only if $P$ is not a commutative plane. This can be easily seen from the formula of the flag curvature in Theorem 5.12.

By Theorem 5.12, we have the following result.

**Theorem 5.14.** *Let $(G/H, F)$ be a globally affine symmetric Berwald space and $\mathfrak{g} = \mathfrak{h} + \mathfrak{m}$ the canonical decomposition of the corresponding Minkowski symmetric*

*Lie algebra. Identifying* $\mathfrak{m}$ *with the tangent space* $T_o(G/H)$ *at the origin, we have* $\mathrm{Ric}(y) = -\frac{1}{2}B(y,y)$, $\forall y(\neq 0) \in \mathfrak{m}$, *where* $B$ *denotes the Killing form of the Lie algebra* $\mathfrak{g}$.

*Proof.* By Theorem 5.12, the curvature tensor of the connection of the Berwald space $(G/H, F)$ is

$$R_o(u,v)w = -[[u,v],w], \quad u,v,w \in \mathfrak{m}.$$

This implies that the Riemann tensor of $(G/H, F)$ satisfies

$$\mathscr{R}_y(u) = -F^2(y)[[y,u],y], \quad y(\neq 0), \ u \in \mathfrak{m}.$$

Taking the trace of the linear endomorphism $R_y$, we find that the Ricci scalar of $(G/H, F)$ is the same as that of the Riemannian manifold $(G/H, Q)$, where $Q$ is a $G$-invariant Riemannian metric on $G/H$. Now, using Corollary 2.1 one can easily deduce that $\mathrm{Ric}(y) = -\frac{1}{2}B(y,y)$. □

## 5.4  Rigidity Theorems

In this section, we present some rigidity results on symmetric Finsler spaces related to flag curvature. We first recall a result of Busemann and Phadke [35] that asserts that a locally symmetric $G$-space has a globally symmetric universal covering space. When the $G$-space is smooth, it is a (complete) Finsler space. Therefore, combining Theorems 5.4, 5.6(d), and 5.8, we have the following theorem.

**Theorem 5.15.** *Let $(M, F)$ be a complete locally symmetric Finsler space. Then for every $x \in M$, there exist a simply connected globally symmetric Finsler space $(\widetilde{M}, \widetilde{F})$, a neighborhood $N$ of $x$ in $M$, and an isometry of $N$ onto an open subset $\widetilde{N}$ of $\widetilde{M}$. In particular, $(M, F)$ is a Berwald space.*

In view of Theorem 5.3, we have the following result.

**Theorem 5.16.** *Let $(M, F)$ be a complete Finsler space. Then $(M, F)$ is locally symmetric if and only if it is a Berwald space and the flag curvature is invariant under all parallel displacements.*

Now we can deduce some global rigidity theorems on symmetric Finsler spaces. First we have the following.

**Theorem 5.17.** *Let $(M, F)$ be a complete locally symmetric Finsler space. If the flag curvature of $(M, F)$ is everywhere nonzero, then $F$ is Riemannian.*

*Proof.* Let $x \in M$. By Theorem 5.15, there exist a neighborhood $N$ of $x$, a simply connected globally symmetric Finsler space $(\widetilde{M}, \widetilde{F})$, and an isometry from $N$ onto an open subset $\widetilde{N}$ of $\widetilde{M}$. By the assumption, the flag curvature of $(\widetilde{M},\widetilde{F})$ is everywhere nonzero on $\widetilde{N}$. Since $(\widetilde{M},\widetilde{F})$ is homogeneous (Theorem 5.6 (c)), the flag curvature of $(\widetilde{M},\widetilde{F})$ is everywhere nonzero. By Theorem 5.7, $\widetilde{M}$ can be written as a coset space of a Riemannian symmetric pair $(G,H)$ and $\widetilde{F}$ corresponds to a $G$-invariant Finsler metric $\widetilde{F}_1$ on $G/H$. Let $\mathfrak{g} = \mathfrak{h} + \mathfrak{m}$ be the canonical decomposition of the Lie algebra. Then by Theorem 5.11, we have $[u,v] \neq 0$ for all vectors $u,v$ in $\mathfrak{m}$ that are linearly independent. That is, $G/H$ is irreducible and of rank one. Fix a $G$-invariant Riemannian metric $Q$ on $G/H$. Then $\mathrm{Ad}(H)$ acts transitively on the unit sphere of the Euclidean space $(T_o(G/H), Q)$. Since both $Q$ and $\widetilde{F}_1$ are invariant under the action of $\mathrm{Ad}(H)$, there exists a positive number $c$ such that $\widetilde{F}_1(y) = c\sqrt{\tilde{g}(y,y)}$ for all $y \in T_o(G/H)$. Hence $\widetilde{F}_1$ is Riemannian. Thus $\widetilde{F}$ is Riemannian. Consequently $F$ is Riemannian on $N$. Since $x$ is arbitrary, we conclude that $F$ is a Riemannian metric.                                                                                         □

Since a compact Finsler space is complete, we have the following corollary.

**Corollary 5.5.** *Let $(M,F)$ be a compact locally symmetric Finsler space. If the flag curvature of $(M,F)$ is everywhere nonzero, then $F$ is Riemannian.*

It should be noted that the negatively curved case in Corollary 5.5 can also be deduced from the main result in [71]. However, the approaches are essentially different. Moreover, the method in [71] is not applicable to the positively curved case here. We mention here that there is a geometric approach to prove the positively curved case of this corollary; see [98].

**Theorem 5.18.** *Let $(M,F)$ be a locally affine symmetric Berwald space. If the flag curvature of $(M,F)$ is everywhere nonzero, then $F$ is Riemannian.*

*Proof.* By Theorem 5.4, a locally symmetric Berwald space is locally isometric to a globally symmetric Berwald space. Therefore the result can be proved in a manner similar to that of Theorem 5.17.                                                                     □

By Theorem 5.4, a Berwald space of constant flag curvature must be locally symmetric. Therefore we have the following corollary.

**Corollary 5.6.** *A Berwald space of nonzero constant flag curvature is Riemannian.*

It should be noted that this result is a special case of Numata's theorem (see [16]). It is interesting that the approaches are very different.

Finally, we mention that recently Matveev and Troyanov proved that a locally symmetric Finsler space must be a Berwald space, without the completeness assumption. This proves a conjecture posed in [61]. See [116] for details. By their results, the assertions of Theorems 5.15–5.18 and Corollary 5.5 are all valid without the completeness assumption.

## 5.5  Complex Structures

In this section we consider homogeneous Finsler spaces with invariant compatible complex structure and give a classification of complex symmetric Finsler spaces. A complex Finsler manifold $(M,I,F)$ is a (connected) complex manifold $(M,I)$ endowed with a complex Finsler metric $F$. Here, by a complex Finsler metric, we mean a continuous function $F : TM \to \mathbb{R}^+$ (where $T_x(M)$, $x \in M$ is viewed as a complex vector space) that satisfies the following conditions:

1. $F$ is smooth on $TM \setminus \{0\}$;
2. $F(u) > 0, \forall u \neq 0$;
3. $F(\lambda u) = |\lambda| F(u)$ for every $\lambda \in \mathbb{C}^*$.

The restriction of $F$ to any $T_xM$ is a complex Minkowski norm, i.e., a function on the complex vector space $T_xM$ that is smooth on $T_xM \setminus \{0\}$ and satisfies conditions (2) and (3). In this section, we will consider only those complex Finsler metrics which are strongly convex as a real Finsler metric. Namely, let $x \in M$ and $v_1, v_2, \ldots, v_{2m}$ be a basis of $T_x(M)$ over the field of real numbers. For $y = y^j v_j \in T_x(M)$, write $F(y) = F(y^1, y^2, \ldots, y^{2m})$. Then the Hessian matrix

$$(g_{ij}) = \left( \frac{1}{2} [F^2_{y^i y^j}] \right)$$

is positive definite at every point in $T_x(M) \setminus \{0\}$. Therefore, $F$ is a real Finsler metric in the usual sense.

We first give some examples of complex Minkowski norms and complex Finsler metrics.

*Example 5.2.* Let $(M,I,Q)$ be a Hermitian manifold. Then it is obvious that the function $F$ defined by

$$F(v) = \sqrt{Q(v,v)}$$

is a complex Finsler metric on $(M,I)$. In this case we say that $F$ is associated with the Hermitian metric $Q$.

*Example 5.3.* Let us give an example that is not associated with any Hermitian metric. Let $n \geq 2$. In the linear vector space $\mathbb{C}^n$, we define

$$F(x) = \sqrt{|x_1|^2 + |x_2|^2 + \cdots + |x_n|^2 + \sqrt[s]{|x_1|^{2s} + |x_2|^{2s} + |x_n|^{2s}}},$$

where $x = (x_1, x_2, \ldots, x_n) \in \mathbb{C}^n$, $|\cdot|$ denotes the modulus of a complex number, and $s \geq 2$ is an integer. Then it is easily seen that $F$ is a complex Minkowski norm that is not the norm of any Hermitian form. In an obvious way, we can view $(\mathbb{C}^n, F)$ as a complex Finsler manifold. It is easily seen that $F$ is not associated with any Hermitian metric on $\mathbb{C}^n$.

Let $(M,I,F)$ be an $n$-dimensional complex Finsler manifold. Then $(M,F)$ is a $(2n)$-dimensional real Finsler space. A differmorphism $\tau$ of $M$ is called holomorphic if $d\tau \circ I = I \circ d\tau$. The set of all holomorphic isometries of $(M,I,F)$ forms a group, denoted by $A(M,I,F)$. It is obvious that $A(M,I,F)$ is a closed subgroup of $I(M,F)$. Therefore, by Theorem 3.4, we have the following result.

**Proposition 5.3.** *The group $A(M,I,F)$ of holomorphic isometries of $(M,I,F)$ is a Lie transformation group of $M$ with respect to the compact-open topology. Moreover, for every $x \in M$, the isotropy subgroup $A_x(M,I,F)$ at $x$ is a compact subgroup of $A(M,I,F)$.*

**Definition 5.6.** A complex Finsler manifold $(M,I,F)$ is called homogeneous if the group of holomorphic isometries $A(M,I,F)$ acts transitively on $M$.

By the above discussion, we have the following.

**Proposition 5.4.** *Let $(M,I,F)$ be a homogeneous complex Finsler manifold. Then $(M,I,F)$ can be written as a coset space $G/H$, where $G = A_0(M,I,F)$ is the unity component of the group $A(M,I,F)$ of holomorphic isometries and $H = (A_0)_x(M,I,F)$ is the isotropy subgroup of $A_0(M,I,F)$ at $x \in M$.*

As an application of the above propositions, we can prove our next theorem.

**Theorem 5.19.** *Let $(M,I,F)$ be a homogeneous complex Finsler manifold. Then there exists a Riemannian metric $Q$ on $M$ such that $(M,I,Q)$ is a homogeneous Hermitian manifold.*

*Proof.* By Proposition 5.4, $M$ can be written as the coset space $G/H$ of a Lie group $G$ with $G$-invariant complex structure $I$ and complex Finsler metric $F$. Let $o = eH$ be the origin of $G/H$ and $V = T_o(G/H)$. Then the group $K$ of linear isometries of the (real) Minkowski space $(V,F)$ is a compact subgroup of $GL(V)$. Fix any inner product $\langle\, ,\, \rangle_0$ on $V$ and define an inner product $\langle\, ,\, \rangle$ by

$$\langle u,v \rangle = \int_K (\langle \mathrm{Ad}(k)u, \mathrm{Ad}(k)v \rangle_0 + \langle I(\mathrm{Ad}(k)u), I(\mathrm{Ad}(k)v) \rangle_0)\, dk,$$

where $dk$ is the standard invariant Haar measure on $K$. It is obvious that $\langle\, ,\, \rangle$ is $K$-invariant. Since $H \subset K$, $\langle\, ,\, \rangle$ is also $H$-invariant. Therefore $\langle\, ,\, \rangle$ can be extended to a $G$-invariant Riemannian metric $Q$ on $G/H$. By definition, it is easy to check that

$$Q(I(u),I(v)) = Q(u,v)$$

for every $u,v \in T_x(G/H)$, $x \in G/H$. Therefore $(G/H,I,Q)$ is a homogeneous Hermitian manifold. $\qquad\square$

By the above results, to study complex homogeneous Finsler spaces, we need only consider invariant Finsler metrics on coset spaces. We have previously introduced several new notions such as Minkowski Lie pairs, Minkowski Lie algebras,

Minkowski symmetric Lie algebras. Now we use a unified method to describe these notions. This will be useful in describing homogeneous complex Finsler spaces. In the following, vector spaces are assumed to be finite-dimensional.

We first give an infinitesimal version of the definition of Minkowski representations of Lie groups.

**Definition 5.7.** Let $\mathfrak{g}$ be a (real or complex) Lie algebra. Then a Minkowski representation of $\mathfrak{g}$ is a representation $(V, \phi)$ of $\mathfrak{g}$ with a (real or complex) Minkowski norm $F$ on the (real or complex) vector space $V$ such that

$$g_y(\phi(w)u, v) + g_y(u, \phi(w)v) + 2C_y(\phi(w)y, u, v) = 0,$$

for every $w \in \mathfrak{g}$, $y(\neq 0)$, $u, v \in V$. We usually denote the Minkowski representation by $(V, \phi, F)$.

We first give some examples of Minkowski representations.

*Example 5.4.* Let $G$ be a Lie group and $H$ a closed connected subgroup of $G$ with Lie algebras $\mathfrak{g}$ and $\mathfrak{h}$, respectively. Suppose $F$ is a Minkowski norm on the quotient space $\mathfrak{g}/\mathfrak{h}$ such that $(\mathfrak{g}, \mathfrak{h}, F)$ is a Minkowski Lie pair. That is,

$$g_y(\mathrm{ad}_{\mathfrak{g}/\mathfrak{h}}(x)(u), v) + g_y(u, \mathrm{ad}_{\mathfrak{g}/\mathfrak{h}}(x)v) + 2C_y(\mathrm{ad}_{\mathfrak{g}/\mathfrak{h}}(x)y, u, v) = 0,$$

for every $y(\neq 0), u, v \in \mathfrak{g}/\mathfrak{h}$, and $w \in \mathfrak{h}$, where $\mathrm{ad}_{\mathfrak{g}/\mathfrak{h}}$ is the representation of $\mathfrak{h}$ on $\mathfrak{g}/\mathfrak{h}$ induced by the adjoint representation. Then it is obvious that $(\mathfrak{g}/\mathfrak{h}, \mathrm{ad}_{\mathfrak{g}/\mathfrak{h}}, F)$ is a Minkowski representation of $\mathfrak{h}$. Recall that by Theorem 4.1, $(\mathfrak{g}/\mathfrak{h}, \mathrm{Ad}_{\mathfrak{g}/\mathfrak{h}}, F)$ is a Minkowski representation of $H$.

*Example 5.5.* Let $(\mathfrak{g}, F)$ be a Minkowski Lie algebra. That is, $\mathfrak{g}$ is a real Lie algebra, $F$ is a Minkowski norm on $\mathfrak{g}$, and the following condition is satisfied:

$$g_y([w, u], v) + g_y(u, [w, v]) + 2C_y([w, y], u, v) = 0,$$

for every $y(\neq 0)$, $w, u, v \in \mathfrak{g}$. Then it is easily seen that $(\mathfrak{g}, \mathrm{ad}, F)$ is a Minkowski representation of $\mathfrak{g}$, where $\mathrm{ad}$ is the adjoint representation of $\mathfrak{g}$.

*Example 5.6.* Let $(\mathfrak{g}, \sigma, F)$ be a Minkowski symmetric Lie algebra. That is, $\mathfrak{g}$ is a real Lie algebra, $\sigma$ is an involutive automorphism of $\mathfrak{g}$ with canonical decomposition $\mathfrak{g} = \mathfrak{h} + \mathfrak{m}$, $F$ is a Minkowski norm on $\mathfrak{p}$, and the following condition is satisfied:

$$g_y([w, u], v) + g_y(u, [w, v]) + 2C_y([w, y], u, v) = 0,$$

for every $y(\neq 0), u, v \in \mathfrak{m}$, $w \in \mathfrak{h}$. Then it is easily seen that $(\mathfrak{m}, \mathrm{ad}, F)$ is a representation of $\mathfrak{h}$, where $\mathrm{ad}$ is the adjoint representation of $\mathfrak{h}$ on $\mathfrak{m}$.

The relation between Minkowski representations of Lie groups and Lie algebras is stated in the following:

**Theorem 5.20.** *Let G be a Lie group with Lie algebra* $\mathfrak{g}$. *If* $(V, \rho, F)$ *is a (real) Minkowski representation of G, then* $(V, d\rho, F)$ *is a Minkowski representation of* $\mathfrak{g}$. *On the other hand, if* $(V, \phi, F)$ *is a Minkowski representation of* $\mathfrak{g}$ *and G is connected, then there exists a Minkwoski representation* $(V, \rho, F)$ *with* $\phi = d\rho$.

The proof of this theorem is similar to that of Theorem 4.2, so we omit it. We also have the following result.

**Theorem 5.21.** *Let G be a Lie group and* $(V, \rho, F)$ *a real (complex) Minkowski representation of G. Then there exists an inner product (Hermitian inner product)* $\langle , \rangle$ *on V such that* $(V, \rho, \langle , \rangle)$ *is an orthogonal (unitary) representation of G.*

*Proof.* The proof is similar to Theorem 5.19. Just note that $\rho(G)$ is contained in the group $K$ of linear isometries of $(V, F)$, which is a compact subgroup of $GL(V)$.  □

Next we will use the notion of Minkowski representations of Lie groups and Lie algebras to study homogeneous complex Finsler spaces. By Proposition 5.4, we need only study the invariant structures on a coset space $G/H$, where $G$ is a connected Lie group and $H$ is a closed subgroup of $G$.

We first consider the special case in which $G$ is a complex Lie group. That is, $G$ is an (abstract) group as well as a complex manifold and the map: $(x, y) \mapsto xy$, resp. $x \mapsto x^{-1}$, is a holomorphic map from $G \times G$, resp. $G$ to $G$. As a manifold, $G$ can be viewed as a real smooth manifold, denoted by $G_\mathbb{R}$. Then $G_\mathbb{R}$ with the original group operation is a real Lie group. The Lie algebra $\mathfrak{g}_\mathbb{R}$ has a natural complex structure $I$. Therefore $\mathfrak{g}_\mathbb{R}$ can be viewed as a complex Lie algebra, which we denote by $\mathfrak{g}$ and call the complex Lie algebra of the complex Lie group $G$.

Let $G$ be a complex Lie group and $H$ a closed complex subgroup of $G$. Then it is easily seen that the coset space $G/H$ has a natural complex structure $I$ such that $(G/H, I)$ is a homogeneous complex manifold. Identifying the tangent space $T_o(G/H)$ of $G/H$ at the origin $o = eH$ with the quotient space $\mathfrak{g}/\mathfrak{h}$, we have the following theorem.

**Theorem 5.22.** *Let G be a complex Lie group and H a closed complex subgroup of G. If F is a complex Finsler metric on* $G/H$ *such that* $(G/H, I, F)$ *is a homogeneous complex Finsler manifold, then* $(\mathfrak{g}/\mathfrak{h}, \mathrm{ad}_{\mathfrak{g}/\mathfrak{h}}, F_o)$ *is a Minkowski representation of* $\mathfrak{h}$. *On the other hand, if* $F_*$ *is a complex Minkowski norm on* $\mathfrak{g}/\mathfrak{h}$ *such that* $(\mathfrak{g}/\mathfrak{h}, \mathrm{ad}_{\mathfrak{g}/\mathfrak{h}}, F_*)$ *is a Minkowski representation of* $\mathfrak{h}$ *and the subgroup H is connected, then there exists a complex Finsler metric F on* $G/H$ *such that* $F_* = F_o$ *and* $(G/H, I, F)$ *is a homogeneous complex Finsler manifold.*

*Proof.* The proof is similar to the real case; see Theorem 4.2.  □

Theorem 5.22 gives a complete description of the structure of invariant complex homogeneous Finsler metrics on the coset spaces of complex Lie groups. However, there exist complex homogeneous manifolds that cannot be written as a coset space of a complex Lie group. Let us give an example.

*Example 5.7.* Consider a bounded domain $D$ in $\mathbb{C}^n$. It is well known that the group of holomorphic diffeomorphisms of $D$ onto itself, denoted by $H(D)$, is a real Lie group (see [83]). If the action of $H(D)$ on $D$ is transitive, then $D$ is a homogeneous complex manifold, called a homogeneous bounded domain. However, in this case, there cannot be any complex structure on $H(D)$ that makes $H(D)$ a complex Lie group. In fact, if $H(D)$ is a complex Lie group, then the orbit of any one-parameter subgroup of $H(D)$ is bounded. According to Liouville's theorem, every bounded holomorphic function on $\mathbb{C}$ is a constant. This means that the orbit of every one-parameter subgroup of $H(D)$ consists of a single point, contracting the fact that $H(D)$ acts transitively on $D$.

Therefore, to obtain a complete algebraic description of all the complex homogeneous Finsler spaces, we have to consider all the coset spaces $G/H$, where $G$ is a *real* Lie group and $H$ is a closed subgroup of $G$. In the following, we will find a sufficient and necessary condition for such a coset space to have a complex structure and an invariant complex Finsler metric simultaneously.

In the following, we will encounter several representations of a Lie algebra $\mathfrak{h}$ that are all induced by the adjoint representation. To keep the notation simple, we sometimes denote such representations simply by ad. Moreover, for a linear transformation $\rho$ on a real vector space $V$, we use $\rho^C$ to denote the induced complex linear transformation on the complexification $V^C$.

**Theorem 5.23.** *Let $G$ be a (real) Lie group, $H$ a closed subgroup of $G$, Lie $G = \mathfrak{g}$, Lie $H = \mathfrak{h}$. If $I$ is a complex structure on $G/H$ and $F$ is a Finsler metric on $G/H$ such that $(G/H, I, F)$ is a homogeneous complex Finsler space, then there exists a complex subalgebra $\mathfrak{a}$ of the complex Lie algebra $\mathfrak{g}^C$ (complexification of $\mathfrak{g}$) such that $\mathfrak{a} \cap \bar{\mathfrak{a}} = \mathfrak{h}^C$, $\mathfrak{a} + \bar{\mathfrak{a}} = \mathfrak{g}^C$, and a complex Finsler metric $F_1$ on $\mathfrak{a}/\mathfrak{h}^C$ such that $(\mathfrak{a}/\mathfrak{h}^C, \mathrm{ad}, F_1)$ is a Minkowski representation of $\mathfrak{h}$. On the other hand, if there exists a complex subalgebra $\mathfrak{a}$ of $\mathfrak{g}^C$ satisfying $\mathfrak{a} \cap \bar{\mathfrak{a}} = \mathfrak{h}^C$, $\mathfrak{a} + \bar{\mathfrak{a}} = \mathfrak{g}^C$ and a complex Minkowski norm $F_1$ on $\mathfrak{a}/\mathfrak{h}^C$ such that $(\mathfrak{a}/\mathfrak{h}^C, \mathrm{ad}, F_1)$ is a Minkowski representation of $\mathfrak{h}$ and if moreover, $H$ is connected, then there exist a complex structure $I$ on $G/H$ and a complex Finsler metric $F$ on $G/H$ such that $(G/H, I, F)$ is a homogeneous complex Finsler space.*

*Proof.* First suppose that $I$ is a $G$-invariant complex structure on $G/H$ and $F$ is a Finsler metric on $G/H$ such that $(G/H, I, F)$ is a homogeneous complex space. Then $I$ corresponds to a Koszul operator $J$ [102, 103] that is a linear transformation of $\mathfrak{g}$ such that there exists a subspace $\mathfrak{m}$ satisfying

$$\mathfrak{g} = \mathfrak{h} + \mathfrak{m} \quad \text{(direct sum of subspaces)},$$

with $J(\mathfrak{h}) = 0$, $J(\mathfrak{m}) \subset \mathfrak{m}$ and $J^2|_\mathfrak{m} = -\mathrm{id}$. Moreover, we also have

$$\pi(JX) = I_o(\pi(X)), \quad X \in \mathfrak{g},$$
$$\mathrm{Ad}(h)J \equiv J\mathrm{Ad}(h) \,(\mathrm{mod}\,\mathfrak{h}), \forall h \in H,$$

where $o$ is the origin of $G/H$ and $\pi$ is the natural projection of $\mathfrak{g}$ onto $\mathfrak{g}/\mathfrak{h}$. Now we extend $J$ to a complex linear transformation $J^C$ of $\mathfrak{g}^C = \mathfrak{h}^C + \mathfrak{m}^C$. Define

$$\mathfrak{n}^\pm = \{X \in \mathfrak{g}^C \mid J^C(X) = \pm\sqrt{-1}X\}.$$

It is easily seen that $\mathfrak{n}^\pm \subset \mathfrak{m}^C$. Set $\mathfrak{a} = \mathfrak{h}^C + \mathfrak{n}^+$. Then $\bar{\mathfrak{a}} = \mathfrak{h}^C + \mathfrak{n}^-$. Therefore we have $\mathfrak{a} \cap \bar{\mathfrak{a}} = \mathfrak{h}^C$, and $\mathfrak{g} = \mathfrak{a} + \bar{\mathfrak{a}}$. Furthermore, by the definition of $\mathfrak{a}$, for every $X \in \mathfrak{h}^C$, $Y \in \mathfrak{a}$, we have

$$I_o^C \pi^C([X,Y]) = I_o^C \pi^C(\mathrm{ad}(X)Y) = I_o^C \mathrm{ad}(X)\pi^C(Y) = \mathrm{ad}(X)I_o^C(\pi^C(Y)),$$

Since $\pi^C(Y) \in \mathfrak{n}^+$, we have $I_o^C(\pi^C(Y)) = \sqrt{-1}\pi^C(Y)$. Thus

$$I_o^C(\pi^C([X,Y])) = \sqrt{-1}\mathrm{ad}(X)\pi^C(Y) = \sqrt{-1}\pi^C([X,Y]).$$

Therefore, $\pi^C([X,Y]) \in \mathfrak{n}^+$. Hence $[X,Y] \in \mathfrak{a}$. That is,

$$[\mathfrak{h}^C, \mathfrak{a}] \subset \mathfrak{a}.$$

Now for any $Y_1, Y_2 \in \mathfrak{a}$, we have

$$J^C[Y_1, Y_2] = \sqrt{-1}[Y_1, Y_2] \quad (\mathrm{mod}\,\mathfrak{h}^C).$$

Combining this fact with $[\mathfrak{h}^C, \mathfrak{a}] \subset \mathfrak{a}$, we see that $\mathfrak{a}$ is a complex subalgebra of $\mathfrak{g}^C$.

Now the Finsler metric $F$ on $G/H$ defines a Minkowski norm on the vector space $T_o(G/H)$. Identifying $\mathfrak{g}/\mathfrak{h}$ with $T_o(G/H)$, we obtain a Minkowski norm $F^*$ on $\mathfrak{g}/\mathfrak{h}$ defined by

$$F^*(au + bI_o(u)) = \sqrt{a^2 + b^2}F(u), \quad u \in \mathfrak{g}/\mathfrak{h}, a, b \in \mathbb{R},$$
$$F^*(\mathrm{Ad}(h)u) = F(u), \quad h \in H, u \in \mathfrak{g}/\mathfrak{h}.$$

Extend $F^*$ to a Minkowski norm on $(\mathfrak{g}/\mathfrak{h})^C = \mathfrak{g}^C/\mathfrak{h}^C$ (denoted by $F_1^*$) by

$$F_1^*(u + \sqrt{-1}v) = \sqrt{(F^*(u))^2 + (F^*(v))^2}, \quad u, v \in \mathfrak{g}/\mathfrak{h}.$$

It is obvious that

$$F_1^*((a\,\mathrm{id} + bI_o^C)u) = \sqrt{a^2 + b^2}F_1^*(u), \quad u \in \mathfrak{g}^C/\mathfrak{h}^C, a, b \in \mathbb{R}.$$

Let $F_1$ be the restriction of $F_1^*$ to $\mathfrak{a}/\mathfrak{h}^C$. Then by definition, we have

$$F_1(\mathrm{Ad}(h)(x)) = F_1(x),$$

for every $x \in \mathfrak{a}/\mathfrak{h}^C$ and $h \in H$. On the other hand, for every $a, b \in \mathbb{R}$ and $u \in \mathfrak{a}$, we have

$$F_1((a+b\sqrt{-1})\pi^C(u)) = F_1(\pi^C((a+bJ^C)u))$$
$$= F_1((a+bI_o^C)\pi^C(u)) = \sqrt{a^2+b^2}F_1(\pi^C(u)).$$

Therefore $F_1$ is a complex Minkowski norm on $\mathfrak{a}/\mathfrak{h}^C$. Since $F_1$ is invariant under $H$, $(\mathfrak{a}/\mathfrak{h}^C, \mathrm{Ad}, F_1)$ is a Minkowski representation of $H$. By Theorem 5.5.4, $(\mathfrak{a}/\mathfrak{h}^C, \mathrm{ad}, F_1)$ is a Minkowski representation of $\mathfrak{h}$.

Conversely, if there exist a complex subalgebra $\mathfrak{a}$ satisfying $\mathfrak{g}^C = \mathfrak{a} + \bar{\mathfrak{a}}$, $\mathfrak{a} \cap \bar{\mathfrak{a}} = \mathfrak{h}^C$ and a complex Minkowski norm $F_1$ on $\mathfrak{a}/\mathfrak{h}^C$ such that $(\mathfrak{a}/\mathfrak{h}^C, \mathrm{ad}, F_1)$ is a Minkowski representation of $\mathfrak{h}$, then we have a direct decomposition

$$(\mathfrak{g}/\mathfrak{h})^C = \pi^C(\mathfrak{a}) + \pi^C(\bar{\mathfrak{a}}).$$

Since $\pi^C(\bar{\mathfrak{a}}) = \overline{\pi^C(\mathfrak{a})}$, we can define a (real) linear transformation $I_o$ on the real vector space $\mathfrak{g}/\mathfrak{h}$ such that $I_o^C|_{\pi^C(\mathfrak{a})} = \sqrt{-1}\,\mathrm{id}$, and $I_o^C|_{\pi^C(\bar{\mathfrak{a}})} = -\sqrt{-1}\,\mathrm{id}$. Then it is known that $I_o$ can be extended to an almost complex structure $I$ on $G/H$ and moreover, $I$ is integrable if and only if $\mathfrak{a}$ is a subalgebra of $\mathfrak{g}^C$ [103]. Thus we have defined a complex structure on $G/H$. Now we extend the Minkowski norm $F_1$ (denoted by $F_1^*$) to $\mathfrak{g}^C/\mathfrak{h}^C = (\mathfrak{g}/\mathfrak{h})^C$ by

$$F_1^*((u+\mathfrak{h}^C) + (\bar{v}+\mathfrak{h}^C)) = \sqrt{(F_1(u+\mathfrak{h}^C))^2 + (F_1(v+\mathfrak{h}^C))^2}, \quad u, v \in \mathfrak{a}.$$

It is easily seen that $F_1^*$ is a complex Minkowski norm on $\mathfrak{g}^C/\mathfrak{h}^C$. Let $F_o$ be the restriction of $F_1^*$ to $\mathfrak{g}/\mathfrak{h}$ and for $w \in \mathfrak{g}$, let

$$\pi^C(w) = (w_1 + \mathfrak{h}^C) + (\bar{w}_2 + \mathfrak{h}^C),$$

where $w_1, w_2 \in \mathfrak{a}$. Then for every $a, b \in \mathbb{R}$, we have (here for simplicity we denote the imaginary unit by $\mathbf{i}$)

$$F_o((a+b\mathbf{i})\pi(w)) = F_o((a+bI_o)\pi(w)) = F_1^*((a+bI_o^C)\pi^C(w))$$
$$= F_1^*((a+bI_o^C)((w_1+\mathfrak{h}^C) + (\bar{w}_2+\mathfrak{h}^C)))$$
$$= F_1^*((a+b\mathbf{i})(w_1+\mathfrak{h}^C) + (a-b\mathbf{i})(\bar{w}_2+\mathfrak{h}^C))$$
$$= \sqrt{(F_1((a+b\mathbf{i})(w_1+\mathfrak{h}^C)))^2 + (F_1((a-b\mathbf{i})(\bar{w}_2+\mathfrak{h}^C))^2}$$
$$= \sqrt{(a^2+b^2)(F_1(w_1+\mathfrak{h}^C))^2 + (a^2+b^2)(F_1(\bar{w}_2+\mathfrak{h}^C))^2}$$
$$= \sqrt{a^2+b^2}\sqrt{(F_1(w_1+\mathfrak{h}^C))^2 + (F_1(\bar{w}_2+\mathfrak{h}^C))^2}$$
$$= \sqrt{a^2+b^2}F_1(\pi^C(w)) = \sqrt{a^2+b^2}F_o(\pi(w)).$$

Therefore, $F_o$ is a complex Minkowski norm on $\mathfrak{g}/\mathfrak{h}$ (with respect to the complex structure $I_o$). On the other hand, since $(\mathfrak{a}/\mathfrak{h}^C, \mathrm{ad}, F_1)$ is a Minkowski representation of $\mathfrak{h}$, it is easily seen that $(\mathfrak{g}/\mathfrak{h}, \mathrm{ad}, F_o)$ is a Minkowski representation of $\mathfrak{h}$. Since $H$ is connected, $(\mathfrak{g}/\mathfrak{h}, \mathrm{Ad}, F_o)$ is a Minkowski representation of the group $H$. By Theorem 4.1, we can define a complex Finsler metric $F$ on $G/H$ that is invariant under the action of $G$. Consequently, $(G/H, I, F)$ is a homogeneous complex Finsler space. $\qquad \square$

*Remark 5.3.* In some special cases, we have a priori an invariant complex structure on $G/H$. In this case, to make $G/H$ a homogeneous complex Finsler space, we need only find an invariant complex Finsler metric on $G/H$.

Finally, we study complex symmetric Finsler spaces. Let $(M, I, F)$ be a complex Finsler space. Then $(M, I, F)$ is called globally symmetric if for every point $x \in M$ there exists an involutive holomorphic isometry $\sigma_x$ of $M$ such that $x$ is an isolated fixed point of $\sigma_x$. It is easily seen that in this case, the group $A(M, I, F)$ acts transitively on $M$. Hence $(M, I, F)$ is a homogeneous complex Finsler space. Let $G = A_0(M, I, F)$ and let $H$ be the isotropy subgroup of $G$ at a fixed point $x$ in $M$. Then $M = G/H$. Define an automorphism $\sigma$ of $G$ by $\sigma(g) = \sigma_x \cdot g \cdot \sigma_x$, $g \in G$. Then $\sigma$ is an involutive automorphism of $G$ and $(K_\sigma)_0 \subset H \subset K_\sigma$, where $K_\sigma$ is the subgroup of $G$ consisting of fixed points of $\sigma$ and $(K_\sigma)_0$ is the identity component of $K_\sigma$. This means that $(G, H)$ is a symmetric pair. Since $H$ is compact, $(G, H)$ is a Riemannian symmetric pair. By standard results on Hermitian symmetric spaces (see [83, Chap. VIII]), we have the following theorem.

**Theorem 5.24.** *Let $(G/H, I, F)$ be a globally symmetric complex Finsler space. Then there exists a $G$-invariant Riemannian metric $Q$ on $G/H$ such that the triple $(G/H, I, Q)$ is a Hermitian symmetric space.*

The Hermitian symmetric spaces were completely classified by Cartan [83]. Hence Theorem 5.24 reduces the problem of classification of symmetric complex Finsler spaces to the problem of determining which Hermitian symmetric space admits an invariant non-Riemannian complex Finsler metric. Now we can prove our next result.

**Theorem 5.25.** *Let $(G_1/H_1, I_1, Q_1)$ and $(G_2/H_2, I_2, Q_2)$ be two Hermitian symmetric spaces. Then on the product manifold $(G_1/H_1) \times (G_2/H_2)$ there exist infinitely many non-Riemannian Finsler metrics $F$ that are invariant under $G_1 \times G_2$ and make $(G_1/H_1 \times G_2/H_2, I_1 \times I_2, F)$ a symmetric complex Finsler space.*

*Proof.* Let $o_1, o_2$ be the origins of $G_1/H_1$, $G_2/H_2$ respectively and $o = (o_1, o_2)$. For every tangent vector $y = (y_1, y_2) \in T_o(G_1/H_1 \times G_2/H_2)$, $y_i \in T_{o_i}(G_i/H_i)$, $i = 1, 2$, and integer $s \geq 2$, define

$$F_o(y) = \sqrt{Q_1(y_1, y_1) + Q_2(y_2, y_2) + \sqrt[s]{Q_1(y_1, y_1)^s + Q_2(y_2, y_2)^s}}.$$

Then $F_o$ is a non-Euclidean Minkowski norm on $T_o(G_1/H_1 \times G_2/H_2)$ that is invariant under $H_1 \times H_2$. Hence it can be extended to a Finsler metric $F$ on

$G_1/H_1 \times G_2/H_2$. It is obvious that $F$ is non-Riemannian. Since $Q_1, Q_2$ are Hermitian metrics, for every $a, b \in \mathbb{R}$, we have

$$F_o((a+bI)y) = F_o((a+bI_1)y_1, (a+bI_2)y_2) = \sqrt{a^2+b^2}F_o(y).$$

Therefore $F$ is a complex Minkowski norm.                                          □

The irreducible cases can also be completely settled. First we prove the following result.

**Theorem 5.26.** *Let $G/H$ be a (compact or noncompact) irreducible Hermitian symmetric space. Then there exists an invariant complex structure $I$ on $G/H$ such that $(G/H, I, F)$ is a symmetric complex Finsler space for every $G$-invariant Finsler metric on $G/H$.*

*Proof.* The only point we need to check is that every invariant Finsler metric $F$ on $G/H$ must be a complex Finsler metric with respect to the complex structure. We treat only the compact case. The noncompact case can be proved using duality. By Theorem 2.38, if $G/H$ is a compact irreducible Hermitian symmetric space, then $G/H$ can also written as $U/K$, where $U$ is a connected compact simple Lie group with center $\{e\}$ and $K$ is a maximal connected proper subgroup of $U$ with nondiscrete center. Moreover, the center $Z_K$ of $K$ is isomorphic to the circle group $S^1$. Therefore there exists an element of order 4, say $j$, in $K$. Let $I$ denote the induced linear transformation of $\mathrm{Ad}(j)$ to $T_o(U/K)$. Then $I^2 = -\mathrm{id}$ and it is true that $I$ induces an invariant complex structure on $U/K$. Now suppose $F$ is a $U$-invariant Finsler metric on $U/K$ and $a, b \in \mathbb{R}$ with $a^2 + b^2 \neq 0$. We assert that the endomorphism

$$\frac{a}{\sqrt{a^2+b^2}}\,\mathrm{id} + \frac{b}{\sqrt{a^2+b^2}}I$$

lies in the image of $Z_K$ under the inducing map (to $T_o(U/K)$). In fact, since $Z_K \simeq S^1$ and $j$ is of order 4, we easily see that there exists $u \in \mathfrak{k}$ such that $Z_K = \{\exp(tu) \mid t \in \mathbb{R}\}$ and $\exp(u) = j$. Since $\mathrm{Ad}(j)$ induces the complex structure $I$ on $T_o(U/K)$, $\mathrm{Ad}(\exp(tu))$ induces the endomorphism $T(t)$ of $T_o(U/K)$, where

$$T(t) = e^{tI} = (\cos t)\,\mathrm{id} + (\sin t)I, \quad t \in \mathbb{R}.$$

This proves our assertion. Now

$$F((a\,\mathrm{id}+bI)y) = F\left(\sqrt{a^2+b^2}\left(\frac{a}{\sqrt{a^2+b^2}}\,\mathrm{id} + \frac{b}{\sqrt{a^2+b^2}}I\right)y\right)$$

$$= \sqrt{a^2+b^2}F\left(\left(\frac{a}{\sqrt{a^2+b^2}}\,\mathrm{id} + \frac{b}{\sqrt{a^2+b^2}}I\right)y\right)$$

$$= \sqrt{a^2+b^2}F(T(\theta)y)$$

$$= \sqrt{a^2+b^2}F(y),$$

**Table 5.1** Irreducible symmetric complex non-Riemannian Finsler spaces. Here $p, q \geq 1$

|      | Noncompact spaces | Compact spaces | Rank | Dimension |
|------|-------------------|----------------|------|-----------|
| AIII | $SU(p,q)/S(U_p \times U_q)$ | $SU(p+q)/S(U_p \times U_q)$ | $\min(p,q)$ | $2pq$ |
| BDI  | $SO_o(p,2)/SO(p) \times SO(2)$ | $SO(p+2)/SO(p) \times SO(2)$ | 2 | $2p$ |
| DIII | $SO^*(2n)/U(n)(n \geq 4)$ | $SO(2n)/U(n)(n \geq 4)$ | $[\frac{1}{2}n]$ | $n(n-1)$ |
| CI   | $Sp(n,\mathbb{R})/U(n)(n \geq 2)$ | $Sp(n)/U(n)(n \geq 2)$ | $n$ | $n(n+1)$ |
| EIII | $(\mathfrak{e}_{6(-14)}, \mathfrak{so}(10) + \mathbb{R})$ | $(\mathfrak{e}_{6(-78)}, \mathfrak{so}(10) + \mathbb{R})$ | 2 | 32 |
| EVII | $(\mathfrak{e}_{7(-25)}, \mathfrak{e}_6 + \mathbb{R})$ | $(\mathfrak{e}_{7(-133)}, \mathfrak{e}_6 + \mathbb{R})$ | 3 | 54 |

where

$$\theta = \arccos \frac{a}{\sqrt{a^2 + b^2}}$$

and we have used the above assertion and the fact that $F$ is invariant under $\mathrm{Ad}(K)$. Thus $F$ is a complex Finsler metric on $U/K$. This proves the theorem.           □

It was proved in Szabó [152] that on each irreducible globally symmetric Riemannian manifold $G/H$ of rank $\geq 2$, there exist infinitely many invariant Finsler metrics on $G/H$, and on a rank-1 symmetric Riemannian manifold $G/H$, there does not exist any non-Riemannian invariant Finsler metric. On the other hand, irreducible Hermitian symmetric spaces were completely classified by Cartan (see [83, p. 518]). Combining these results with Theorem 5.26, we get a complete classification of irreducible symmetric complex non-Riemannian Finsler spaces.

**Theorem 5.27.** *Let $G/H$ be an irreducible Riemannian symmetric space. If $G/H$ admits an invariant complex structure $I$ and a non-Riemannian Finsler metric $F$ such that $(G/H, I, F)$ is a symmetric complex Finsler space, then $G/H$ must be one of the manifolds in Table 5.1. Furthermore, on each manifold in the table, there exist infinitely many invariant symmetric complex non-Riemannian Finsler metrics.*

# Chapter 6
# Weakly Symmetric Finsler Spaces

The notion of Riemannian weakly symmetric spaces was introduced by Selberg in 1956. In the paper [135], he generalized the Poisson summation formula to the celebrated Selberg trace formula in his study of harmonic analysis on such manifolds. Every Riemannian symmetric space is weakly symmetric. But the converse is not true, as pointed out by Selberg by constructing some explicit examples. Moreover, a Riemannian weakly symmetric space is a commutative space, in the sense that the differential operators that are invariant under the action of the full group of isometries form a commutative algebra. Recently the study of Riemannian weakly symmetric spaces has become rather active. An excellent survey of related results in this field can be found in Wolf's recent book [176].

In recent years, we have generalized the notion of a weakly symmetric space to the Finslerian case and studied weakly symmetric Finsler spaces extensively. One of the most important applications of this study is that we have found many examples of reversible non-Berwald spaces with vanishing S-curvature, such spaces as have been sought in an open problem posed by Shen in the problem section [146].

In this chapter we will introduce our results on weakly symmetric Finsler spaces. Here is an outline of the main results. We first study the geometric properties of weakly symmetric Finsler spaces and particularly prove that every weakly symmetric Finsler space must have vanishing S-curvature. Then we introduce a new algebraic notion of a weakly symmetric Lie algebra to describe weakly symmetric Finsler spaces. As an application, we obtain a global classification of all weakly symmetric Finsler spaces with dimension less than or equal to 3. Finally, we classify weakly symmetric spaces with a reductive transitive group of isometries, as well as admissible weakly symmetric metrics on H-type Lie groups. The last section is devoted to constructing examples of non-Berwaldian reversible Finsler spaces with vanishing S-curvature.

Here we point out some possible developments in related topics in this chapter. First, Finslerian g.o. spaces (see Sect. 6.2 for the definition) deserve to be studied thoroughly. Up to now, the only known examples of Finslerian g.o. spaces are the Riemannian ones and weakly symmetric Finsler spaces studied in this chapter. It would be interesting to find some special types of Finslerian g.o. spaces.

S. Deng, *Homogeneous Finsler Spaces*, Springer Monographs in Mathematics, DOI 10.1007/978-1-4614-4244-8_6, © Springer Science+Business Media New York 2012

For example, the problem of under what condition a homogeneous Randers space is a g.o. space deserves to be considered. For the Riemannian case of this problem, see [74]. Another interesting problem concerns the notion of weakly symmetric Lie algebras introduced in Sect. 6.3 of this chapter. It would be very interesting to establish a theory of such algebraic structures from a purely algebraic point of view. For example, if we can classify all weakly symmetric Lie algebras with the underlying Lie algebra being semisimple or reductive, then an intrinsic proof of the classification result of Yakimova and Wolf (see Sect. 6.6) can be achieved.

The standard references of this chapter are [49–51, 66, 179]; see also [54, 65, 104, 105] for further information.

## 6.1   Definitions and Fundamental Properties

We first recall the original definition of Selberg.

**Definition 6.1.** A connected Riemannian manifold $(M, Q)$ is called weakly symmetric if there exists a subgroup $G$ of the full group $I(M, Q)$ of isometries such that $G$ acts transitively on $M$ and there exists an isometry $f$ of $(M, Q)$ with $f^2 \in G$ and $fGf^{-1} = G$ such that for every two points $p, q \in M$, there exists an isometry $g$ of $(M, Q)$ satisfying $g(p) = f(q)$ and $g(q) = f(p)$.

This definition seems somehow complicated and roundabout. In [25], Berndt and Vanhecke proved the following simple geometric characterization of weakly symmetric spaces.

**Proposition 6.1.** *A Riemannian manifold $(M, Q)$ is weakly symmetric if and only if for every two points $p, q \in M$ there exists an isometry $f$ of $(M, Q)$ such that $f(p) = q$ and $f(q) = p$, or in short, $f$ interchanges $p$ and $q$.*

*Proof.* Suppose $(M, Q)$ is a weakly symmetric Riemannian manifold and $G$ is the subgroup of $I(M, Q)$ satisfying the condition in Definition 6.1, along with a specific isometry $f_1$. Then for every $p, q$, there is $g_1 \in I(M, Q)$ such that $f_1(p) = g_1(q)$ and $f_1(q) = g_1(p)$. Hence the isometry $f = g_1^{-1}f_1$ interchanges $p$ and $q$. Conversely, suppose for every two points $p, q$ in $M$ there is an isometry $f_1$ that interchanges $p$ and $q$. Set $G = I(M, Q)$ and $f = \mathrm{id}$. Then it is easy to check that $G$ and $f$ satisfy the conditions in Definition 6.1.                                                                  □

Now we give the definition of a weakly symmetric Finsler space.

**Definition 6.2.** Let $(M, F)$ be a connected Finsler space and $I(M, F)$ the full group of isometries. Then $(M, F)$ is called weakly symmetric if for every two points $p, q$ in $M$ there exists an isometry $\sigma \in I(M, F)$ such that $\sigma(p) = q$ and $\sigma(q) = p$.

From the definition, we see that a weakly symmetric Finsler space is homogeneous, i.e., the group $I(M, F)$ of isometries acts transitively on $M$. Therefore, a weakly symmetric Finsler space is necessarily (both forward and backward) complete. Moreover, a weakly symmetric Finsler space $(M, F)$ must be reversible.

In fact, for any $p, q \in M$, let $\sigma$ be an isometry such that $\sigma(p) = q$, $\sigma(q) = p$. Then we have $d(p,q) = d(\sigma(p), \sigma(q)) = d(q,p)$ (Since an isometry is distance-preserving). Therefore the distance function of $(M,F)$ is reversible. Hence $F$ is reversible.

The following result is an alternative characterization of weakly symmetric Finsler spaces. In the Riemannian case, this is given by Szabó; see [153].

**Proposition 6.2.** *A Finsler space* $(M,F)$ *is a weakly symmetric space if and only if for every maximal geodesic* $\gamma$ *in M and every point* $m \in \gamma$ *there exists an isometry* $\sigma \in I(M,F)$ *such that* $\sigma(\gamma) \subset \gamma$, $\sigma(m) = m$, $\sigma|_\gamma \neq \mathrm{id}$, *and* $\sigma^2|_\gamma = \mathrm{id}$ *(in other words, $\sigma$ is a nontrivial involution along $\gamma$ fixing $m$).*

*Proof.* Let $(M,F)$ be a weakly symmetric Finsler space and $\gamma$ a maximal geodesic in $M$ containing $m$. Without loss of generality, we can assume that $\gamma$ is of unit speed and $m = \gamma(0)$. Let $U$ be a neighborhood of $m$ such that each pair of points in $U$ can be connected by a unique (minimal) geodesic in $U$. Select $p$, $q$ in $\gamma \cap U$ such that $m$ is the midpoint along the geodesic segment between $p$, $q$, i.e., $m$ is the unique point in $\gamma$ with $d(p,m) = d(m,q)$ (note that $d$ is reversible), or equivalently $p = \gamma(\varepsilon)$, $q = \gamma(-\varepsilon)$, where $\varepsilon > 0$. Then by assumption, there exists an isometry $\sigma$ of $(M,F)$ such that $\sigma(p) = q$ and $\sigma(q) = p$. Therefore the curve $\sigma(\gamma(t))$, $-\varepsilon \leq t \leq \varepsilon$, as another distance-minimizing curve connecting $p$ and $q$, must coincide with the curve $\gamma(-t)$, $-\varepsilon \leq t \leq \varepsilon$. Thus $\sigma(\gamma(t)) = \gamma(-t)$ for $-\varepsilon \leq t \leq \varepsilon$. In particular, $\sigma(m) = m$. Now consider the maximal geodesics $\sigma(\gamma(t))$ and $\gamma(-t)$, $t \in \mathbb{R}$. By the above argument and the uniqueness of the geodesics, we see that these two geodesics must coincide. Hence $\sigma(\gamma(t)) = \gamma(-t)$, $\forall t \in \mathbb{R}$. Therefore $\sigma$ is the desired involution on $\gamma$.

On the other hand, suppose for every geodesic $\gamma$ and $m \in \gamma$ there exists an involution along $\gamma$ fixing $m$. Given $p_1, p_2 \in M$, we can find a series of (constant-speed) distance-minimizing geodesic segments $\gamma_i$, $i = 1, 2, \ldots, s$, such that $\gamma_1(0) = p_1$, $\gamma_i(0) = \gamma_{i-1}(1)$, $i = 2, 3, \ldots, s$, and $\gamma_s(1) = p_2$. Let $m_i$ be the unique point in $\gamma_i$ satisfying $d(m_i, \gamma_i(0)) = d(m_i, \gamma_i(1))$ and let $\sigma_i$ be the involution along $\gamma_i$ fixing $m_i$. Then it is easily seen that the isometry $\sigma_i$ sends $\gamma_i(o)$ to $\gamma_i(1)$ and $\gamma_i(1)$ to $\gamma_i(0)$. Therefore the composition map $\sigma_s \sigma_{s-1} \cdots \sigma_1$ sends $p_1$ to $p_2$. Thus $(M,F)$ is homogeneous. In particular, $(M,F)$ is (both forward and backward) complete. Now for every two points $p, q \in M$, let $\gamma(t)$, $t \in \mathbb{R}$, be a maximal geodesic containing $p$ and $q$ such that $\gamma$ realizes the distance of $p$ and $q$. Let $m \in \gamma$ be the unique point in $\gamma$ that lies between $p$ and $q$ and satisfies $d(m,p) = d(m,q)$. Then the involutive isometry along $\sigma$ that keeps $m$ fixed interchanges $p$ and $q$. Hence $(M,F)$ is weakly symmetric. $\square$

Next we give another geometric description of weakly symmetric Finsler spaces.

**Proposition 6.3.** *A Finsler space* $(M,F)$ *is weakly symmetric if and only if for every* $m \in M$ *and* $u \in T_m(M)$ *there exists an isometry* $\sigma$ *of* $(M,F)$ *such that* $\sigma(m) = m$ *and* $d\sigma(u) = -u$.

*Proof.* The proof of the "only if" part is contained in the first part the proof of Proposition 6.2. In fact, the isometry $\sigma$ constructed there must reverse the geodesic $\gamma$. Therefore the differential map of the isometry must send the initial vector of the geodesic to its negative. Now we prove the "if" part. First we prove that such a space must be complete. We prove only the forward completeness. The proof of backward completeness is similar. Let $\gamma(t)$, $0 \leq t < b$, be a (forward) maximal unit-speed geodesic. Suppose conversely that $b < +\infty$. Choose an isometry $\sigma$ such that $\sigma(\gamma(\frac{3}{4}b)) = \gamma(\frac{3}{4}b)$ and $d\sigma(v) = -v$, where $v$ is the initial vector of the geodesic $\gamma_1(t) = \gamma(\frac{3}{4}b - t)$, $0 \leq t \leq \frac{3}{4}b$. Then $\sigma(\gamma_1(t))$, $0 \leq t \leq \frac{3}{4}b$, is a geodesic. It is easily seen that the curve $\tau(t)$, $0 \leq t \leq \frac{3}{2}b$, defined by

$$
\tau(t) = \begin{cases} \gamma(t) & 0 \leq t \leq \frac{3}{4}b \\ \sigma(\gamma_1(t - \frac{3}{4}b)) & \frac{3}{4}b \leq t \leq \frac{3}{2}b, \end{cases}
$$

is a geodesic on $[0, \frac{3}{4}b]$ and $[\frac{3}{4}b, \frac{3}{2}b]$. Furthermore, it is $C^1$ smooth at $\gamma(\frac{3}{4}b)$. Hence it is a smooth geodesic on the interval $[0, \frac{3}{2}b]$, which is an extension of $\gamma$. This is a contradiction. Thus $(M, F)$ is complete. Moreover, it is obvious that $F$ is reversible. Now for every $p, q \in M$, we select a unit-speed geodesic connecting $p$ and $q$. Let $m$ be the midpoint of the geodesic and let $\sigma$ be an isometry such that $\sigma(m) = m$ and $d\sigma(w) = -w$, where $w$ is the tangent vector of the geodesic at $m$. Then it is easily seen that $\sigma(p) = q$ and $\sigma(q) = p$. Therefore $(M, F)$ is weakly symmetric.   $\square$

In the following, we will adopt some definitions in [124]; see also [176].

**Definition 6.3.** Let $G$ a Lie group and $H$ be a closed subgroup of $G$. The coset space $G/H$ is called a homogeneous weakly symmetric space if there exists an analytic diffeomorphism $\mu$ of $M = G/H$ such that

(i) $\mu G \mu^{-1} = G$, $\mu(eH) = eH$, and $\mu^2 \in H$,
(ii) Given any two points $x$ and $y$ in $M$, there exists $g \in G$ such that $gx = \mu y$ and $gy = \mu x$.

The proof of the following lemma is similar to the Riemannian case; see [124].

**Lemma 6.1.** *Let $(M, F)$ be a weakly symmetric Finsler space. Fix $x \in M$ and let $H$ be the isotropy group of $G = I(M, F)_0$ at $x$. Then there exists an isometry $\tilde{\mu}$ of $M = G/H$ that makes $G/H$ a homogeneous weakly symmetric space.*

**Definition 6.4.** Let $G$ be a Lie group and $H$ a closed subgroup of $G$. The pair $(G, H)$ is called a weakly symmetric pair if there exists an automorphism $\theta$ of $G$ such that

(i) $\theta(H) \subset H$, and there exists $h \in H$ such that $\theta^2 = \mathrm{Ad}(h)$ (i.e., $\theta^2(g) = hgh^{-1}$, $g \in G$).
(ii) $H\theta(g)H = Hg^{-1}H$ for all $g \in G$.

Furthermore, $(G, H)$ is called a Riemannian symmetric pair if $\mathrm{Ad}_G(H)$ is compact.

**Lemma 6.2.** *A pair $(G, H)$ is weakly symmetric if and only if $G/H$ is a homogeneous weakly symmetric space.*

For the proof, see [124].

The following theorem will be very useful for studying weakly symmetric Finsler spaces.

**Theorem 6.1.** *Let $(M,F)$ be a weakly symmetric Finsler space. Fix $x \in M$ and let $H$ be the isotropy subgroup of $G = I(M,F)_0$ at $x$. Then $(G,H)$ is a Riemannian weakly symmetric pair. On the other hand, if $(G,H)$ is a weakly symmetric pair, then every G-invariant reversible Finsler metric (if it exists) on $G/H$ makes $G/H$ a weakly symmetric Finsler space. In particular, if $(G,H)$ is a Riemannian weakly symmetric pair, then there exists a G-invariant Finsler metric $F$ on $G/H$ such that $(G/H,F)$ is a weakly symmetric Finsler space.*

*Proof.* By Theorem 3.2, the isotropic subgroup $H$ of the full group of isometries of $(M,F)$ is a compact subgroup. Then by Lemma 6.1, $G/H$ is a homogeneous weakly symmetric manifold. Hence $(G,H)$ is a weakly symmetric pair. Now the compactness of $H$ implies that $\mathrm{Ad}_G(H)$ is compact. Therefore $(G,H)$ is a Riemannian weakly symmetric pair. On the other hand, if $(G,H)$ is a weakly symmetric pair and $F$ is a $G$-invariant Finsler metric on $M = G/H$, then by Lemma 6.1, $G/H$ is a homogeneous weakly symmetric space. Hence there exists an analytic diffeomorphism $\mu$ of the homogeneous manifold $M = G/H$ such that $\mu g \mu^{-1} = G$, $\mu(eH) = eH$, and $\mu^2 \in H$. Moreover, for every two points $p, q \in M$, there exists an element $g \in G$ such that $g(p) = \mu(q)$ and $g(q) = \mu(p)$. Since $F$ is $G$-invariant, $g$ is an isometry of $(M,F)$. Therefore $\mathrm{d}(\mu(p),\mu(q)) = \mathrm{d}(g(q),g(p)) = \mathrm{d}(q,p) = \mathrm{d}(p,q)$ (note that $F$ is reversible; hence $d$ is symmetric). Hence $\mu$ is an isometry. Consequently, $(M,F,G,\mu)$ is a weakly symmetric Finsler space. Finally, if $(G,H)$ is a Riemannian weakly symmetric pair, then there exists a $G$-invariant reversible Finsler metric (for example, a Riemannian metric) on $G/H$. Hence it can be made into a weakly symmetric Finsler space. $\qquad\square$

## 6.2 Geodesics and S-Curvature

To illustrate the geodesics in weakly symmetric Finsler spaces, we introduce the following definition.

**Definition 6.5.** Let $(M,F)$ be a Finsler space and $G = I(M,F)$ the (full) group of isometries. The space $(M,F)$ is called a Finsler g.o. space if every maximal (constant-speed) geodesic of $(M,F)$ is the orbit of a one-parameter subgroup of $G$. That is, if $\gamma$ is a maximal geodesic, then there exist $w \in \mathfrak{g} = \mathrm{Lie}\, G$ and $o \in M$ such that $\gamma(t) = \exp(tw) \cdot o, t \in \mathbb{R}$.

By definition, a Finsler g.o. space is necessarily (forward and backward) complete. Moreover, it is easily seen that such spaces must be homogeneous. For the general properties of Riemannian g.o. spaces, we refer to [74]. The following theorem gives an important property of a Finsler g.o. space.

**Theorem 6.2.** *Let $(M,F)$ be a Finsler g.o. space. Then the S-curvature of $(M,F)$ is vanishing.*

*Proof.* Let $(x,y) \in TM \setminus \{0\}$ and let $\gamma(t)$ be a geodesic with $\gamma(0) = x$, $\dot{\gamma}(0) = y$. Then there exists $u \in \mathfrak{g}$, the Lie algebra of the full group of isometries, such that

$$\gamma(t) = (\exp tu) \cdot x.$$

we assert that

$$\dot{\gamma}(t) = d(\exp tu)|_x(y).$$

In fact, for every $t_0 \in \mathbb{R}$, we have

$$\gamma(t_0 + t) = \exp(t_0 + t)u \cdot x = (\exp t_0 u)(\exp tu) \cdot x = (\exp t_0 u) \cdot \gamma(t).$$

Thus the curve $\gamma(t_0 + t)$ is just the image of the curve $\gamma(t)$ under the isometry $\exp t_0 u$. From this our assertion follows. Now let $\{b_i(0)\}$ be an arbitrary basis of $T_x(M)$. Define $b_i(t) = d(\exp tu)|_x(b_i(0))$ $(i = 1,2,,\ldots,n)$. Since $\exp(tu)$ are isometries, $\{b_i(t)\}$ is a basis of $T_{\gamma(t)}(M)$ and

$$\begin{aligned}
g_{ij}(t) &= g_{\dot{\gamma}(t)}(b_i(t), b_j(t)) \\
&= g_{d(\exp tu)|_x(y)}(d(\exp tX)|_x(b_i(0)), d(\exp tu)|(b_j(0))) \\
&= g_y(b_i(0), b_j(0)).
\end{aligned}$$

Thus $\det(g_{ij}(t))$ is constant along $\sigma$.

On the other hand, for $(y^i) \in \mathbb{R}^n$ and the vector field $U(t) = y^i b_i(t)$ along $\gamma(t)$, we have

$$F(\sigma(t), y^i b_i(t)) = F(\sigma(t), y^i d(\exp tu)|_x(b_i(0))) = F(\sigma(0), y^i b_i(0)).$$

Thus $\sigma_F$ is constant along $\gamma$. Therefore the distortion $\tau$ of $(M,F)$ is constant along $\gamma(t)$. Hence the S-curvature of $(M,F)$ vanishes.                                    □

Finally we have the following result.

**Theorem 6.3.** *A weakly symmetric Finsler space must be a Finsler g.o. space.*

*Proof.* The proof of the theorem for the Riemannian case was provided by Berndt et al. in [26]. Their proof can be applied to the Finslerian case after some changes. First recall that a submanifold $M_1$ of a manifold $M$ is called quasiregular if $M_1$ is an immersed submanifold of $M$ such that every smooth map from every manifold $M'$ to $M$ whose image is contained in $M_1$ is also a smooth map when considered as a map from $M'$ to $M_1$.

Let $(M,F)$ be a connected weakly symmetric Finsler space and $G = I(M,F)$ the full group of isometries. Then $(M,F)$ is homogeneous hence complete. Now we consider the sphere bundle

$$S(M,F) = \{(x,y) \in TM \,|\, x \in M, \, y \in T_x(M), \, F(x,y) = 1\}.$$

Consider the action of $G$ on $S(M,F)$ defined by

$$g(x,y) = (g(x), dg(y)),$$

where $dg$ is the differential of the isometry $g$. Let $\gamma$ be any maximal constant-speed geodesic. Without loss of generality, we assume that $\gamma$ is of unit speed. Then as pointed out in [26], the orbit of $(\gamma(0), \dot{\gamma}(0))$ under $G$, denoted by $N$, is a quasiregular submanifold of $S(M,F)$. Let $K$ be the isotropy subgroup of $G$ at $(\gamma(0), \dot{\gamma}(0))$. Then $K$ is a closed subgroup of $G$ and we have $N = G/K$. Note that each element of $K$ keeps the whole geodesic $\gamma$ pointwise fixed. In fact, if $k \in K$, then $k \cdot \gamma$ is a geodesic defined on the whole of $\mathbb{R}$, since an isometry sends every geodesic to a geodesic. Since the initial points and the initial directions of $k \cdot \gamma$ and $\gamma$ coincide, by the uniqueness of the geodesics in Finsler geometry (see [16]), they must coincide everywhere.

By Proposition 6.2, for every $t \in \mathbb{R}$, there exists an isometry $\tau_t$ whose restriction to $\gamma$ is a nontrivial involution along $\gamma$ fixing $\gamma(t)$. Since $\gamma$ is a constant-speed geodesic, this means that $\tau_t(\gamma(t+s)) = \gamma(t-s)$. It is easily seen that $\tau_{\frac{t}{2}} \cdot \tau_0$ $(\gamma(0), \dot{\gamma}(0)) = (\gamma(t), \dot{\gamma}(t))$. Thus $(\gamma(t), \dot{\gamma}(t)) \in N$, for every $t$. Since $N$ is a quasiregular submanifold of $S(M,F)$ and $\dot{\gamma} : \mathbb{R} \to S(M,F)$ is an immersed submanifold whose image is contained in $N$, $\dot{\gamma}(\mathbb{R})$ is a smooth submanifold of $N$.

By definition it is obvious that $K$ is a closed subgroup of $G$. Similarly as in [26], we have the following:

- The set

$$G_\gamma = \{g \in G \,|\, dg(\dot{\gamma}(0)) = \dot{\gamma}(t) \text{ for some } t \in \mathbb{R}\}$$

  is a submanifold as well as an abstract subgroup of $G$.
- The submanifolds $g \cdot G_\gamma$, $g \in G$, form a foliation $\mathfrak{F}$ of $G$. Denote by $\mathfrak{F}_0$ its restriction to the unity component $G_0$ of $G$. Then $\mathfrak{F}_0$ gives rise to a left-invariant distribution $\Delta = T\mathfrak{F}_0$. Moreover, $\Delta$ is an involutive distribution. The maximal integral submanifold $H$ of $\Delta$ through the identity element $e \in G_0$ is a Lie subgroup of $G_0$.
- Let $(G_\gamma)_0$ be the unity component of $G_\gamma$. Then $H$ is generated by $(G_\gamma)_0$ as a group, and $\dim H = \dim G_\gamma = 1 + \dim K$.

Given $u \in \mathfrak{g}$, we can define a curve $\alpha_u : \mathbb{R} \to M$ by $\alpha_u(t) = \exp(tu) \cdot \gamma(0)$, where exp is the exponential map of $G$. By the dimension condition, there is $w \in \mathfrak{h}$, the Lie algebra of $H$, such that $\dot{\alpha}_w(0) = \dot{\gamma}(0)$. Then we have $\alpha_w(\mathbb{R}) \subset \gamma(\mathbb{R})$. Now consider the set

$$D = \{t \in \mathbb{R} \mid \alpha_w(t) = \gamma(t), \dot{\alpha}_w(t) = \dot{\gamma}(t)\}.$$

Then $0 \in D$. Moreover, by continuity, $D$ is a closed subset of $\mathbb{R}$.

We now prove that $D$ is also an open subset of $\mathbb{R}$. Suppose $t_o \in D$. Then we have $\alpha_w(t_0) = \gamma(t_0)$ and also $\dot{\alpha}_w(t_0) = \dot{\gamma}(t_0) \neq 0$. Since $\alpha_w(\mathbb{R}) \subset \gamma(\mathbb{R})$ and both $\alpha_w$ and $\gamma$ are immersed curves on $M$, there is a positive number $\varepsilon$ such that $\alpha_w$ is just a reparametrization curve of $\gamma$ in the interval $(t_0 - \varepsilon, t_0 + \varepsilon)$. That is, there is a continuous function $\lambda(t)$, with $\dot{\lambda}(t) > 0$ in $(t_0 - \varepsilon, t_0 + \varepsilon)$, such that $\alpha_w(t) = \gamma(\lambda(t))$, for $t_0 - \varepsilon < t < t_0 + \varepsilon$. Then we have $\lambda(t_0) = t_0$. Moreover, considering the length of the tangent vectors of the two curves, we get

$$F(\dot{\alpha}_u(t)) = \lambda'(t)F(\dot{\gamma}(\lambda(t))) = 1,$$

for $t_0 - \varepsilon < t < t_0 + \varepsilon$. By the condition that $F(\dot{\alpha}_w(t)) = F(\dot{\gamma}(s)) = 1$, for every $t, s \in \mathbb{R}$, we get that $\lambda'(t) \equiv 1$, for $t_0 - \varepsilon < t < t_0 + \varepsilon$. Thus in the interval $(t_0 - \varepsilon, t_0 + \varepsilon)$ we have $\lambda(t) = t$. Therefore $D$ is an open subset of $\mathbb{R}$. Hence $D = \mathbb{R}$ and $\gamma$ is the orbit of the one-parameter subgroup $\exp(tw)$ of $G$. This completes the proof of the theorem.                                                                                        $\square$

## 6.3  Weakly Symmetric Lie Algebras

In this section, we introduce the notion of a weakly symmetric Lie algebra to give an algebraic description of simply connected weakly symmetric Finsler spaces.

**Definition 6.6.** Let $\mathfrak{g}$ be a real Lie algebra and $\mathfrak{h}$ a subalgebra of $\mathfrak{g}$. Suppose there exists a subspace $\mathfrak{p}$ of $\mathfrak{g}$ such that $\mathfrak{g} = \mathfrak{h} + \mathfrak{p}$ (direct sum) and $[\mathfrak{h}, \mathfrak{p}] \subset \mathfrak{p}$. Then the pair $(\mathfrak{g}, \mathfrak{h})$ is called a weakly symmetric Lie algebra if there exists a finite set of automorphisms of $\mathfrak{g}$, $\{\sigma_0, \sigma_1, \sigma_2, \ldots, \sigma_s\}$ ($\sigma_0 = \mathrm{id}$), satisfying the following conditions:

W1:  Every $\sigma_i$, $0 \leq i \leq s$, preserves the subspaces $\mathfrak{h}$ and $\mathfrak{p}$, i.e., $\sigma_i(\mathfrak{h}) = \mathfrak{h}$, $\sigma_i(\mathfrak{p}) = \mathfrak{p}$.
W2:  For every pair $i, j$, $0 \leq i, j \leq s$, there exist $k$, $0 \leq k \leq s$, and an vector $u \in \mathfrak{h}$ such that $\sigma_i \sigma_j = e^{\mathrm{ad}(u)} \sigma_k$.
W3:  Given $v \in \mathfrak{p}$, there exist $v' \in \mathfrak{h}$ and an index $k$ such that $e^{\mathrm{ad}(v')} \cdot \sigma_k(v) = -v$.

We usually say that $(\mathfrak{g}, \mathfrak{h})$ is weakly symmetric with respect to $\{\sigma_0, \sigma_1, \ldots, \sigma_s\}$. Moreover, a weakly symmetric Lie algebra $(\mathfrak{g}, \mathfrak{h})$ is called Riemannian if in addition, $\mathfrak{h}$ is a compactly embedded subalgebra of $\mathfrak{g}$.

We now give two remarks concerning the above definition.

*Remark 6.1.* The notion of Riemannian weakly symmetric Lie algebras is a natural generalization of that of orthogonal symmetric Lie algebras. In fact, if $(\mathfrak{g}, \sigma)$ is an orthogonal symmetric Lie algebra, then the fixed points of the involution $\sigma$, denoted by $\mathfrak{h}$, is a compactly embedded subalgebra of $\mathfrak{g}$. Let $\mathfrak{p}$ be the eigenspace of $\sigma$ with eigenvalue $-1$. Then $(\mathfrak{g}, \mathfrak{p})$ satisfies the condition in the above definition with respect to the set $\{\sigma_0, \sigma_1\}$, where $\sigma_0 = \mathrm{id}$ and $\sigma_1 = \sigma$. In the next section, we will construct a series of examples of Riemannian weakly symmetric Lie algebras that are not orthogonal symmetric.

*Remark 6.2.* In general, the set of automorphisms $\{\sigma_0, \sigma_1, \ldots, \sigma_s\}$ in the above definition is not unique. However, if $(\mathfrak{g}, \mathfrak{h})$ is a weakly symmetric Lie algebra, then we can choose a set of automorphisms $\{\tau_0(= \text{id}), \tau_1, \tau_2, \ldots, \tau_l\}$ that satisfy W1, W2, W3, and the following additional condition:

W4: For every $j \neq k$, $\tau_j(\tau_k)^{-1}$ cannot be written in the form $e^{\text{ad}(u)}$, $u \in \mathfrak{h}$.

In this case, we say that the set $\{\tau_0(= \text{id}), \tau_1, \tau_2, \ldots, \tau_l\}$ is a reduced set of automorphisms. It is easily seen that each set of automorphisms in the definition can be reduced to a reduced one. In the following, we will usually select a reduced set of automorphisms to study weakly symmetric Lie algebras.

Now we can give the algebraic description of weakly symmetric Finsler spaces.

**Theorem 6.4.** *Let $(M, F)$ be a weakly symmetric Finsler space. Then there exist a Lie group $G$ and a closed subgroup $H$ of $G$ such that $M = G/H$ and $F$ is $G$-invariant. Furthermore, the Lie algebra pair $(\mathfrak{g}, \mathfrak{h})$ of $(G, H)$ is a Riemannian weakly symmetric Lie algebra.*

*Proof.* Fix $x \in M$. By Proposition 6.3, for every $u \in T_x(M)$, there exists an isometry $\tau$ of $(M, F)$ such that $\tau(x) = x$ and $d\tau|_x(u) = -u$. Let $G$ be the full group of isometries of $(M, F)$ and $H$ the isotropy subgroup of $G$ at $x$. Since a weakly symmetric Finsler spaces is homogeneous, $G$ acts transitively on $M$. Hence $M$ is diffeomorphic to the coset space $G/H$ and $F$ is $G$-invariant. Now we prove that the Lie algebra pair $(\mathfrak{g}, \mathfrak{h})$ of $(G, H)$ is a weakly symmetric Lie algebra. Since $H$ is compact, $G/H$ is a reductive homogeneous manifold. Hence there exists a subspace $\mathfrak{p}$ of $\mathfrak{g}$ such that $\mathfrak{g} = \mathfrak{h} + \mathfrak{p}$ (direct sum) and $\text{Ad}(h)(\mathfrak{p}) \subset \mathfrak{p}$, $\forall h \in H$. In particular, $[\mathfrak{h}, \mathfrak{p}] \subset \mathfrak{p}$. Now we identify the space $\mathfrak{p}$ with the tangent space $T_x(M)$ through the map $v \mapsto \frac{d}{dt}(\exp tv) \cdot x|_{t=0}$, $v \in \mathfrak{p}$. Then the isotropic action of $H$ on $T_x(M)$ corresponds to the adjoint action of $H$ on $\mathfrak{p}$. Let $e$ be the unit element and $H_e$ the identity component of $H$. Then $H_e$ is a normal subgroup of $H$ and the quotient group $H/H_e$ is finite, since $H$, as a compact Lie group, has at most a finite number of components. Let $\{e, h_1, \ldots, h_s\}$ be a set of elements of $H$ such that $\{eH, h_1 H, \ldots, h_s H\}$ are all the (distinct) elements of the quotient group.

Let $\sigma_0$ be the identity transformation of $\mathfrak{g}$ and $\sigma_j = \text{Ad}(h_j)$, $j = 1, 2, \ldots, s$. Then we can prove that the set $\{\sigma_0, \sigma_1, \ldots, \sigma_s\}$ satisfies the conditions W1, W2, W3. In fact, W1 is obviously satisfied. Now we prove W2. Given a pair $i, j$, suppose in the quotient group $H/H_e$ we have $h_i H_e \cdot h_j H_e = h_k H_e$. Then there exist $m_1, m_2, m_3 \in H_e$ such that $h_i m_1 h_j m_2 = h_k m_3$, i.e.,

$$h_i h_j = h_k (m_3 m_2^{-1} (h_j m_1^{-1} h_j^{-1})).$$

Since $h_j m_1^{-1} h_j^{-1} \in H_e$, we have $m = m_3 m_2^{-1} (h_j m_1^{-1} h_j^{-1}) \in H_e$. Since $H_e$ is a connected compact Lie group, the exponential map is surjective. Hence there exists $u_{ij} \in \mathfrak{h}$ such that $\exp(u_{ij}) = m$. Then $\text{Ad}(m) = e^{\text{ad}(u_{ij})}$. Therefore $\sigma_i \sigma_j = \sigma_k e^{\text{ad}(u_{ij})} = \sigma_k e^{\text{ad}(u_{ij})} \sigma_k^{-1} \cdot \sigma_k = e^{\text{ad}(\sigma_k(u_{ij}))} \sigma_k$, i.e., W2 is satisfied. Finally, we prove W3. By Proposition 6.3, for every $v \in \mathfrak{p}$ we can select $h \in H$ such that $\text{Ad}(h)(v) = -v$. Suppose $h$ lies in the component $h_i H_e$. Then there exists $h_0 \in H_e$ such that $h = h_i h_0 = h_i h_0 h_i^{-1} h_i$. Since $h_i h_0 h_i^{-1} \in H_e$, we can write $h = \exp(v') h_i$, for some $v' \in \mathfrak{h}$. From this we easily see that W3 is satisfied. This completes the proof. $\square$

Now we show that every Riemannian weakly symmetric Lie algebra gives rise to a weakly symmetric Finsler space, although in general the spaces constructed from a weakly symmetric Lie algebra are not unique.

**Theorem 6.5.** *Let* $(\mathfrak{g},\mathfrak{h})$ *be a Riemannian weakly symmetric Lie algebra. Suppose* $G$ *is a connected simply connected Lie group with Lie algebra* $\mathfrak{g}$ *and* $H$ *is the (unique) connected Lie subgroup of* $G$ *with Lie algebra* $\mathfrak{h}$. *If* $H$ *is closed in* $G$ *(this is the case if* $C(\mathfrak{g}) = 0$*), then there exists a* $G$*-invariant Riemannian metric* $Q$ *on the coset space* $G/H$ *such that* $(G/H,Q)$ *is a Riemannian weakly symmetric space. In particular, there exists an invariant Finsler space* $F$ *on* $G/H$ *such that* $(G/H,F)$ *is a weakly symmetric Finsler space.*

*Proof.* Since $\mathfrak{h}$ is a compactly embedded subalgebra of $\mathfrak{g}$, the Lie algebra $\mathrm{ad}_{\mathfrak{g}}(\mathfrak{h})$ is compact. If we identify the tangent space $T_o(G/H)$ with $\mathfrak{p}$, where $\mathfrak{p}$ is as in Definition 6.6 and $o = eH$ is the origin of the coset space $G/H$, then $\mathrm{Ad}(H)$ is a compact group of linear transformations of $\mathfrak{p}$. Hence there exists an $\mathrm{Ad}(H)$-invariant inner product $\langle\,,\,\rangle_1$ on $\mathfrak{p}$. Fix a reduced set of automorphisms $\{\tau_0,\tau_1,\ldots,\tau_l\}$ ($\tau_0 = \mathrm{id}$) as in Remark 6.2 after Definition 6.6, and define an inner product $\langle\,,\,\rangle$ on $\mathfrak{p}$ as follows:

$$\langle u,v\rangle = \sum_{j=0}^{l}\langle\tau_j u,\tau_j v\rangle_1,\quad u,v\in\mathfrak{p}.$$

By condition W2, $\langle\,,\,\rangle$ is invariant under the action of each $\tau_i$, $i = 1,2,\ldots,l$. We assert that $\langle\,,\,\rangle$ is also $\mathrm{Ad}(H)$-invariant. In fact, for every $h\in H$ and every index $j$, we have $\mathrm{Ad}(h)\cdot\tau_j = \tau_j\cdot(\tau_j^{-1}\cdot\mathrm{Ad}(h)\cdot\tau_j)$. Since $H$ is connected, it is generated by the elements $\exp(u)$, $u\in\mathfrak{h}$. Hence $h$ can be written as $(\exp u_1\exp u_2\cdots\exp u_k)$, where $u_i\in\mathfrak{h}$, $i = 1,2,\ldots,k$. Then we have

$$\tau_j^{-1}\cdot\mathrm{Ad}(h)\cdot\tau_j = e^{\mathrm{ad}(\tau_j u_1)}e^{\mathrm{ad}(\tau_j u_2)}\ldots e^{\mathrm{ad}(\tau_j u_k)}.$$

From this our assertion follows. Now by Proposition 2.8, $\langle\,,\,\rangle$ can be extended to a $G$-invariant Riemannian metric $Q$ on the coset space $G/H$ (here the condition that $H$ be closed is required, for otherwise it may happen that there is no differentiable structure on $G/H$) whose restriction to $T_o(G/H) = \mathfrak{p}$ is equal to $\langle\,,\,\rangle$. We assert that the homogeneous Riemannian manifold constructed above is weakly symmetric. Since $G$ is connected and simply connected, each automorphism $\tau_j$ of $\mathfrak{g}$ can be lifted to an automorphism of $G$. We denote the corresponding automorphisms by $\widetilde{\tau}_j$, $j = 0,1,2,\ldots$. Since $\tau_j(\mathfrak{h}) = \mathfrak{h}$, $\widetilde{\tau}_j(H)\subset H$. Hence $\widetilde{\tau}_j$ induces a diffeomorphism of $G/H$, denoted by $\hat{\tau}_j$, by sending $gH$ to $\widetilde{\tau}_j(g)H$. The diffeomorphism $\hat{\tau}_j$ keeps the origin $o = eH$ invariant, and its differential at $o$ is just the restriction of $\tau_j$ to $\mathfrak{p}$. From this we see that $\hat{\tau}_j$ keeps the Riemannian metric $Q$ invariant, or in other words, $\hat{\tau}_j$ lies in the isotropic subgroup (at $o$) of the full group of isometries of $(G/H,Q)$. By W3, for every $v\in\mathfrak{p} = T_o(G/H)$, we can choose $v'\in\mathfrak{h}$ and an index $j$ such that $e^{\mathrm{ad}(v')}\tau_j(v) = -v$. This means that the isometry $\tau_{\exp(v')}\cdot\hat{\tau}_j$ of the Riemannian manifold $(G/H,Q)$, where $\tau_h(gH) = hgH$, $h\in H$, reverses the tangent vector $v$.

Since $(G/H,Q)$ is homogeneous, for every $x \in G/H$ and $w \in T_x(G/H)$, there exists an isometry $f$ such that $f(x) = x$ and $df|_x(w) = -w$. By Proposition 6.3, $(G/H,Q)$ is weakly symmetric.

Finally, if $C(\mathfrak{g}) = 0$, then $H$ is closed in $G$; see [83, pp. 213–214]. □

By Theorem 4.8, for almost all weakly symmetric Lie algebras $(\mathfrak{g},\mathfrak{h})$, one can find invariant non-Riemannian Finsler metrics on the coset space $G/H$ in Theorem 6.5 that make $G/H$ a weakly symmetric Finsler space.

## 6.4 Examples of Weakly Symmetric Lie Algebras

In this section, we present some examples of weakly symmetric Lie algebras. They will be used in the classification of 3-dimensional weakly symmetric Finsler spaces in the next section.

*Example 6.1.* Let $\mathfrak{h}$ be the 1-dimensional real Lie algebra and

$$\mathfrak{g} = \mathfrak{h} + \mathfrak{su}(2) \quad \text{(direct sum of subspaces)}.$$

Now we define Lie brackets on $\mathfrak{g}$. The Lie brackets between the elements of $\mathfrak{su}(2)$ are defined as usual. Let $u$ be a nonzero element in $\mathfrak{h}$ and

$$\varepsilon_1 = \begin{pmatrix} 0 & 1 \\ -1 & 0 \end{pmatrix}, \quad \varepsilon_2 = \begin{pmatrix} 0 & \sqrt{-1} \\ \sqrt{-1} & 0 \end{pmatrix}, \quad \varepsilon_3 = \begin{pmatrix} \sqrt{-1} & 0 \\ 0 & -\sqrt{-1} \end{pmatrix}.$$

Then we define

$$[u,\varepsilon_1] = 0, \quad [u,\varepsilon_2] = -\varepsilon_3, \quad [u,\varepsilon_3] = \varepsilon_2.$$

These brackets can be extended linearly to a skew-symmetric binary operation on $\mathfrak{g}$. It is easy to check that the Jacobian identities hold for this operation. Hence $\mathfrak{g}$ is a Lie algebra. Obviously $[\mathfrak{h},\mathfrak{su}(2)] \subset \mathfrak{su}(2)$. Now we define an endomorphism $\tau$ on $\mathfrak{g}$ by

$$\tau(u) = -u, \quad \tau(\varepsilon_1) = -\varepsilon_1, \quad \tau(\varepsilon_2) = -\varepsilon_2, \quad \tau(\varepsilon_3) = \varepsilon_3.$$

Then it is easy to check that $\tau$ is a Lie algebra automorphism, and $\tau^2 = \text{id}$. Now we prove that $(\mathfrak{g},\mathfrak{h})$ is a weakly symmetric Lie algebra with respect to $S = \{\text{id}, \tau\}$. Since W1, W2 are obviously satisfied, we need only check W3. Note that the action of $e^{\text{ad}(tu)}$ on $\mathfrak{p} = \mathfrak{su}(2)$ keeps the subspaces $V_1 = \text{span}(\varepsilon_1)$ and $V_2 = \text{span}(\varepsilon_2,\varepsilon_3)$ invariant. Moreover, the restriction of $e^{\text{ad}(tu)}$ to $V_1$ is equal to the identity transformation, and its restriction on $V_2$ has the matrix

$$\begin{pmatrix} \cos t & \sin t \\ -\sin t & \cos t \end{pmatrix}$$

with respect to the basis $\varepsilon_2, \varepsilon_3$. Thus $e^{\mathrm{ad}(tu)}$ is a rotation through the angle $t$ if we define an inner product on $V_2$ by requiring that $\varepsilon_2$, $\varepsilon_3$ form an orthonormal basis. Now given an element $\varepsilon = a\varepsilon_1 + b\varepsilon_2 + c\varepsilon_3$ in $\mathfrak{p}$, we have $\tau(\varepsilon) = -a\varepsilon_1 - b\varepsilon_2 + c\varepsilon_3$. Since $-b\varepsilon_2 + c\varepsilon_3$ is an element in $V_2$ with the same length as $-b\varepsilon_2 - c\varepsilon_3$, there exists an appropriate $t_0 \in \mathbb{R}$ such that

$$e^{\mathrm{ad}(t_0 u)}(-b\varepsilon_2 + c\varepsilon_3) = -b\varepsilon_2 - c\varepsilon_3.$$

Then

$$e^{(t_0 \mathrm{ad}(u))}\tau(\varepsilon) = -a\varepsilon_1 - b\varepsilon_2 - c\varepsilon_3 = -\varepsilon.$$

Therefore W3 is satisfied. Hence $(\mathfrak{g}, \mathfrak{h})$ is a weakly symmetric Lie algebra. Note that the action of $\mathrm{ad}(u)$ on $\mathfrak{p}$ has a skew-symmetric matrix with respect to the basis $\varepsilon_1, \varepsilon_2, \varepsilon_3$. Thus $\mathfrak{h}$ is a compactly embedded subalgebra of $\mathfrak{h}$. Hence $(\mathfrak{g}, \mathfrak{h})$ is a Riemannian weakly symmetric Lie algebra.

*Example 6.2.* This example is similar to Example 6.1. Here we let

$$\mathfrak{g} = \mathfrak{h} + \mathfrak{sl}(2, \mathbb{R}) \quad \text{(direct sum of subspaces).}$$

The Lie brackets are defined similarly. In $\mathfrak{sl}(2, \mathbb{R})$ we use the usual Lie operations. Let

$$\varepsilon_1 = \begin{pmatrix} 0 & 1 \\ -1 & 0 \end{pmatrix}, \quad \varepsilon_2 = \begin{pmatrix} 1 & 0 \\ 0 & -1 \end{pmatrix}, \quad \varepsilon_3 = \begin{pmatrix} 0 & 1 \\ 1 & 0 \end{pmatrix}.$$

Then we define

$$[u, \varepsilon_1] = 0, \quad [u, \varepsilon_2] = -\varepsilon_3, \quad [u, \varepsilon_3] = \varepsilon_2.$$

Moreover, we define an endomorphism $\tau$ on $\mathfrak{g}$ by

$$\tau(u) = -u, \quad \tau(\varepsilon_1) = -\varepsilon_1, \quad \tau(\varepsilon_2) = -\varepsilon_2, \quad \tau(\varepsilon_3) = \varepsilon_3.$$

Then it can be checked directly that $\tau$ is an automorphism of $\mathfrak{g}$. Similarly as in Example 6.1, we can prove that $(\mathfrak{g}, \mathfrak{h})$ is a Riemannian weakly symmetric Lie algebra with respect to $\{\mathrm{id}, \tau\}$.

*Example 6.3.* In this example we consider Heisenberg Lie algebras. Let $\mathfrak{n}$ be a $(2n+1)$-dimensional real Lie algebra with basis $x_1, x_2, \ldots, x_n, y_1, y_2, \ldots, y_n, z$ and brackets

$$[x_i, y_j] = \delta_{ij} z, \quad [x_i, x_j] = [y_i, y_j] = 0, [x_i, z] = [y_i, z] = 0, \quad i, j = 1, 2, \ldots, n.$$

Then $\mathfrak{n}$ is a 2-step nilpotent Lie algebra. Let $\mathfrak{g} = \mathfrak{u}(n) + \mathfrak{n}$ (direct sum of subspaces) and define the brackets as follows. The brackets among the elements in $\mathfrak{u}(n)$ are the usual operations. For $A \in \mathfrak{u}(n)$ we define $[A, z] = 0$, and for the element

$$w = \sum_{i=1}^{n}(a_i x_i + b_i y_i), \quad a_i, b_i \in \mathbb{R},$$

we set

$$z_i = a_i + \sqrt{-1}b_i, \quad i = 1, 2, \ldots, n. \tag{6.1}$$

Let

$$(z_1', z_2', \ldots, z_n') = (z_1, z_2, \ldots, z_n)A$$

and write $z_i' = a_i' + \sqrt{-1}b_i'$, $a_i', b_i' \in \mathbb{R}$. Then we define

$$[A, w] = w' = \sum_{i=1}^{n} (a_i' x_i + b_i' y_i).$$

It is easy to check that the Jacobian identities hold among these brackets. Therefore these brackets together with the brackets of $\mathfrak{n}$ define a Lie algebra structure on $\mathfrak{g}$. By definition, we have $[\mathfrak{u}(n), \mathfrak{n}] \subset \mathfrak{n}$. Now we define an endomorphism $\tau$ of $\mathfrak{g}$ by

$$\tau(A) = \bar{A}, \quad \tau(x_i) = -x_i, \quad \tau(y_i) = y_i, \quad \tau(z) = -z,$$

where $A \in \mathfrak{u}(n)$ and $\bar{A}$ is the complex conjugate matrix of $A$. It is easy to check that $\tau$ is a (real) automorphism of the real Lie algebra $\mathfrak{g}$ and $\tau^2 = \mathrm{id}$.

Now we prove that $(\mathfrak{g}, \mathfrak{u}(n))$ is a weakly symmetric Lie algebra with respect to $\{\mathrm{id}, \tau\}$. The conditions W1, W2 are obviously satisfied, so we need only check W3. Note that the action of $\mathfrak{u}(n)$ in the subspace $V_2 = \mathrm{span}(x_1, y_1, \ldots, x_n, y_n)$ is just the usual operation of $\mathfrak{u}(n)$ on the complex linear space $\mathbb{C}^n$ if we write the corresponding coordinates as complex numbers as in (6.1). Hence the action of $e^{\mathrm{ad}(tA)}$, $A \in \mathfrak{u}(n)$, on $V_2$ is the operation $w \mapsto w \times \exp(tA)$ (matrix multiplication), i.e., it is just the usual action of the unitary Lie group $U(n)$ on $\mathbb{C}^n$. Now for every $w + cz$, $w \in V_2$, $c \in \mathbb{R}$, we have

$$\tau(w + cz) = \tau(w) - cz,$$

where $\tau(w)$ lies in $V_2$ and has the same length as $w$ when we identify $V_2$ with $\mathbb{C}^n$ (with the usual Hermitian metric). Since the group $U(n)$ acts as the identity transformation on $\mathrm{span}(z)$ and acts transitively on the unit sphere in $\mathbb{C}^n$, there exists $A_0$ in $\mathfrak{u}(n)$ such that

$$e^{\mathrm{ad}(A_0)}(\tau(w)) = -w, \quad e^{\mathrm{ad}(A_0)}(z) = z.$$

Thus $e^{\mathrm{ad}(A_0)} \cdot \tau(w + cz) = -(w + cz)$. This proves our assertion.

When $n = 1$, the Heisenberg Lie algebra has dimension 3. The Lie algebra $\mathfrak{u}(1)$ is 1-dimensional. Let $x, y, z$ be the basis of $\mathfrak{n}$ such that $[x, y] = z$, $[x, z] = [y, z] = 0$. Then we can choose an element $u$ of $\mathfrak{u}(1)$ such that

$$[u, z] = 0, \quad [u, x] = -y, \quad [u, y] = x.$$

The automorphism $\tau$ satisfies $\tau(u) = -u$, $\tau(x) = -x$, $\tau(y) = y$. This weakly symmetric Lie algebra will appear in our classification of 3-dimensional weakly symmetric Finsler spaces in the next section.

## 6.5 Spaces of Low Dimension

As an application of the notion of a weakly symmetric Lie algebra, we now give a classification of weakly symmetric Finsler spaces of dimension $\leq 3$.

We begin with some general observations. Suppose $(M,F)$ is a connected weakly symmetric Finsler space. Let $\widetilde{G} = I(M,F)$ be the full group of isometries and let $\widetilde{H}$ be the isotropic subgroup at a point $x \in M$. Then $M$ can be written as $\widetilde{G}/\widetilde{H}$ and $F$ can be viewed as a $\widetilde{G}$-invariant metric on $\widetilde{G}/\widetilde{H}$. Note that as a connected manifold, the identity component $G$ of $\widetilde{G}$ is also transitive on $M$. Denote by $H$ the isotropy subgroup of $G$ at $x$. Then $M$ can also be written as the coset space $G/H$.

**Lemma 6.3.** *The isotropic representation $\rho$ of $\widetilde{H}$ on $T_x(M)$ is faithful. In particular, the isotropic representation of $H$ on $M$ is faithful.*

The proof is similar to the Riemannian case. Just note that an isometry sends a geodesic to a geodesic and that $M$ is homogeneous, hence complete.

The dimension-2 case is settled by the following theorem.

**Theorem 6.6.** *Let $(M,F)$ be a 2-dimensional connected simply connected weakly symmetric Finsler space. Then $(M,F)$ must be one of the following:*

1. *$(M,F)$ is a reversible Minkowski space;*
2. *$F$ is Riemannian and $(M,F)$ is isometric to a rank-one globally symmetric Riemannian space.*

*In each case, $(M,F)$ must be a globally symmetric Finsler space.*

*Proof.* Let $G$, $H$, $\rho$ be as above. Then by Lemma 6.3 we have $\dim H = \dim \rho(H)$. Since $\rho(H)$ is a compact group of linear transformations of $T_x(M)$, we have $\dim \rho(H) \leq 1$. Thus there are only two cases:

*Case (i)*: $\dim \rho(H) = 0$. Then $H = \{e\}$. Hence $M$ is diffeomorphic to the Lie group $G$. Then by Theorem 6.2, the Lie algebra $\mathfrak{g}$ admits finitely many automorphisms $\tau_0(= \mathrm{id})$, $\tau_1$, $\tau_2,\ldots,\tau_s$ such that for every $Y \in \mathfrak{g}$ there exists an index $j_Y$ such that $\tau_{j_Y}(Y) = -Y$. Let $V_j = \{Y \in \mathfrak{g} \mid \tau_j(Y) = -Y\}$. Then $V_j$ are subspaces of $\mathfrak{g}$ with $\mathfrak{g} = \cup V_j$. Therefore there must be a $j_0$ such that $V_{j_0} = \mathfrak{g}$. Thus for every $X,Y \in \mathfrak{g}$, we have

$$-[X,Y] = \tau_{j_0}([X,Y]) = [\tau_{j_0}(X), \tau_{j_0}(Y)] = [-X,-Y] = [X,Y].$$

This means that $\mathfrak{g}$ is abelian. Hence $G$ is a two-dimensional connected simply connected commutative Lie group, i.e., $G = \mathbb{R}^2$ (here $\mathbb{R}^2$ is viewed as an additive group). Since $F$ is invariant under $G$, $F$ is defined by a Minkowski norm in the canonical way. Hence in this case, $(M,F)$ is a reversible Minkowski space.

*Case (ii)*: $\dim \rho(H) = 1$. Since $\rho(H)$ is connected and compact, $\rho(H)$ is isomorphic to $S^1$. Hence $\rho(H)$ acts transitively on the indicatrix

$$\mathscr{I}_x = \{u \in T_x(M) \mid F(u) = 1\}$$

of $F$ at $T_x(M)$. On the other hand, by the compactness of $\rho(H)$ there exists a $\rho(H)$-invariant inner product on $T_x(M)$. The group $\rho(H)$ also acts transitively on the unit circle of $T_x(M)$ with respect to this inner product. Therefore $F|_{T_x(M)}$ is a Euclidean norm. Since $(M, F)$ is homogeneous, $F$ is Riemannian. Furthermore, if $p_1, p_2$ are two points in $M$ such that $\mathrm{d}(x, p_1) = \mathrm{d}(x, p_2)$, then we can select two unit vectors $u_1, u_2$ in $T_x(M)$ such that

$$p_1 = \mathrm{Exp}\,(tu_1), \quad p_2 = \mathrm{Exp}(tu_2),$$

where Exp is the exponential of $M$ with the Riemannian metric $F$, and $t = d(x, p_1)$. Since $\rho(H)$ acts transitively on the unit circle of $T_x(M)$, we can choose $h \in H$ such that $\rho(h)(u_1) = u_2$. Hence $h(p_1) = p_2$. This argument and the fact that $(M, F)$ is homogeneous imply that $(M, F)$ is actually a two-point homogeneous Riemannian manifold. By the classification results of Wang [164] and Tits [154] on two-point homogeneous Riemannian manifolds, if $(M, F)$ is not a Euclidean space, then it must be a globally symmetric Riemannian manifold of rank 1. This proves the theorem.                                                                    □

Next we consider the 3-dimensional case. The situation here is much more complicated. Let $(M, F)$ be a 3-dimensional connected simply connected weakly symmetric Finsler space and let $\widetilde{G}$, $\widetilde{H}$, $G$, $H$, $\rho$ be as above. Then $\rho(H)$ is a connected compact Lie subgroup of $\mathrm{GL}(T_x(M))$. Hence it is a connected compact Lie subgroup of $\mathrm{SL}(T_x(M))$. By the conjugacy of the maximal compact subgroup of semisimple Lie groups (Theorem 2.36), there exists an element $g \in \mathrm{SL}(T_x(M))$ such that $g\rho(H)g^{-1} \subset \mathrm{SO}(T_x(M))$, where we have fixed an inner product in $T_x(M)$ and $\mathrm{SO}(T_x(M))$ is defined as usual. Without loss of generality, we can assume that $\rho(H) \subset \mathrm{SO}(T_x(M))$. Then there are only two cases:

*Case (i)*: $\rho(H) = \mathrm{SO}(T_x(M))$. In this case, $H$ acts transitively on the indicatrix of $F$ at $T_x(M)$. Similarly as in the 2-dimensional case, we can prove that $F$ is Riemannian and $(M, F)$ is isometric to a two-point homogeneous Riemannian manifold.

*Case (ii)*: $\rho(H) \neq \mathrm{SO}(T_x(M))$. According to a result of Montgomery and Samelson [121], $\mathrm{O}(N)$ has no proper subgroup of dimension greater than $\frac{1}{2}(N-1)\,(N-2)$, when $N > 2$ and $N \neq 4$. In particular, $\mathrm{SO}(T_x(M))$ has no proper subgroup of dimension greater than $\frac{1}{2} \times 2 \times 1 = 1$. Therefore we have either $\dim\rho(H) = 0$ or $\dim\rho(H) = 1$. Next we tackle the problem case by case.

The case of $\dim\rho(H) = 0$ is easy. In fact, we can proceed in exactly the same way as in Theorem 6.6 to prove that in this case, $(M, F)$ is just a reversible Minkowski space. So we are left with the case of $\dim\rho(H) = 1$. In this case, we need some complicated reasoning and computation.

Since $\rho$ is a faithful representation, the dimension of $\dim H$ must be 1. So in the weakly symmetric Lie algebra $(\mathfrak{g}, \mathfrak{h})$ we have $\dim\mathfrak{h} = 1$. Let $\mathfrak{g} = \mathfrak{h} + \mathfrak{p}$ be the corresponding decomposition. As pointed out in the proof of Theorem 6.5, there exists an inner product in $\mathfrak{p}$ that is invariant under the actions of $\rho(H)$ and $\tau_j$,

$j = 0, 1, 2, \ldots, s$, where $T = \{\tau_0, \tau_1, \tau_2, \ldots, \tau_s\}$ is a reduced set of automorphisms in Definition 6.5. Then for every $v \in \mathfrak{h}$, the endomorphism $\mathrm{ad}(v)$ is skew-symmetric with respect to this inner product. Fix a nonzero element $u$ in $\mathfrak{h}$. Then $\mathrm{ad}(u)|_{\mathfrak{p}}$ must have a zero eigenvalue, and the corresponding eigenspace $V_0$ is either 1-dimensional or 3-dimensional. If $\dim V_0 = 3$, then $\mathfrak{g}$ is the direct sum of the ideals $\mathfrak{h}$ and $\mathfrak{p}$. Since $e^{\mathrm{ad}(tu)}$ acts as the identity transformation on $\mathfrak{p}$ for every $t \in \mathbb{R}$, there must be a $\tau_i$, $1 \leq i \leq s$, such that $\tau_i(v) = -v$ for every $v \in \mathfrak{p}$. This means that for every $v_1, v_2 \in \mathfrak{p}$, we have

$$-[v_1, v_2] = \tau_i([v_1, v_2]) = [\tau_i(v_1), \tau_i(v_2)] = [-v_1, -v_2] = [v_1, v_2].$$

Hence $[\mathfrak{p}, \mathfrak{p}] = 0$, i.e., $\mathfrak{p}$ is an abelian ideal of $\mathfrak{g}$. If $\dim V_0 = 1$, then it is easily seen that there exists a basis $\varepsilon_1, \varepsilon_2, \varepsilon_3$ of $\mathfrak{p}$ such that $\mathrm{ad}\, u$ has the matrix

$$\begin{pmatrix} 0 & 0 & 0 \\ 0 & 0 & 1 \\ 0 & -1 & 0 \end{pmatrix}$$

with respect to this basis.

**Lemma 6.4.** *The reduced set $T$ can be selected to consist of only two automorphisms $\tau_0$, $\tau_1$ such that $\tau_0 = \mathrm{id}$ and $\tau_1(\varepsilon_1) = -\varepsilon_1$.*

*Proof.* First, there exist an index $j_0$ and $t \in \mathbb{R}$ such that

$$e^{t\,\mathrm{ad}(u)} \tau_{j_0}(\varepsilon_1) = -\varepsilon_1. \tag{6.2}$$

We assert that $\tau_{j_0}(\varepsilon_1) = -\varepsilon_1$. In fact, if

$$\tau_{j_0}(\varepsilon_1) = a\varepsilon_1 + b\varepsilon_2 + c\varepsilon_3,$$

where $\varepsilon = b\varepsilon_2 + c\varepsilon_3$ is a nonzero vector in $V = \mathrm{span}\,(\varepsilon_2, \varepsilon_3)$, then $\varepsilon' = e^{t\,\mathrm{ad}(u)}(\varepsilon)$ is also a nonzero vector in $V$. But

$$-\varepsilon_1 = e^{t\,\mathrm{ad}(u)} \tau_{j_0}(\varepsilon_1) = a\varepsilon_1 + \varepsilon'.$$

This contradicts the assumption that $\varepsilon_1, \varepsilon_2, \varepsilon_3$ is a basis of $\mathfrak{p}$. Hence $\tau_{j_0}(\varepsilon_1) = a\varepsilon_1$. Substituting this into (6.2) yields $a = -1$. This proves the lemma.  □

Now we will proceed to give a classification of the weakly symmetric Lie algebras corresponding to 3-dimensional weakly symmetric Finsler spaces in the case of $\dim H = 1$ and $\dim V_0 = 1$.

The Lie algebra $\mathfrak{g}$ has a basis $u, \varepsilon_1, \varepsilon_2, \varepsilon_3$, where $u$ spans $\mathfrak{h}$ and $\varepsilon_1, \varepsilon_2, \varepsilon_3$ span $\mathfrak{p}$. Moreover, we have determined the following brackets:

$$[u, \varepsilon_1] = 0, \quad [u, \varepsilon_2] = -\varepsilon_3, \quad [u, \varepsilon_3] = \varepsilon_2.$$

It is also known that the automorphism $\tau_1$ keeps the subspaces $\mathfrak{h}$ and $\mathfrak{p}$ invariant. Thus

$$\tau_1(u) = du, \quad d \in \mathbb{R}.$$

Since $\tau_1$ keeps the inner product invariant and $\tau_1(\varepsilon_1) = -\varepsilon_1$, $\tau_1$ must keep the space $V = \text{span}(\varepsilon_2, \varepsilon_3)$ invariant, and $\tau_1|_V$ is an orthogonal linear transformation with respect to the inner product. Hence with respect to the basis $\varepsilon_2, \varepsilon_3$ of $V$, the linear transformation $\tau_1|_V$ has matrix

$$\begin{pmatrix} \cos\theta & \sin\theta \\ -\sin\theta & \cos\theta \end{pmatrix}$$

or

$$\begin{pmatrix} -\cos\theta & -\sin\theta \\ -\sin\theta & \cos\theta \end{pmatrix},$$

where $\theta \in \mathbb{R}$.

To determine the structure of the Lie algebra $\mathfrak{g}$, we need only determine the Lie brackets $[\varepsilon_i, \varepsilon_j]$, $i, j = 1, 2, 3$. Taking into account the skew-symmetry of the Lie bracket, we are to determine $[\varepsilon_1, \varepsilon_2]$, $[\varepsilon_1, \varepsilon_3]$, and $[\varepsilon_1, \varepsilon_3]$.

By the Jacobi identity, we have

$$[u, [\varepsilon_1, \varepsilon_2]] = [[u, \varepsilon_1], \varepsilon_2] + [\varepsilon_1, [u, \varepsilon_2]].$$

Thus

$$[u, [\varepsilon_1, \varepsilon_2]] = -[\varepsilon_1, \varepsilon_3]. \tag{6.3}$$

Similarly,

$$[u, [\varepsilon_1, \varepsilon_3]] = [\varepsilon_1, \varepsilon_2]. \tag{6.4}$$

Since $[\mathfrak{h}, \mathfrak{g}] \subset \mathfrak{p}$, we have

$$[\varepsilon_1, \varepsilon_2] \in \mathfrak{p}, \quad [\varepsilon_1, \varepsilon_3] \in \mathfrak{p}.$$

Suppose

$$[\varepsilon_1, \varepsilon_2] = a\varepsilon_1 + b\varepsilon_2 + c\varepsilon_3.$$

Then by (6.3) we have

$$[\varepsilon_1, \varepsilon_3] = -c\varepsilon_2 + b\varepsilon_3.$$

Substituting this into (6.4) yields

$$[\varepsilon_1, \varepsilon_2] = b\varepsilon_2 + c\varepsilon_3.$$

Therefore $a = 0$. Now from

$$[u, [\varepsilon_2, \varepsilon_3]] = [[u, \varepsilon_2], \varepsilon_3] + [\varepsilon_2, [u, \varepsilon_3]] = [-\varepsilon_3, \varepsilon_3] + [\varepsilon_2, \varepsilon_2] = 0,$$

it follows that that

$$[\varepsilon_2, \varepsilon_3] \in \mathrm{span}(u, \varepsilon_1).$$

Therefore we can write $[\varepsilon_2, \varepsilon_3] = a_1 u + a_2 \varepsilon$, where $a_1, a_2 \in \mathbb{R}$.

First we consider the case in which $\tau_1|_V$ has the matrix

$$\begin{pmatrix} -\cos\theta & -\sin\theta \\ -\sin\theta & \cos\theta \end{pmatrix}, \quad \theta \in \mathbb{R},$$

with respect to the basis $\varepsilon_2, \varepsilon_3$. Since $\tau_1$ is an automorphism, we have

$$-\tau_1(\varepsilon_3) = \tau_1([u, \varepsilon_2]) = [\tau_1(u), \tau_1(\varepsilon_2)] = d[u, \tau_1(\varepsilon_2)].$$

Thus

$$(\sin\theta)\varepsilon_2 - (\cos\theta)\varepsilon_3 = -d(\sin\theta)\varepsilon_2 + d(\cos\theta)\varepsilon_3.$$

This implies that

$$d\cos\theta = -\cos\theta, \quad d\sin\theta = -\sin\theta.$$

Thus $d = -1$.

Next applying $\tau_1$ to both sides of

$$[\varepsilon_1, \varepsilon_2] = b\varepsilon_2 + c\varepsilon_3,$$

we get

$$(b\cos\theta - c\sin\theta)\varepsilon_2 + (c\cos\theta + b\sin\theta)\varepsilon_3$$
$$= (-b\cos\theta - c\sin\theta)\varepsilon_2 + (c\cos\theta - b\sin\theta)\varepsilon_3.$$

Therefore we have

$$b\cos\theta = 0, \quad b\sin\theta = 0.$$

Thus $b = 0$.

What we have obtained about the Lie brackets of the Lie algebra $\mathfrak{g}$ can be summarized as follows:

$$[u, \varepsilon_1] = 0, \quad [u, \varepsilon_2] = -\varepsilon_3, \quad [u, \varepsilon_3] = \varepsilon_2,$$
$$[\varepsilon_1, \varepsilon_2] = c\varepsilon_3, \quad [\varepsilon_1, \varepsilon_3] = -c\varepsilon_2, \quad [\varepsilon_2, \varepsilon_3] = a_1 u + a_2 \varepsilon_1,$$

where $c, a_1, a_2$ are real numbers. If $a_1 \neq 0$, we set

$$\varepsilon_1' = X + \frac{a_2}{a_1}\varepsilon_1,$$

and

$$\mathfrak{p}' = \operatorname{span}(\varepsilon_1', \varepsilon_2, \varepsilon_3).$$

Then we have $\mathfrak{g} = \mathfrak{h} + \mathfrak{p}'$, $[\mathfrak{h}, \mathfrak{p}'] \subset \mathfrak{p}'$, and the above brackets are still valid with respect to the new basis $u, \varepsilon_1', \varepsilon_2, \varepsilon_3$. Moreover, the matrix of $\tau_1$ on $\mathfrak{p}'$ with respect to the basis $\varepsilon_1', \varepsilon_2, \varepsilon_3$ is the same as that of $\tau_1$ on $\mathfrak{p}$ with respect to the basis $\varepsilon_1, \varepsilon_2, \varepsilon_3$. Thus we can assume that $a_1 = 0$. Now we have the following cases:

1. $a_2 = 0$. In this case, we set $\varepsilon_1'' = cu + \varepsilon_1$, $\mathfrak{p}'' = \operatorname{span}(\varepsilon_1'', \varepsilon_2, \varepsilon_3)$. Then $\mathfrak{g} = \mathfrak{h} + \mathfrak{p}''$. It is easy to check that $\mathfrak{p}''$ is an abelian ideal of $\mathfrak{g}$, and the action of $u$ on $\mathfrak{p}''$ is

$$[u, \varepsilon_1''] = 0, \quad [u, \varepsilon_2] = -\varepsilon_3, \quad [u, \varepsilon_3] = \varepsilon_2.$$

2. $a_2 \neq 0$ and $c = 0$. Without loss of generality, we can assume that $a_2 = 1$ (otherwise, we use $a_2\varepsilon_1$ to substitute $\varepsilon_1$). Then $\mathfrak{p}$ is a 3-dimensional Heisenberg Lie algebra, since $[\varepsilon_1, \varepsilon_2] = [\varepsilon_1, \varepsilon_3] = 0$, $[\varepsilon_2, \varepsilon_3] = \varepsilon_1$. The action of $u$ is

$$[u, \varepsilon_1] = 0, \quad [u, \varepsilon_2] = -\varepsilon_3, \quad [u, \varepsilon_3] = \varepsilon_2.$$

3. $a_2 c > 0$. Setting $\varepsilon_1' = \frac{2}{c}\varepsilon_1$, $\varepsilon_2' = \frac{2}{\sqrt{a_2 c}}\varepsilon_2$, $\frac{2}{\sqrt{a_2 c}}\varepsilon_3$, we have

$$[\varepsilon_1', \varepsilon_2'] = 2\varepsilon_3', \quad [\varepsilon_1', \varepsilon_3'] = -2\varepsilon_2', \quad [\varepsilon_2', \varepsilon_3'] = 2\varepsilon_1'.$$

Thus $\mathfrak{p}$ is an ideal of $\mathfrak{g}$ isomorphic to the (real) compact simple Lie algebra $\mathfrak{su}(2)$ of skew-Hermitian traceless complex matrices. For simplicity, we just suppose $\mathfrak{p} = \mathfrak{su}(2)$. Let

$$H_1 = \begin{pmatrix} 0 & 1 \\ -1 & 0 \end{pmatrix}, \quad X_1 = \begin{pmatrix} 0 & \sqrt{-1} \\ \sqrt{-1} & 0 \end{pmatrix}, \quad Y_1 = \begin{pmatrix} \sqrt{-1} & 0 \\ 0 & -\sqrt{-1} \end{pmatrix}$$

be the standard basis of $\mathfrak{su}(2)$. Then

$$[H_1, X_1] = 2Y_1, \quad [H_1, Y_1] = -2X_1, \quad [X_1, Y_1] = 2H_1.$$

The action of $u$ on $\mathfrak{p}$ is

$$[u, H_1] = 0, \quad [u, X_1] = -Y_1, \quad [u, Y_1] = X_1.$$

4. $a_2 c < 0$. Similarly as in case (3), we can prove that $\mathfrak{p}$ is an ideal of $\mathfrak{g}$ that is isomorphic to the real simple Lie algebra $\mathfrak{sl}(2, \mathbb{R})$ of traceless $2 \times 2$ real matrices. Let

$$H_2 = \begin{pmatrix} 0 & 1 \\ -1 & 0 \end{pmatrix}, \quad X_2 = \begin{pmatrix} 1 & 0 \\ 0 & -1 \end{pmatrix}, \quad Y_2 = \begin{pmatrix} 0 & 1 \\ 1 & 0 \end{pmatrix}.$$

Then $H_2, X_2, Y_2$ form a basis of $\mathfrak{p}$ and

$$[H_2, X_2] = -2Y_2, \quad [H_2, Y_2] = 2X_2, \quad [X_2, Y_2] = 2H_2.$$

The action of $u$ on $\mathfrak{p}$ is

$$[u, H_2] = 0, \quad [u, X_2] = -Y_2, \quad [u, Y_2] = X_2.$$

It remains to consider the case in which the restriction to $V = \mathrm{span}(\varepsilon_2, \varepsilon_3)$ of the automorphism $\tau_1$ has the matrix

$$\begin{pmatrix} \cos\theta & \sin\theta \\ -\sin\theta & \cos\theta \end{pmatrix}, \quad \theta \in \mathbb{R},$$

with respect to the basis $\varepsilon_2, \varepsilon_3$. In this case, applying $\tau_1$ to both sides of

$$[\varepsilon_1, \varepsilon_2] = b\varepsilon_2 + \varepsilon_3,$$

we get

$$b\cos\theta + c\sin\theta = 0,$$

$$-b\sin\theta + c\cos\theta = 0.$$

Thus $b = c = 0$. This means that $[\varepsilon_1, \varepsilon_2] = [\varepsilon_1, \varepsilon_3] = 0$. If $[\varepsilon_2, \varepsilon_3] = 0$, then $\mathfrak{g}$ is abelian. Otherwise, we can assume that (see case (2) above) $[\varepsilon_2, \varepsilon_3] = \varepsilon_1$, so $\mathfrak{p}$ is a Heisenberg Lie algebra. From this we see that no new structure appears in this case.

Combining the above arguments with the examples in Sect. 6.4, we have the following result.

**Theorem 6.7.** *Let $(\mathfrak{g}, \mathfrak{h})$ be a Riemannian weakly symmetric Lie algebra such that $\dim \mathfrak{h} = 1$ and $\dim \mathfrak{g} = 4$. Then $(\mathfrak{g}, \mathfrak{h})$ must be one of the following:*

1. *$\mathfrak{g}$ is an abelian Lie algebra. In this case the reduced set of automorphisms can be chosen to be $\{\mathrm{id}, -\mathrm{id}\}$.*
2. *$\mathfrak{g}$ has a decomposition $\mathfrak{g} = \mathfrak{h} + \mathfrak{p}$ (direct sum), where $\mathfrak{p}$ is an abelian ideal of $\mathfrak{g}$ and we can select a basis $u$ of $\mathfrak{h}$, and $\varepsilon_1, \varepsilon_2, \varepsilon_3$ of $\mathfrak{p}$, such that*

$$[u, \varepsilon_1] = 0, \quad [u, \varepsilon_2] = -\varepsilon_3, \quad [u, \varepsilon_3] = \varepsilon_2.$$

*In this case, the reduced set of automorphisms can be chosen to be $\{\mathrm{id}, \tau_1\}$, where $\tau_1$ is defined by*

$$\tau_1(u) = -u, \quad \tau_1(\varepsilon_1) = -\varepsilon_1, \quad \tau_1(\varepsilon_2) = -\varepsilon_2, \quad \tau_1(\varepsilon_3) = \varepsilon_3.$$

3.  $\mathfrak{g}$ *has a decomposition* $\mathfrak{g} = \mathfrak{h} + \mathfrak{su}(2)$ *(direct sum), where* $\mathfrak{su}(2)$ *is an ideal of* $\mathfrak{g}$.
    *In this case, let*

$$\varepsilon_1 = \begin{pmatrix} 0 & 1 \\ -1 & 0 \end{pmatrix}, \quad \varepsilon_2 = \begin{pmatrix} 0 & \sqrt{-1} \\ \sqrt{-1} & 0 \end{pmatrix}, \quad \varepsilon_3 = \begin{pmatrix} \sqrt{-1} & 0 \\ 0 & -\sqrt{-1} \end{pmatrix}$$

*be the standard basis of* $\mathfrak{su}(2)$. *Then the action of a nonzero vector* $u$ *of* $\mathfrak{h}$ *on* $\mathfrak{su}(2)$ *is*

$$[u, \varepsilon_1] = 0, \quad [u, \varepsilon_2] = -\varepsilon_3, \quad [u, \varepsilon_3] = \varepsilon_2.$$

*The reduced set of automorphisms can be chosen the same as in case (2).*

4.  $\mathfrak{g}$ *has a decomposition* $\mathfrak{g} = \mathfrak{h} + \mathfrak{sl}(2, \mathbb{R})$ *(direct sum), where* $\mathfrak{sl}(2, \mathbb{R})$ *is an ideal of*
    $\mathfrak{g}$. *In this case, let*

$$\varepsilon_1 = \begin{pmatrix} 0 & 1 \\ -1 & 0 \end{pmatrix}, \quad \varepsilon_2 = \begin{pmatrix} 1 & 0 \\ 0 & -1 \end{pmatrix}, \quad \varepsilon_3 = \begin{pmatrix} 0 & 1 \\ 1 & 0 \end{pmatrix}$$

*be the standard basis of* $\mathfrak{sl}(2, \mathbb{R})$. *Then the action of a nonzero vector* $u$ *of* $\mathfrak{h}$ *on*
$\mathfrak{sl}(2, \mathbb{R})$ *is the same as in case (3). The reduced set of automorphisms can be*
*chosen the same as in case (2).*

5.  $\mathfrak{g}$ *has a decomposition* $\mathfrak{g} = \mathfrak{h} + \mathfrak{n}$ *(direct sum), where* $\mathfrak{n}$ *is an ideal of* $\mathfrak{g}$ *and can be*
    *identified with the 3-dimensional Heisenberg Lie algebra consisting of the real*
    *matrices*

$$\begin{pmatrix} 0 & a & c \\ 0 & 0 & b \\ 0 & 0 & 0 \end{pmatrix}, \quad a, b, c \in \mathbb{R}.$$

*Let*

$$\varepsilon_1 = \begin{pmatrix} 0 & 0 & 1 \\ 0 & 0 & 0 \\ 0 & 0 & 0 \end{pmatrix}, \quad \varepsilon_2 = \begin{pmatrix} 0 & 1 & 0 \\ 0 & 0 & 0 \\ 0 & 0 & 0 \end{pmatrix}, \quad \varepsilon_3 = \begin{pmatrix} 0 & 0 & 0 \\ 0 & 0 & 1 \\ 0 & 0 & 0 \end{pmatrix}$$

*be the standard basis of* $\mathfrak{n}$. *Then the action of a nonzero vector* $u$ *of* $\mathfrak{h}$ *on* $\mathfrak{n}$ *is the*
*same as in case (3). The reduced set of automorphisms can be chosen the same*
*as in case (2).*

Using Theorem 6.7, we can give a complete classification of 3-dimensional
weakly symmetric Finsler spaces.

**Theorem 6.8.** *Let* $(M, F)$ *be a 3-dimensional connected simply connected weakly*
*symmetric Finsler space. Then* $(M, F)$ *must be one of the following:*

1.  $(M, F)$ *is a (reversible) globally symmetric Finsler space.*
2.  *M is the compact simple Lie group* SU(2) *and F is a left-invariant Finsler metric*
    *on M. The restriction of F at the unit element of M is a reversible Minkowski*
    *norm on the Lie algebra* $\mathfrak{su}(2)$ *satisfying*

$$F(a\varepsilon_1 + b\varepsilon_2 + c\varepsilon_3) = F(a\varepsilon_1 + b_1\varepsilon_2 + c_1\varepsilon_3), \tag{$**$}$$

$$F(y, y) \neq d\sqrt{-B(y, y)}, \quad \forall d > 0,$$

where $\varepsilon_1, \varepsilon_2, \varepsilon_3$ constitute the basis as in (3) of Theorem 6.7, $a, b, c, \theta$ are arbitrary real numbers, and $b_1 = b\cos\theta + c\sin\theta, c_1 = c\cos\theta - b\sin\theta$. In this case, there are infinitely many Riemannian metrics as well as infinitely many non-Riemannian Finsler metrics. All these metrics are nonsymmetric.

3. $M$ is the universal covering group of $\mathrm{SL}(2,\mathbb{R})$ and $F$ is a left-invariant Finsler metric on the Lie group $M$. The restriction of $F$ at the unit element $e$ of $M$ is a reversible Minkowski norm on the Lie algebra $\mathfrak{sl}(2,\mathbb{R})$ satisfying

$$F(a\varepsilon_1 + b\varepsilon_2 + c\varepsilon_3) = F(a\varepsilon_1 + (b\cos\theta + c\sin\theta)\varepsilon_2 + (c\cos\theta - b\sin\theta)\varepsilon_3),$$

where $\varepsilon_1, \varepsilon_2, \varepsilon_3$ is the basis as in (4) of Theorem 6.7 and $a, b, c, \theta$ are arbitrary real numbers. In this case, there are infinitely many Riemannian metrics as well as infinitely many non-Riemannian Finsler metrics. Further, all these metrics are nonsymmetric.

4. $M$ is the 3-dimensional Heisenberg Lie group

$$G = \left\{ \begin{pmatrix} 1 & x & z \\ 0 & 1 & y \\ 0 & 0 & 1 \end{pmatrix} \mid x, y, z \in \mathbb{R} \right\},$$

and $F$ is a left-invariant Finsler metric on $M$. The restriction of $F$ at the unit element $e$ of $M$ is a reversible Minkowski norm on the 3-dimensional Heisenberg Lie algebra satisfying

$$F(a\varepsilon_1 + b\varepsilon_2 + c\varepsilon_3) = F(a\varepsilon_1 + (b\cos\theta + c\sin\theta)\varepsilon_2 + (c\cos\theta - b\sin\theta)\varepsilon_3),$$

where $\varepsilon_1, \varepsilon_2, \varepsilon_3$ is the basis as in (5) of Theorem 6.7 and $a, b, c, \theta$ are arbitrary real numbers. In this case, in the sense of isometric diffeomorphism, there exists a unique (up to a positive scalar) Riemannian metric, but there are infinitely many non-Riemannian Finsler metrics. All these metrics are nonsymmetric.

*Proof.* Let $(M, F)$ be a 3-dimensional connected and simply connected weakly symmetric Finsler space. As before, let $\widetilde{G}$ be the full group of isometries and $\widetilde{H}$ the isotropic subgroup at a fixed point in $M$. Let $H$, $G$ be the identity components of $\widetilde{H}$, $\widetilde{G}$, respectively. Then we have pointed out that $\dim H = 0, 1$, or 3. If $\dim H = 3$, then $(M, F)$ is a two-point homogeneous Riemannian manifold. If $\dim H = 0$, then $M$ itself is a commutative Lie group. Hence $(M, F)$ is just a reversible Minkowski space. If $\dim H = 1$, then by Theorem 6.4, we see that $(\mathfrak{g}, \mathfrak{h})$, where $\mathfrak{g} = \mathrm{Lie}\ G$, $\mathfrak{h} = \mathrm{Lie}\ H$, is a Riemannian weakly symmetric Lie algebra. By Theorem 6.7, there are only five kinds of structures for $(\mathfrak{g}, \mathfrak{h})$. In case (1), $(M, F)$ is obviously a reversible Minkowski space. Therefore we need consider only cases (2)–(5) of Theorem 6.7.

We first consider case (2). Let $P$ be the connected Lie subgroup of $G$ with Lie algebra $\mathfrak{p}$. Then $P$ is a commutative normal subgroup of $G$ and $G$ is the semiproduct of $H$ and $P$. Hence the coset space $M = G/H$ is diffeomorphic to the Lie group $P$. This means that $P$ is a connected simply connected commutative Lie group, i.e., $P = \mathbb{R}^3$ (as an additive group). Thus $(M, F)$ must be the Euclidean space $\mathbb{R}^3$ endowed with a reversible Minkowski norm that is invariant under the actions of $\widetilde{H}$ and $\tau_1$. Hence it is a globally symmetric Finsler space.

Next we consider case (3). Similarly to case (2), $(M, F)$ must be a left-invariant Finsler metric on the Lie group $SU(2)$ whose restriction to the tangent space at the unit element $(= \mathfrak{su}(2))$ is invariant under the actions of $\widetilde{H}$ and $\tau_1$. In particular, it is invariant under the action of $H$. Hence $F$ satisfies $(**)$. On the other hand, by Theorem 6.5, every Minkowski norm on $\mathfrak{su}(2)$ satisfying $(**)$ defines a weakly symmetric left-invariant Finsler metric on $SU(2)$. However, if $F(y) = d_1 \sqrt{-B(y,y)}$ for some positive number $d_1$, then this metric is globally symmetric and the isotropic subgroup at $e$ contains $\mathrm{Ad}(SU(2))$, which is 3-dimensional. Therefore the corresponding weakly symmetric Lie algebra cannot be of type (3). Hence $F(y) \neq d\sqrt{-B(y,y)}$ for every positive number $d$. It is easily seen that among these metrics there are infinitely many Riemannian ones as well as non-Riemannian ones. Finally, if one such metric is (globally) symmetric, then the Lie group pair $(\widetilde{G}, \widetilde{H})$ is a Riemannian symmetric pair. Thus by Theorems 2.29 and 2.30, the Lie algebra pair $(\mathfrak{g}, \mathfrak{h})$ is an orthogonal symmetric Lie algebra. But this is impossible because in an orthogonal Lie algebra we have $[\mathfrak{p}, \mathfrak{p}] \subset \mathfrak{h}$, but in our case we have $[\mathfrak{p}, \mathfrak{p}] \subset \mathfrak{p}$. Thus all the metrics in this case are nonsymmetric.

The above arguments are all valid in case (4) except the proof that such metrics are nonsymmetric. Now we proceed as follows. If one such metric on the universal covering of $SL(2, \mathbb{R})$ is symmetric, then as in case (3), the isotropic subgroup at the unit element must be 3-dimensional, for otherwise it will contradict the fact that $(\mathfrak{g}, \mathfrak{h})$ cannot be an orthogonal symmetric Lie algebra. If the isotropic group is 3-dimensional, then the space $(M, F)$ is a two-point homogeneous Riemannian manifold. Hence it is a simply connected noncompact symmetric space of rank 1, i.e., it is 3-dimensional hyperbolic space. In particular, it is a homogeneous Riemannian space of negative constant curvature $-1$. Now we can deduce a contraction as follows. Note that the Lie group $SL(2, \mathbb{R})$, endowed with the left-invariant Riemannian metric $Q$ whose restriction at $e$ satisfying $F(Y) = \sqrt{Q(Y,Y)}$, $Y = \mathfrak{sl}(2, \mathbb{R})$, is locally isometric to $(M, F)$. Hence $(SL(2, \mathbb{R}), Q)$ is also of constant curvature $-1$. This is impossible because a classical result of Kobayashi asserts that a homogeneous Riemannian manifold of strictly negative curvature is necessarily simply connected [100]. This completes the proof for case (4).

Finally, in case (5), the assertion that the weakly symmetric Riemannian metric is unique up to a positive factor follows from the fact that on the 3-dimensional Heisenberg Lie group all left-invariant Riemannian metrics are isometric (up to a positive factor); see [183]. It is also proved in [183] that this metric is not symmetric. But then none of the non-Riemannian ones can be symmetric. Otherwise, a contraction would arise in both cases in which the dimension of isotropic subgroup is 2 (the Lie algebra pair $(\mathfrak{g}, \mathfrak{h})$ cannot be an orthogonal symmetric Lie algebra), or is 3 (Riemannian metric). That there are infinitely many non-Riemannian weakly symmetric Finsler metrics in this case follows from the facts that the Minkowskian norms

$$F_\lambda(a\varepsilon_1 + b\varepsilon_2 + c\varepsilon_3) = \sqrt{a^2 + b^2 + c^2 + \lambda \sqrt[4]{a^4 + (b^2 + c^2)^2}}, \quad \lambda > 0,$$

satisfy condition of (4) and that $F_{\lambda_1}$ is not linearly isometric to $F_{\lambda_2}$ for $\lambda_1 \neq \lambda_2$ (see [16, p. 21]).                                                                                      □

## 6.6   The Classification

In this section we give a classification of non-Riemannian weakly symmetric Finsler spaces with reductive isometry groups. First we present a general principle to classify weakly symmetric Finsler spaces.

**Theorem 6.9.** *Let $(M, F)$ be a connected weakly symmetric Finsler space and $G$ the full group of isometries of $(M, F)$. Then there exists a Riemannian metric $Q$ on $M$ that is invariant under the action of $G$ such that $(M, Q)$ is a weakly symmetric Riemannian manifold. Moreover, every $G$-invariant Finsler metric on $M$ must be weakly symmetric.*

*Proof.* Since a connected weakly symmetric Finsler space must be homogeneous, $G$ acts transitively on $M$. Hence the identity component $G_0$ of $G$ acts transitively on $M$. Fix $x \in M$ and denote the isotropy subgroup of $G$ at $x$ by $H$. Then the isotropy subgroup of $G_0$ at $x$ is $H \cap G_0$. Since the Lie algebras of $G$ and $G_0$ coincide, it is easily seen that the identity component $H_0$ of $H$ is also the identity component of $H \cap G_0$. Since $H$ is compact, it has at most finite components. Therefore the index $[H : H_0]$ is finite. Hence the index of $H \cap G_0$ in $H$ is also finite. Suppose

$$H = (H \cap G_0) \cup \mu_1(H \cap G_0) \cup \cdots \cup \mu_s(H \cap G_0)$$

is the decomposition of $H$ into left cosets of $H \cap G_0$. Since $M = G_o/(H \cap G_0)$, we have

$$M = (G_0 \cup \mu_1 G_0 \cup \cdots \cup \mu_s G_0)/H. \tag{6.5}$$

It is obvious that $G = G_0 \cup \mu_1 G_0 \cup \cdots \cup \mu_s G_0$. Since $H$ is compact, by Weyl's unitary trick there exists an $H$-invariant inner product $\langle , \rangle$ on $T_x(M)$. By (6.5), we easily see that $\langle , \rangle$ can induce a Riemannian metric $Q$ on $M$ that is invariant under the action of $G = G_0 \cup \mu_1 G_0 \cup \cdots \cup \mu_s G_0$. Now we assert that $(M, Q)$ is a weakly symmetric space. In fact, since $(M, F)$ is weakly symmetric and $H$ is the isotropic subgroup (at $x$) of the full group of isometries of $(M, F)$, for every fixed $u \in T_x(M)$, there exists $h \in H$ such that $dh|_x(u) = -u$. Thus the isotropy representation of the homogeneous Riemannian manifold on the right-hand side of (6.5) satisfies the condition of Proposition 6.3. Hence $(M, Q)$ is weakly symmetric. The last assertion is clear.                                                                                      □

By Theorem 6.9, the strategy to classify weakly symmetric Finsler spaces can be reduced to two steps: the first step is to classify weakly symmetric Riemannian manifolds and the second step is to find all the Finsler metrics on a weakly symmetric Riemannian manifold that make a weakly symmetric Finsler space. In particular, if there is a non-Riemannian Finsler metric that is invariant under the full group of isometries of a weakly symmetric Riemannian metric, then this manifold admits non-Riemannian weakly symmetric Finsler metrics. Combining this argument with Theorem 4.8, we have the following.

**Theorem 6.10.** *Let $(M, Q)$ be a connected weakly symmetric Riemannian manifold. If $M$ is not diffeomorphic to a rank-one Riemannian symmetric space, then there exists a non-Riemannian Finsler metric $F$ on $M$ that is invariant under the full group of isometries of $Q$ such that $(M, F)$ is a weakly symmetric Finsler space.*

Now we consider the case of rank-one symmetric Riemannian spaces. By duality, we need treat only the compact ones. These manifolds include the spheres $S^n$ $(n \geq 2)$, the real projective spaces $\mathbb{R}P^n$ $(n \geq 2)$, the complex projective spaces $\mathbb{C}P^n$ $(n \geq 2)$, the quaternion projective spaces $\mathbb{H}P^n$ $(n \geq 2)$, and the Cayley projective plane $\mathrm{Cay}P^2$ (see [83]). Suppose $M$ is one of the manifolds above and $F$ is a weakly symmetric Finsler metric on $M$. Then the full group of isometries of $F$ acts transitively on $M$. The compact connected groups that admit effective transitive action on spheres have been classified by Montgomery–Samelson [121] and Borel [31]; see also [28]. Furthermore, compact Lie groups that admit transitive action on rank-one Riemannian symmetric spaces were classified by Onishchik [126]. These results lead to the following theorem.

**Theorem 6.11.** *On the even-dimensional sphere $S^{2n}$, the even-dimensional real projective space $\mathbb{R}P^{2n}$, the $4n$-dimensional complex projective space $\mathbb{C}P^{2n}$, the quaternion projective space $\mathbb{H}P^n$, and the Cayley projective plane $\mathrm{Cay}\,P^2$, every weakly symmetric Finsler metric must be Riemannian and globally symmetric. On the odd-dimensional sphere $S^{2n-1}$ $(n \geq 2)$, the odd-dimensional real projective space $\mathbb{R}P^{2n-1}$ $(n \geq 2)$, and the $(4n-2)$-dimensional complex projective space $\mathbb{C}P^{2n-1}$, there exist infinitely many non-Riemannian weakly symmetric Finsler metrics.*

*Proof.* As pointed out in [126], if a compact connected Lie group $G$ acts transitively on $S^{2n}$ $(n \neq 3)$, $\mathbb{R}P^{2n}$, $\mathbb{C}P^{2n}$, $\mathbb{H}P^n$, or $\mathrm{Cay}\,P^2$, then $G$ must be the the identity component of the full group of isometries of the canonical Riemannian metric $Q_c$ on the corresponding manifold $M$. Fix $x \in M$ and denote the isotropy subgroup of $G$ at $x$ by $H$. Since $(M, Q_c)$ is a rank-one Riemannian symmetric space, $(G, H)$ must be a rank-one Riemannian symmetric pair. This means that $H$ acts transitively on the unit sphere (with respect to $Q_c$) of $T_x(M)$. Therefore every $H$-invariant Minkowski norm on $T_x(M)$ must be Euclidean and must be a positive multiple of $Q_c|_{T_x(M)}$. Hence every $G$-invariant Finsler metric on $G/H$ must be a positive multiple of $Q_c$.

On the other hand, the group $G_2$ has an effective transitive action on the sphere $S^6$ as isometries. The isotropy subgroup at the point $p = (1, 0, \ldots, 0)$ is $\mathrm{SU}(3)$. However, in this case, the linear isotropic representation of $\mathrm{SU}(3)$ on $T_p(S^6)$ is also transitive on the unit sphere with respect to the standard metric. Hence every $G_2$-invariant Finsler metric on $S^6$ must be the standard Riemannian metric. This proves the first assertion.

Now we prove the second assertion. We first consider the sphere $S^{2n-1}$ as the unit sphere of $\mathbb{C}^n$ with respect to the standard Hermitian metric. The unitary group $U(n)$ then acts transitively on $S^{2n-1}$ and the isotropy group at the point $p = (1,0,0,\ldots,0)$ is $U(n-1)$. The tangent space $T_p(M)$ can be viewed as $\mathbb{C}^{n-1} \oplus \mathbb{R}\sqrt{-1}p$ and the isotropy representation is given by $\chi(A)(v,a) = (Av,a)$. Now we consider another specific isometry $\sigma$ of $S^{2n-1}$, given by taking the conjugate on each coordinate. Note that $\sigma$ leaves $p$ fixed. If we take $G = U(n) \cup \sigma \circ U(n-1)$ and $H_p = U(n-1) \cup \sigma \circ U(n-1)$, then we have $S^{2n-1} = G/H_p$. We assert that $H_p$ satisfies the condition of Proposition 6.3. In fact, for every $(v,a) \in T_p(M)$, we first apply $\sigma$ to send $(v,a)$ to $(\bar{v},-a)$. Then by the transitivity of $U(n-1)$ on the unit sphere of $\mathbb{C}^{n-1}$ we can select an element of $U(n-1)$ to send $(\bar{v},-a)$ to $(-v,-a)$. This proves our assertion. This means that every reversible $G$-invariant Finsler metric on $S^{2n-1}$ is weakly symmetric. Next we consider the isotropy representation of $H_p$ on $T_p(M)$. It is easily seen that both the subspaces $\mathbb{C}^{n-1}$ and $\mathbb{R}\sqrt{-1}p$ are invariant under $H_p$. By Theorem 4.6, this implies that there exist infinitely many $H_p$-invariant non-Euclidean Minkowski norms on $T_x(M)$. Each such Minkowski norm will then induce a $G$-invariant Finsler metric on $S^{2n-1}$ that is weakly symmetric. This proves the second assertion for the case of the sphere. A similar argument $\mathbb{R}P^{2n-1} = U(n)/(U(n-1) \times \{I_n, -I_n\})$ shows that there also exist infinitely many non-Riemannian Finsler metrics on it that are weakly symmetric. Now we consider $\mathbb{C}P^{2n-1}$. According to [183], the group $G = Sp(n)$ acts transitively on it with the isotropic subgroup $H = Sp(n-1)U(1)$. The isotropy representation can be described as follows: the tangent space $T_o(G/H)$ can be identified with $\mathbb{H}^{n-1} \oplus \mathbb{R}^2$, the group $H$ acts on $\mathbb{R}^2$ by the usual action (as rotations) of $U(1)$, and it acts on $\mathbb{H}^{n-1}$ by

$$(A,z)(v) = A(v)\bar{z}.$$

From these we can easily deduce that $H$ satisfies the condition of Proposition 6.3. In fact, for $(v,w) \in T_o(M)$, we can first use $e^{\sqrt{-1}\pi/2}$ to send it to $(-\sqrt{-1}v,-w)$. Then by the transitivity of the action of $Sp(n-1)$ on the unit sphere of $\mathbb{H}^{n-1}$ we can select $h \in Sp(n-1)$ that sends $-\sqrt{-1}v$ to $-v$. This proves our assertion. The above argument shows that every reversible $G$-invariant Finsler metric on $M$ must be weakly symmetric. Since the action of $H$ leaves the subspaces $\mathbb{H}^{n-1}$ and $\mathbb{R}^2$ invariant, there exist infinitely many such metrics that are non-Riemannian. This completes the proof of the theorem.                                                                    □

Now we consider homogeneous manifolds that admit non-Riemannian weakly symmetric Finsler metrics with reductive isometry groups. We first recall some results concerning the classification of weakly symmetric Riemannian manifolds. Let $G$ be a connected reductive algebraic group over the field of complex numbers $\mathbb{C}$ acting on an algebraic variety $X$. Then $X$ is called spherical (with respect to $G$)

if there exists a Borel subgroup $B \subset G$ such that $B$ has a Zariski dense orbit in $X$. É. Cartan proved that any complexified symmetric space is spherical with respect to the identity component of the full group of isometries [38]. An algebraic subgroup $H$ of $G$ is called spherical if $G/H$ is spherical with respect to $G$. In this case the pair $(G,H)$ is usually called a spherical pair. D.N. Akhiezer and E.B. Vinberg proved in [3] that if $(G,H)$ is a Lie group pair with $G$ reductive, with the property that every $G$-invariant Riemannian metric is weakly symmetric, then the complexification $(G^C, H^C)$ must be a spherical pair. This reduces the classification of weakly symmetric Riemannian manifolds to the classification of the real forms of the spherical pairs. The classification of spherical pairs was achieved by Akhiezer and Vinberg [3], Brion [33], Krämer [106], and Mikityuk [119]. Using these results and some reduction, Yakimova and Wolf independently obtained a complete classification of weakly symmetric Riemannian manifolds with reductive isometry groups [176, 180]. The result is presented in Table 6.1 (note that the last three types of manifolds were missed in [180]).

We will adopt the approaches of [180]. Let us recall some terminology related to Yakimova's classification. Let $G$ be a reductive Lie group and $K$ a closed subgroup of $G$. Let $Z(G)$ denote the center of $G$ and $K_r = K/(K \cap Z(G))$. The coset space $G'/K_r$, where $G'$ denotes the commutator subgroup of $G$, is called the central reduction of $G/K$. A Lie group pair $(G,K)$, where $G$ is a semisimple Lie group and $K$ is a closed subgroup of $G$, is called principal if $Z(K)^0 = Z = (G^1 \cap Z) \times (G^2 \cap Z) \times \cdots \times (Z \cap G^s)$, where $G^1, \ldots, G^s$ are all the simple factors of $G$.

**Theorem 6.12 (Yakimova [180]).** *Let $M = G/K$ be a simply connected compact irreducible weakly symmetric homogeneous space with $G$ reductive. If $(G,K)$ is principal, then it must be one of the pairs in Table 6.1.*

Now we give some explanation of the notation in Table 6.1. If $M = G/H$ is a weakly symmetric homogeneous manifold such that there exists a Riemannian symmetric pair $(P,Q)$ with $G \subset P$, $H \subset Q$, then $(P,Q)$ is called a symmetric extension of $G/H$. In this case, on the homogeneous manifold $M = G/H$ there exists some $G$-invariant Riemannian metric that is globally symmetric. We use $\mathbb{B}$ to denote the set of all $G$-invariant Riemannian metrics on $M$ and $\mathfrak{m}$ to denote the set of all $P$-invariant Riemanian metrics on $M$. The dimensions of $\mathbb{B}$ and $\mathfrak{m}$ are given in the cases in which there exists a symmetric extension of $G/H$. If $K_1$ and $K_2$ are two closed subgroups of a Lie group $G$, then we use $K_1 \cdot K_2$ to denote the quotient group of $K_1 \times K_2$ with respect to the central subgroup of $K_1 \times K_2$. The notation $\times^2 G$ means the direct product group $G \times G$.

Let us consider invariant Finsler metrics on the manifolds in Table 6.1. We first find out which one admits weakly symmetric non-Riemannian Finsler metrics. Note that there are some manifolds that are diffeomorphic to rank-one Riemannian symmetric manifolds or the product of some rank-one symmetric Riemannian manifolds. In fact, we have the following relations:

**Table 6.1** Weakly symmetric homogeneous manifolds and symmetric extensions

| | $M = G/K$ | $P$ | $Q$ | $\dim \mathbb{B}(m)$ |
|---|---|---|---|---|
| 1a | $SU(n)/SU(n-1)$ | $SO(2n)$ | $SO(2n-1)$ | $2(1)$ |
| 1b | $SU(n)/(SU(n-k) \times SU(k))$ | | | |
| | $(n \neq n-k)$ | | | |
| 1c | $U(2n)/(U(n) \times SU(n))$ | | | |
| 2 | $SU(2n+1)/(Sp(n) \cdot U(1))$ | | | |
| 3 | $SU(2n+1)/Sp(n)$ | $SU(2n+2)$ | $Sp(n+1)$ | $3(1)$ |
| 4 | $(U(1) \cdot Sp(n))/U(n)$ | | | |
| 5 | $Sp(n)/(Sp(n-1) \cdot U(1))$ | $SU(2n)$ | $U(2n-1)$ | $2(1)$ |
| 6 | $(Sp(n) \cdot U(1))/(Sp(n-1) \cdot U(1))$ | | | |
| 7 | $SO(2n+1)/U(n)$ | $SO(2n+2)$ | $U(n+1)$ | $2(1)$ |
| 8 | $U(1) \cdot SO(2n+1)/U(n)$ | | | |
| 9 | $U(1) \cdot SO(4n+2)/U(2n+1)$ | | | |
| 10 | $SO(10)/Spin(7) \times SO(7)$ | | | |
| 11 | $U(1) \cdot SO(10)/Spin(7) \times SO(7)$ | | | |
| 12 | $SO(9)/Spin(7)$ | $SO(16)$ | $SO(15)$ | $2(1)$ |
| 13 | $Spin(8)/G_2$ | $SO(8) \times SO(8)$ | $SO(7) \times SO(7)$ | $3(2)$ |
| 14 | $Spin(7)/G_2$ | $SO(8)$ | $SO(7)$ | $1(1)$ |
| 15 | $E_6/Spin(10)$ | | | |
| 16 | $G_2/SU(3)$ | $SO(7)$ | $SO(6)$ | $1(1)$ |
| 17 | $U(n+1) \times SU(n)/U(n)$ | $\times^2 SU(n+1)$ | $SU(n+1)$ | $2(1)$ |
| 18 | $SU(n) \times Sp(m)/(U(n-2)$ | | | |
| | $\times SU(2) \times Sp(m-1))$ | | | |
| 19a | $Sp(n) \times Sp(l) \times Sp(m)/Sp(n-1)$ | | | |
| | $\times Sp(1) \times Sp(l-1) \times Sp(m-1)$ | | | |
| 19b | $Sp(n) \times Sp(1) \times Sp(m)/Sp(n-1)$ | $\times^2 SO(4n)$ | $\times^2 SO(4n-1)$ | $6(2)$ |
| | $\times Sp(1) \times Sp(m-1)$ | | | |
| 19b' | $Sp(1) \times Sp(n) \times Sp(1)/Sp(1)$ | $\times^2 SO(4n)$ | $\times^2 SO(4n-1)$ | $5(2)$ |
| | $\times Sp(n-1) \times Sp(1)$ | | | |
| 20 | $(Sp(n) \times Sp(2))/(Sp(n-2) \times Sp(2))$ | | | |
| 21 | $SU(n) \times Sp(m)/(SU(n-2) \times Sp(2)$ | | | |
| | $\times Sp(m-1))$ | | | |
| 22a | $Sp(n) \times Sp(2) \times Sp(m)/(Sp(n-1)$ | | | |
| | $\times Sp(1) \times Sp(1) \times Sp(m-1))$ | | | |
| 22b | $Sp(1) \times Sp(2) \times Sp(1)/(Sp(1) \times Sp(1))$ | $Sp(2) \times Sp(2)$ | $Sp(2)$ | $3(1)$ |
| 23 | $SO(n+1) \times SO(n)/SO(n)$ | $\times^2 SO(n+1)$ | $SO(n+1)$ | $2(1)$ |
| 24a | $Sp(n) \times Sp(m)/(Sp(n-1) \times Sp(1)$ | | | |
| | $\times Sp(m-1))$ | | | |
| 24b | $Sp(n) \times Sp(1)/(Sp(n-1) \times Sp(1))$ | $SO(4n)$ | $SO(4n-1)$ | $3\,(1)$ |

Type 1a:   $SU(n)/SU(n-1) = S^{2n-1}$,
Type 5:     $Sp(n)/(Sp(n-1) \cdot U(1)) = \mathbb{C}P^{2n-1}$,
Type 6:     $(U_1 \cdot SP(n))/(Sp(n-1) \cdot U(1)) = S^{4n-1}$,
Type 12:   $SO(9)/Spin(7) = S^{15}$,
Type 13:   $Spin(8)/G_2 = S^7 \times S^7$,
Type 14:   $Spin(7)/G_2 = S^7$,

Type 16: $G_2/\mathrm{SU}(3) = S^6$,

Type 19b: $\mathrm{Sp}(n) \times \mathrm{Sp}(l) \times \mathrm{Sp}(m)/\mathrm{Sp}(n-1) \times \mathrm{Sp}(1) \times \mathrm{Sp}(l-1) \times \mathrm{Sp}(m-1) = S^{4n-1} \times S^{4m-1}$,

Type 19b': $\mathrm{Sp}(1) \times \mathrm{Sp}(n) \times \mathrm{Sp}(1)/\mathrm{Sp}(1) \times \mathrm{Sp}(n-1) \times \mathrm{Sp}(1) = S^{4n-1} \times S^{4n-1}$,

Type 24b: $\mathrm{Sp}(n) \times \mathrm{Sp}(1)/(\mathrm{Sp}(n-1) \times \mathrm{Sp}(1)) = S^{4n-1}$.

As we have explained above, all these manifolds except type 16, $G_2/\mathrm{SU}(3)$, admit weakly symmetric non-Riemannian Finsler metrics. Note also that in types 14 and 16, every $G$-invariant Finsler metric on $M = G/H$ must be Riemannian. This means that although $S^7 = \mathrm{Spin}(7)/G_2$ admits weakly symmetric non-Riemannian Finsler metrics, such metrics cannot be $\mathrm{Spin}(7)$-invariant. To get such a metric, we must write $S^7$ as $\mathrm{SU}(4)/\mathrm{SU}(3)$ or $\mathrm{Sp}(2) \times \mathrm{Sp}(1)/\mathrm{Sp}(1) \times \mathrm{Sp}(1)$ and find invariant Finsler metrics on these coset spaces.

Let us summarize the above as the following:

**Theorem 6.13.** *Let $M$ be a connected homogeneous manifold. If $M$ can be written as $M = G/K$, where $(G,K)$ is a principal pair and $M$ admits $G$-invariant weakly symmetric non-Riemannian Finsler metrics, then $M$ must be diffeomorphic either to one manifold or to the product of some manifolds from the following list:*

1. *The odd-dimensional spheres $S^{2n-1}$ ($n \geq 2$),*
2. $\mathrm{SU}(n)/(SU(n-k) \times \mathrm{SU}(k))$ *($n \neq n-k$),*
3. $\mathrm{U}(2n)/(\mathrm{U}(n) \times \mathrm{SU}(n))$,
4. $\mathrm{SU}(2n+1)/(\mathrm{Sp}(n) \cdot \mathrm{U}(1))$,
5. $\mathrm{SU}(2n+1)/\mathrm{Sp}(n)$,
6. $(\mathrm{U}(1) \cdot \mathrm{Sp}(n))/\mathrm{U}(n)$,
7. *The $(4n-2)$-dimensional complex projective spaces $\mathbb{C}P^{2n-1}$,*
8. $\mathrm{SO}(2n+1)/\mathrm{U}(n)$,
9. $\mathrm{U}(1) \cdot \mathrm{SO}(2n+1)/\mathrm{U}(n)$,
10. $\mathrm{U}(1) \cdot \mathrm{SO}(4n+2)/\mathrm{U}(2n+1)$,
11. $E_6/\mathrm{Spin}(10)$,
12. $\mathrm{SU}(n+1)$,
13. $\mathrm{SU}(n) \times \mathrm{Sp}(m)/(\mathrm{U}(n-2) \times \mathrm{SU}(2) \times \mathrm{Sp}(m-1))$,
14. $\mathrm{Sp}(n) \times \mathrm{Sp}(l) \times \mathrm{Sp}(m)/\mathrm{Sp}(n-1) \times \mathrm{Sp}(1) \times \mathrm{Sp}(l-1) \times \mathrm{Sp}(m-1)$,
15. $(\mathrm{Sp}(n) \times \mathrm{Sp}(2))/(\mathrm{Sp}(n-2) \times \mathrm{Sp}(2))$,
16. $\mathrm{SU}(n) \times \mathrm{Sp}(m)/(\mathrm{SU}(n-2) \times \mathrm{Sp}(2) \times \mathrm{Sp}(m-1))$,
17. $\mathrm{Sp}(n) \times \mathrm{Sp}(2) \times \mathrm{Sp}(m)/(\mathrm{Sp}(n-1) \times \mathrm{Sp}(1) \times \mathrm{Sp}(1) \times \mathrm{Sp}(m-1))$,
18. $\mathrm{Sp}(2)$,
19. $\mathrm{SO}(n+1)$ *($n \geq 2$),*
20. $\mathrm{Sp}(n) \times \mathrm{Sp}(m)/(\mathrm{Sp}(n-1) \times \mathrm{Sp}(1) \times \mathrm{Sp}(m-1))$,

*Moreover, on each of the above manifolds there exists a non-Riemannian weakly symmetric Finsler metric.*

## 6.7   Weakly Symmetric Metrics on H-Type Groups

We now consider left-invariant weakly symmetric Finsler metrics on nilpotent Lie groups. A Lie algebra $\mathfrak{g}$ is called two-step nilpotent if $[\mathfrak{g}, [\mathfrak{g}, \mathfrak{g}]] = 0$. A connected Lie group is called two-step nilpotent if its Lie algebra is two-step nilpotent. A result of C. Gordon asserts that if a connected nilpotent Lie group $N$ admits a left-invariant weakly symmetric Riemannian metric, then $N$ must be two-step nilpotent [73, 74]. Taking into account Theorem 6.2, we have the following result.

**Proposition 6.4.** *Let $N$ be a connected nilpotent Lie group. If $N$ admits a left-invariant weakly symmetric Finsler metric, then it must be two-step nilpotent.*

Now we turn to find the conditions for a connected two-step nilpotent Lie group to admit a left-invariant weakly symmetric non-Riemannian Finsler metric. The following theorem reduces the problem to the Riemannian case.

**Theorem 6.14.** *If a connected simply connected two-step nilpotent Lie group admits a left-invariant weakly symmetric Riemannian metric, then it admits a left-invariant weakly symmetric non-Riemannian Finsler metric.*

*Proof.* Without loss of generality, we can assume that $N$ is not abelian. Suppose $Q$ is a left-invariant weakly symmetric Riemannian metric on $N$. Let $G$ be the full group of isometries of $Q$ and let $H$ be the isotropy subgroup of $G$ at the identity element $e$. Then $N = G/H$. N. Wilson proved that $N$ is a normal subgroup of the identity component $G_0$ of $G$ [170]. Now we assert that the action of $H$ on $T_e(N) = T_o(G/H)$ $(o = H)$ cannot be transitive on the unit sphere of $T_e(N)$ (with respect to the restriction of $Q$ to $T_e(N)$). In fact, otherwise, $(N, Q)$ would an isotropy Riemannian manifold hence must be a rank-one Riemannian symmetric space (see [83, p. 535]). Since $N$ is noncompact, the sectional curvature of $(N, Q)$ must be $< 0$ everywhere. But a result of Wolf [173] asserts that for every nonabelian connected nilpotent Lie group $N_1$, and every left-invariant Riemannian metric $Q_1$ on $N_1$, there exist three tangent planes $\Pi_0$, $\Pi_+$, and $\Pi_-$ such that $\Pi_0$ has sectional curvature 0, $\Pi_+$ has sectional curvature $> 0$, and $\Pi_-$ has sectional curvature $< 0$. This is a contradiction. Therefore $H$ cannot be transitive on the unit sphere of $T_e(N)$. By Theorem 4.8, there must be a non-Euclidean Minkowski norm $F_0$ on $T_e(N)$ that is invariant under $H$. Then $F_0$ induces a $G$-invariant Finsler metric $F$ on $N = G/H$. By Theorem 6.2, $(N, F)$ must be weakly symmetric. On the other hand, since $N$ is a normal subgroup of $G_0$, $F$ is left-invariant. This completes the proof of the theorem. $\qquad\square$

By the results above, to determine all the left-invariant weakly symmetric Finsler metrics on nilpotent Lie groups, we must first find all the two-step nilpotent Lie groups that admit left-invariant weakly symmetric Riemannian metrics, and then classify all the left-invariant Finsler metrics that are weakly symmetric (there must be non-Riemannian ones among them). However, no sufficient and necessary condition for a two-step nilpotent Lie group to admit left-invariant weakly symmetric Riemannian metrics is known. So in the following we consider a special type of two-step nilpotent Lie group, namely, the Lie groups of Heisenberg type.

We first recall the definition of Lie groups of Heisenberg type, or simply H-type Lie groups. This type of Lie groups were introduced by Kaplan in [95]. After that, several open problems in Riemannian geometry were settled using H-type Lie groups. We begin with the notion of an H-type Lie algebra. An H-type Lie algebra is a two-step nilpotent Lie algebra $\mathfrak{n}$ with an inner product $\langle , \rangle$ such that the following property holds: if $w$ is any element in the center $\mathfrak{z}$ of $\mathfrak{n}$, and $\mathfrak{a}$ is the orthogonal complement of $\mathfrak{z}$, then the linear operator $J_w \colon \mathfrak{a} \to \mathfrak{a}$ defined by

$$\langle J_w(u), v \rangle = \langle w, [u,v] \rangle, \tag{6.6}$$

for $u, v \in \mathfrak{a}$, satisfies the identity

$$|J_w(u)| = |w| \cdot |u|, \tag{6.7}$$

where $|\cdot|$ is the length with respect to $\langle \cdot, \cdot \rangle$. A connected simply connected two-step nilpotent Lie group $N$ is called of H-type if its Lie algebra $\mathfrak{n}$ is of H-type.

H-type groups are closely related to the representations of Clifford algebras and can be classified using this point of view. Let $\mathfrak{n} = \mathfrak{a} \oplus \mathfrak{z}$ be the decomposition of an H-type Lie algebra and let $J_w$ be the maps defined in (6.6). Then $J_w$ induce an action of the Clifford algebra $C(\mathfrak{z}, -|\cdot|^2)$ on $\mathfrak{a}$. Conversely, given a Euclidean space $\mathfrak{z}$ and a $C(\mathfrak{z}, -|\cdot|^2)$-module $\mathfrak{a}$, denote by $J_w$ the action of $w \in \mathfrak{z}$ on $\mathfrak{a}$. Then one can define an inner product on $\mathfrak{a}$ such that (6.7) holds, as well as a Lie algebra structure on $\mathfrak{a} \oplus \mathfrak{z}$ through (6.6). Obviously different choices of the modules may lead to isomorphic H-type Lie algebras. Therefore the classification of H-type Lie algebras up to isomorphism can be achieved by a detailed analysis of modules of Clifford algebras. The result can be summarized as follows (see [95]):

1. If $\dim \mathfrak{z} \not\equiv 3 \,(\mathrm{mod}\,4)$, then up to isomorphism, there is only one irreducible $C(\mathfrak{z}, -|\cdot|^2)$ module, denoted by $\mathfrak{a}_0$. In this case, all the H-type Lie algebras with center $\mathfrak{z}$ are of the form $\mathfrak{z} \oplus (\mathfrak{a}_0)^p$ with $p \geq 1$.
2. If $\dim \mathfrak{z} \equiv 3 \,(\mathrm{mod}\,4)$, then up to isomorphism there are two nonisomorphic irreducible $C(\mathfrak{z}, -|\cdot|^2)$-modules, denoted by $\mathfrak{a}_1$ and $\mathfrak{a}_2$. In this case, all the H-type Lie algebras with center $\mathfrak{z}$ are of the form $\mathfrak{z} \oplus ((\mathfrak{a}_1)^p \oplus (\mathfrak{a}_2)^q)$, where $p, q \geq 0$, $p+q \geq 1$. The H-type Lie algebras $\mathfrak{z} \oplus ((\mathfrak{a}_1)^p \oplus (\mathfrak{a}_2)^q)$ and $\mathfrak{z} \oplus ((\mathfrak{a}_1)^q \oplus (\mathfrak{a}_2)^p)$ are isomorphic. In the special case $p = 0$ or $q = 0$, $\mathfrak{a}$ is called isotypic.

Now we consider weakly symmetric Finsler metrics on H-type Lie groups. We first need to generalize the above-mentioned result of Wilson on the group of isometries of a left-invariant Riemannian metric to the case of a left-invariant Finsler metric.

**Proposition 6.5.** *Let $N$ be a connected simply connected nilpotent Lie group with Lie algebra $\mathfrak{n}$. Suppose $F$ is a left-invariant Finsler metric on $N$ and denote the full group of isometries of $(M, F)$ by $G$. Then $N$ (viewed as the group of left translations) is a normal subgroup of $G$. Moreover, $G$ is the semiproduct of $N$ with the isotropy subgroup $G_e$ of $G$ at the unity element $e$ and $G_e$ is equal to the group of automorphisms of $N$ whose differential of $F$ at $e$ is a linear isometry of the Minkowski space $(\mathfrak{n}, F|_\mathfrak{n})$.*

*Proof.* We first prove that $N$ is a normal subgroup of $G$. This follows from a simple observation. Note that $N = G/G_e$ and $F$ is a $G$-invariant Finsler metric on $N$, where $G_e$ is compact. It is easily seen that there exists a $G$-invariant Riemannian metric $Q$ on $G/G_e$. Since $N \subset G$, $Q$ is left-invariant on $N$. Wilson's result implies that $N$ is normal in the full group $I(M,Q)$ of $Q$. Since $G \subset I(M,Q)$, $N$ is normal in $G$. From the facts that each element of $G$ can be written as a product of an element in $N$ and an element of $G_e$ and that $N \cap G_e = \{e\}$, one easily deduces that $G$ is the semiproduct of $N$ and $G_e$. Now we prove the last assertion. If $k \in G_e$, then it is obvious that $dk|_e \in L(\mathfrak{n}, F)$ (the group of linear isometries of the Minkowski space $(\mathfrak{n}, F)$). So we need only prove that $k$ is an automorphism of $N$. Suppose $x, y \in N$. Then we have

$$k(xy) = k(L_x(y)) = kL_x k^{-1}(k(y)).$$

Since $N$ is normal in $G$, there exists $x' \in N$ such that $kL_x k^{-1} = L_{x'}$. Hence $k(xy) = L_{x'}k(y)$. Considering the value at $y = e$, we find that $x' = k(x)$. Thus $k(xy) = L_{k(x)}K(y) = k(x)k(y)$. This completes the proof of the proposition. $\qquad\square$

Now we can give a classification of a special class of left-invariant weakly symmetric Finsler metrics on H-type Lie groups. We call a left-invariant Finsler metric $F$ on an H-type nilpotent Lie group $N$ admissible if the full group of isometries $(N, F)$ is equal to the full group of isometries $A(N)$ (where $N$ is endowed with the left-invariant Riemannian metric induced by the inner product $\langle \cdot, \cdot \rangle$ in (6.6). To prove our result, we need the following lemma.

**Lemma 6.5.** *Let $V$ be a real vector space and let $\langle , \rangle_1$, $\langle , \rangle_2$ be two inner products on $V$. Denote the groups of orthogonal transformations with respect to $\langle , \rangle_1$ and $\langle , \rangle_2$ by $O_1$ and $O_2$, respectively. Suppose $K$ is a compact subgroup of $GL(V)$ that is contained both in $O_1$ and $O_2$. Then there is a linear transformation $g \in GL(V)$ such that $g O_1 g^{-1} = O_2$ and $gkg^{-1} = k, \forall k \in K$.*

*Proof.* The proof uses some elementary results from representation theory. Since $K$ is compact and $\langle , \rangle_1$ is invariant under $K$, $V$ has a decomposition

$$V = V_1 \oplus V_2 \oplus \cdots \oplus V_m,$$

where the $V_i$ are irreducible subspaces of $K$ and the decomposition is orthogonal with respect to $\langle , \rangle_1$. Similarly, $V$ also has a decomposition

$$V = V_1' \oplus V_2' \oplus \cdots \oplus V_{m'}',$$

where the $V_i'$ are irreducible subspaces of $K$ and the decomposition is orthogonal with respect to $\langle , \rangle_2$. By elementary representation theory (see [34]), $m = m'$ and we can assume that $V_i \equiv V_i'$ as $K$-modules. Let $g_i$ be an isomorphism of $K$-modules from $V_i$ to $V_i'$. Since $\langle , \rangle_1|_{V_i}$ is $K$-invariant, the inner product

$$\langle x, y \rangle' = \langle g_i(x), g_i(y) \rangle_2, \quad x, y \in V_i,$$

is $K$-invariant. Since $V_i$ is $K$-irreducible, by Schur's lemma there is a positive constant $c_i$ such that

$$\langle\,,\,\rangle' = c_i\langle\,,\,\rangle_1.$$

Without loss of generality we can assume that $c_i = 1$. Then we have

$$\langle g(x), g(y)\rangle' = \langle x, y\rangle_1, \quad \forall x, y \in V_i.$$

Now we define a linear transformation $g$ such that $g|_{V_i} = g_i$. Then it is easily seen that $g^{-1}O_1 g = O_2$. Moreover, for every $k \in K$, since $g_i$ is an automorphism between the $K$-modules $V_i$ and $V_i'$, we have

$$g_i(k|_{V_i}) = (k|_{V_i'})g_i.$$

Therefore $gkg^{-1} = k$. This completes the proof of the lemma. $\qquad\square$

**Theorem 6.15.** *An H-type Lie group $N$ admits admissible weakly symmetric non-Riemannian Finsler metrics if and only if it falls into the following cases:*

(i) $\dim\mathfrak{z} = 1, 2, 3$.
(ii) $\dim\mathfrak{z} = 5, 6, 7$ *and* $\dim\mathfrak{a} = 8$.
(iii) $\dim\mathfrak{z} = 7$, $\dim\mathfrak{a} = 16$ *and* $\mathfrak{a}$ *is isotypic.*

*Proof.* Bernt et al. proved in [27] that an H-type Lie group $N$ with the left-invariant Riemannian metric induced by the inner product on $\mathfrak{n}$ is weakly symmetric if and only if it falls into the above three cases. Hence we need only prove that the above Lie groups admit admissible weakly symmetric non-Riemannian Finsler metrics. Let $N$ be one of the H-type Lie groups in the above three cases. Then $N$ is weakly symmetric. From the proof of Theorem 6.10, we see that there exists a non-Euclidean Minkowski norm $F_1$ on $\mathfrak{n}$ that is invariant under the isotropy group $A_0(N)$. Let $L(F_1)$ be the group of linear isometries of $F$. Then $L(F_1)$ is a compact Lie group. Hence it preserves an inner product $\langle\,,\,\rangle'$ on $\mathfrak{n}$. Let $O$ and $O'$ be respectively the groups of orthogonal transformations of $\mathfrak{n}$ with respect to $\langle\,,\,\rangle$ and $\langle\,,\,\rangle'$. Then $A_0(N)$ can be viewed as a closed subgroup of $O$ as well as of $O'$. By Lemma 6.5, there exists an invertible linear transformation $\tau$ of $\mathfrak{n}$ such that $\tau O'\tau^{-1} = O$ and $\tau k\tau^{-1} = k$, $\forall k \in A_0(N)$. Now we define a Minkowski norm $F$ on $\mathfrak{n}$ by

$$F(X) = F_1(\tau(X)).$$

It is easily seen that $L(F) = \tau L(F_1)\tau^{-1}$. Since $L(F_1) \subset O'$, we have $L(F) \subset O$. Moreover, since $\tau k\tau^{-1} = k$, $\forall k \in A_0(N)$, we also have $A_0(N) \subset L(F)$. Now we extend $F$ to a left-invariant Finsler metric on $N$, still denoted by $F$. By Proposition 6.5, we have

$$I(N, F) = NI_0(N, F) \quad \text{(semiproduct)},$$

where $I_0(N,F)$ consists of the automorphisms of $N$ whose differential at the identity belongs to $L(F)$. Since $L(F) \subset O$, we have $I(N,F) \subset A(n)$. On the other hand, since $A_0(N) \subset L(F)$, we also have $A(n) \subset I(N,F)$. Therefore $I(N,F) = A(N)$. This means that $F$ is an admissible weakly symmetric non-Riemannian Finsler metric on $N$.  $\square$

## 6.8   Reversible Non-Berwald Finsler Spaces with Vanishing S-Curvature

In this section we will solve an open problem posed by Shen in [146] concerning S-curvature. The problem can be stated as follows.

**Problem 6.1.** Is there any reversible non-Berwaldian Finsler space with vanishing S-curvature?

The reason to ask this question is that in [138], Shen proved that the Bishop–Gromov volume comparison theorem holds for a Finsler space with vanishing S-curvature. Therefore it is important to determine how large the class of Finsler spaces with vanishing S-curvature is. By Theorem 1.12, every Berwald space has vanishing S-curvature. It is also known that there are some non-Berwald spaces (e.g., some Randers spaces) whose S-curvature vanishes. However, all the known examples of such spaces are not reversible. Therefore it is natural to ask whether there is a reversible non-Berwald space with vanishing S-curvature.

In this section we shall show that many of the non-Riemannian weakly symmetric Finsler metrics we have constructed in the previous sections are non-Berwaldian. Since a weakly symmetric Finsler space must be reversible and with vanishing S-curvature, this presents a large number of examples that give a positive solution to Z. Shen's problem. First we prove the following result.

**Theorem 6.16.** *Let $M$ be a connected simply connected manifold and $F$ a weakly symmetric Finsler metric on $M$. If $(M,F)$ is a Berwald space, then $(M,F)$ can be decomposed into a Berwald product as*

$$(M,F) = (\mathbb{R}^m, F_0) \times (M_1, Q_1) \times (M_s, Q_s) \times (N_1, F_1) \times \cdots \times (N_t, F_t),$$

*where $(\mathbb{R}^m, F_0)$ is a reversible Minkowski space, $(M_i, Q_i)$ $(i = 1, \ldots, s)$ are holonomy irreducible weakly symmetric Riemannian manifolds, and $(N_j, F_j)$ $(j = 1, \ldots, t)$ are holonomy irreducible globally symmetric non-Riemannian Finsler spaces of rank $\geq 2$.*

*Proof.* Note that a weakly symmetric Finsler space must be complete. By the generalized de Rham decomposition theorem for Berwald spaces due to Szabó [152], $(M,F)$ can be decomposed into a Berwald product as

$$(M,F) = (\mathbb{R}^m, F_0) \times (M_1, Q_1) \times (M_s, Q_s) \times (N_1, F_1) \times \cdots \times (N_t, F_t), \qquad (6.8)$$

where $F_0$ is a Minkowski metric on $\mathbb{R}^m$, $(M_i, Q_i)$ $(i = 1, \ldots, s)$ are holonomy irreducible Riemannian manifolds, and $(N_j, F_j)$ $(j = 1, \ldots, t)$ are holonomy irreducible globally affine symmetric non-Riemannian Berwald spaces of rank $\geq 2$. Since a totally geodesic submanifold of a weakly symmetric space must also be weakly symmetric (see [180]), all the factors in (6.8) must be weakly symmetric. In particular, it is true that $F_0$ must be reversible, that $Q_i$ must be weakly symmetric, and that $F_j$ must be reversible. It is obvious that a reversible affine symmetric Berwald space must be globally symmetric. This proves the theorem. $\qquad \square$

**Theorem 6.17.** *Let $M$ be a connected simply connected compact manifold of dimension $\geq 2$ and suppose that every homogeneous Riemannian metric on it is holonomy irreducible. Then a weakly symmetric Finsler metric on $M$ is Berwaldian if and only if it is either Riemannian or globally symmetric. In particular, a weakly symmetric Finsler metric on a connected simply connected compact rank-one symmetric Riemannian manifold is Berwaldian if and only if it is Riemannian.*

*Proof.* For the first assertion we need only prove the "only if" part. Suppose $F$ is a Berwlad metric on $M$ that is also weakly symmetric. Then by Theorem 1.9, there exists a Riemannian metric $Q$ whose Levi-Civita connection coincides with the linear connection of $F$. Since $M$ is compact, the identity component $I_0(M, g)$ of the full group $I(M, Q)$ of isometries coincides with the identity component $A_0(M, Q)$ of the full group of affine transformations $A(M, Q)$ of the Levi-Civita connection of $(M, Q)$ (see [102, Vol. 1, p. 244]). Since $(M, F)$ is weakly symmetric, the full group of isometries $I(M, F)$ acts transitively on it. Hence the identity component $I_0(M, F)$ also acts transitively on $M$. By Theorem 5.1, an isometry of a Berwald space must be an affine transformation with respect to its linear transformation. Therefore we have $I_0(M, F) \subset A_0(M, Q) = I_0(M, Q)$. Hence $(M, Q)$ is a homogeneous Riemannian manifold. By the assumption, $(M, Q)$ must be holonomy irreducible. Then $(M, F)$ is holonomy irreducible. From this the first assertion follows directly from Theorem 6.9. Now we prove the second assertion. We need only prove that on a connected simply connected compact rank-one symmetric Riemannian manifold $M$ every homogeneous Riemannian metric must be holonomy irreducible. Suppose conversely that $Q$ is a homogeneous Riemannian metric on $M$ that is holonomy reducible. Then we have the de Rham decomposition

$$(M, Q) = (M_1, Q_1) \times \cdots \times (M_s, Q_s), \quad s \geq 2,$$

where $(M_i, Q_i)$ $(i = 1, \ldots, s)$ are irreducible. According to Hano's theorem [102], we have

$$I_0(M, Q) = I_0(M_1, Q_1) \times \cdots \times I_0(M, Q_s).$$

However, from the list of groups that admit transitive action on a compact rank-one Riemannian symmetric space (see [28, p. 179], and in particular Table 7.1 in this book), we can easily see that no such group can be decomposed as the product of two connected (nontrivial) subgroups. This is a contradiction. $\qquad \square$

**Corollary 6.1.** *On the real projective space $\mathbb{R}P^n$, a weakly symmetric Finsler metric is Berwaldian if and only if it is Riemannian.*

*Proof.* Let $\pi : S^n \to \mathbb{R}P^n$ be the covering projection. Suppose $F$ is a weakly symmetric Finsler space on $\mathbb{R}P^n$. Then $\pi^*F$ is a weakly symmetric Finsler metric on $S^n$. Since being a Berwald space is a local property, $\pi^*F$ is Berwaldian if and only if $F$ is Berwaldian. From this the corollary follows.

Combining the above results with Theorem 6.2, we get the following result.

**Proposition 6.6.** *Every weakly symmetric non-Riemannian Finsler metric on the odd-dimensional sphere $S^{2n-1}$ ($n \geq 2$), the odd-dimensional real projective space $\mathbb{R}P^{2n-1}$ ($n \geq 2$), and the $(4n-2)$-dimensional complex projective space $\mathbb{C}P^{2n-1}$ must be non-Berwaldian. In particular, on each of the above manifolds, there exist infinitely many Finsler metrics that are reversible, non-Berwaldian, and with vanishing S-curvature.*

Besides the above examples, there are still many examples that possess the above properties. In particular, we can prove that all the admissible weakly symmetric non-Riemanian Finsler metrics on H-type nilpotent Lie groups are of this type. We first prove the following.

**Proposition 6.7.** *Let $N$ ($\dim N \geq 2$) be a connected simply connected nilpotent Lie group that is indecomposable, i.e., $N$ cannot be written as the direct product of two nontrivial normal subgroups. Then a left-invariant Finsler metric $F$ on $N$ is Berwaldian if and only if it is Riemannian.*

*Proof.* Suppose conversely that there is a left-invariant Finsler metric $F$ on $N$ that is non-Riemannian but Berwaldian. Let $L$ be the group of linear isometries of the Minkowski space $(T_e(N), F)$, where $e$ is the unity element of $N$. Then $L$ is a compact Lie group. Therefore there exists an inner product on $T_e(N)$ that is invariant under the action of $L$. Let $Q$ be the left-invariant Riemannian metric on $N$ whose restriction to $T_e(N)$ is $\langle , \rangle$. We assert that $Q$ is affinely equivalent to $F$, or equivalently, the Levi-Civita connection of $Q$ coincides with the linear connection of $F$. Let $K$ be the holonomy group of the Berwald space. By Theorem 1.11, every parallel displacement of $F$ is a linear isometry. Thus $K \subset L$. On the other hand, for every piecewise smooth curve $c$ from $e$ to $g \in N$, let $P_c$ be the parallel displacement along $c$. Then we have

$$F(P_c(X)) = F(X), \quad \forall X \in T_e(N).$$

Moreover, $F(L_g^*(X)) = F(X)$. Thus $F((L_g^*)^{-1}P_c(X)) = F(X)$, that is, $(L_g^*)^{-1}P_c \in L$. Hence $P_c$ keeps the Riemannian metric $Q$ invariant. This proves the assertion. Next we assert that $Q$ is holonomy irreducible. For this we use again N. Wilson's result on the full group of isometries of $(N, Q)$. If $Q$ is holonomy reducible, then we have a de Rham decomposition

$$N = N_0 \times N_1 \times \cdots \times N_s,$$

where $N_0$ is a Euclidean space and $N_1, \ldots, N_s$, are connected simply connected holonomy irreducible Riemannian manifolds with dimension $\geq 1$. Then by J. Hano's theorem the identity component of the full group of $(N, Q)$ can be written as the product of the identity components of full groups of isometries of $N_i$ $(i = 0, 1, \ldots, s)$. From this one can easily deduce a contradiction to the assumption that $N$ is indecomposable. This proves our second assertion. Now $(N, Q)$ is a holonomy irreducible Riemannian manifold whose Levi-Civita connection coincides with the linear connection of a non-Riemannian Berwald space. Then by Theorem 1.10, $(M, Q)$ must be an (irreducible) globally symmetric Riemannian manifold of rank $\geq 2$. This implies that the sectional curvature of $(N, Q)$ is either everywhere $\geq 0$ (compact type) or everywhere $\leq 0$ (noncompact type). This contradicts again the result of J. Wolf [173] asserting that every left-invariant Riemannian metric on a nonabelian nilpotent Lie group must have tangent planes $\Pi_0$, $\Pi_1$, and $\Pi_{-1}$ such that $\Pi_0$ has sectional curvature 0, $\Pi_1$ has sectional curvature $> 0$, and $\Pi_{-1}$ has sectional curvature $< 0$, respectively. □

Now we conclude this chapter with the following corollary.

**Corollary 6.2.** *An admissible weakly symmetric Finsler metric on an H-type Lie group is Berwaldian if and only if it is Riemannian. In particular, on any of the H-type Lie groups in Theorem 6.15, there exist infinitely many left-invariant Finsler metrics that are reversible, non-Berwaldian, and with vanishing S-curvature.*

*Proof.* By Proposition 6.7, we need only prove that every H-type Lie group is indecomposable. For this it suffices to prove that an H-type Lie algebra cannot be written as the direct sum of two nontrivial ideals. Suppose conversely that an H-type Lie algebra $\mathfrak{n}$ has a decomposition

$$\mathfrak{n} = \mathfrak{n}_1 \oplus \mathfrak{n}_2,$$

where $\mathfrak{n}_1$ and $\mathfrak{n}_2$ are nontrivial ideals. Without loss of generality, we can assume that $\mathfrak{n}_1$ contains an element $u$ that is not in the center $\mathfrak{z}$ of $\mathfrak{n}$. Then it is easily seen (see [95]) that the map $\mathrm{ad}(u) : \mathfrak{n} \to \mathfrak{z}$ is surjective. This means that $\mathfrak{z} \subset \mathfrak{n}_1$. On the other hand, we can also select a nonzero element $v \in \mathfrak{n}_2$. Then $v \notin \mathfrak{z}$. The above argument then shows that we should also have $\mathfrak{z} \subset \mathfrak{n}_2$, which is a contradiction. □

# Chapter 7
# Homogeneous Randers Spaces

In this chapter we study homogeneous Randers spaces. Randers spaces were first introduced by Randers in 1941 [131], in his study of general relativity. Thus Randers metrics have important applications in the theory of relativity. Moreover, They occur naturally in other physical applications, especially in electron optics. In fact, as explained by Ingarden, the Lagrangian of the relativistic electrons gives rise to a Finsler metric of Randers type. Randers metrics are also used as unifying model for gravitation and electromagnetism; see [10, 93]. In geometry, Randers metrics provide a rich source of explicit examples of $y$-global Berwald spaces, particularly those that are neither Riemannian nor locally Minkowskian. Since they are most closely related to Riemannian metrics among the class of Finsler spaces, many new geometric invariants are first computed for them; see for example [17] for the computation of Laplacians of Randers metrics by Bao and Lackey.

In this chapter, we will present an exposition on recent results on homogeneous Randers spaces. This is the most fruitful subject in the very general field of homogeneous Finsler spaces, as can be expected from the definition of a Randers space. We now give an outline of the contents of the individual sections. In Sect. 7.1, we give a method to construct invariant Randers metrics on a coset space of a Lie group, through a thorough study of invariant vector fields on homogeneous manifolds. In Sect. 7.2, we use the formula of the Levi-Civita connection of an invariant Riemannian metric on a coset space to deduce an explicit simple formula for the S-curvature of invariant Randers metrics. In Sects. 7.3 and 7.4, we study homogeneous Einstein–Randers metrics. The main results are a rigidity result that every homogeneous Einstein–Randers metric with negative Ricci scalar must be Riemannian, and a complete classification of all homogeneous Einstein–Randers metrics on spheres. As a result, we find many new homogeneous Einstein metrics on spheres. In Sect. 7.5, we prove that a homogeneous Randers metric is Ricci quadratic if and only if it is Berwald. Finally, in Sects. 7.6 and 7.7, we study homogeneous Randers metrics with positive flag curvature or negative flag curvature. This results in an isometric classification of homogeneous Randers spaces with positive flag

S. Deng, *Homogeneous Finsler Spaces*, Springer Monographs in Mathematics,
DOI 10.1007/978-1-4614-4244-8_7, © Springer Science+Business Media New York 2012

curvature and almost isotropic S-curvature, and a rigidity result asserting that a homogeneous Randers metric with negative flag curvature and almost isotropic S-curvature must be Riemannian.

There are some problems deserving to be studied in the future. The classification of homogeneous Randers spaces of positive flag curvature, without the restriction of S-curvature, would be an important achievement. It would also be interesting to study other curvature tensors of homogeneous Randers spaces. For example, up to now, we have not obtained an explicit formula of the Landsberg tensor and Berwald tensor.

The standard references of this chapter are [48, 60, 64, 87–89, 147, 162, 168]; see [67, 120, 136, 145, 161] for further information on related topics.

## 7.1  General Description

Let $M$ be a smooth $n$-dimensional manifold. Recall that a Randers metric on $M$ consists of a Riemannian metric $\alpha = \sqrt{a_{ij}dx^i \otimes dx^j}$ on $M$ and a 1-form $\beta := b_i dx^i$. Using $\alpha$ and $\beta$ we define a function $F$ on $TM$:

$$F(x,y) = \alpha(x,y) + \beta(x,y), x \in M, y \in T_x(M),$$

where $F$ is a Finsler structure if and only if

$$\|\beta\| := \sqrt{b_i b^i} < 1, \tag{7.1}$$

where

$$b^i := a^{ij}b_j,$$

and $(a^{ij})$ is the inverse of the matrix $(a_{ij})$.

There is a global way to express a Randers metric on a Riemannian manifold, which is convenient when we consider such structures on homogeneous Riemannian manifolds. Let $x \in M$. Then the Riemannian metric induces an inner product in the cotangent space $T_x^*(M)$ in a standard way. An easy computation shows that $\langle dx^i, dx^j \rangle = a^{ij}$. This inner product defines a linear isomorphism between $T_x^*(M)$ and $T_x(M)$. Through this inner product the 1-form $\beta$ corresponds to a smooth vector field $U$ on $M$. Let

$$U = u^i \partial/\partial x_i.$$

Then we have

$$u^i = \sum_{j=1}^n a^{ij}b_j = b^i,$$

and for every $y \in T_x(M)$ we have

$$\langle y, U \rangle = \left\langle y, \left(\sum_{j=1}^n a^{ij}(x)b_j\right)\partial\Big/\partial x_i \right\rangle = b_i(x)y^i = \beta(x,y).$$

It is obvious that $\|\beta\| = \|U\|$. Thus (7.1) holds if and only if

$$\|U\| < 1.$$

Therefore we have the following result.

**Lemma 7.1.** *The Randers metric on a manifold consisting of a Riemannian metric*

$$\alpha = \sqrt{a_{ij}dx^i \otimes dx^j}$$

*together with a smooth vector field $U$ with $\alpha(U|_x) < 1, \forall x \in M$, is defined by*

$$F(x,y) = \alpha(y) + \langle U, y \rangle, x \in M, y \in T_x(M),$$

*where $\langle \, , \, \rangle$ is the inner product induced by the Riemannian metric $\alpha$.*

Recall that the Randers metric defined by Riemannian metric $\alpha$ and 1-form $\beta$ is of Berwald type if and only if $\beta$ is parallel with respect to $\alpha$. It is obvious that $\beta$ is parallel if and only if the corresponding vector field $U$ is parallel with respect to $\alpha$. Therefore we have the following.

**Lemma 7.2.** *Let $F$ be a Randers metric on $M$ defined by the Riemannian metric $\alpha$ and the vector field $U$. Then $(M,F)$ is a Berwald space if and only if $U$ is parallel with respect to $\alpha$.*

Now we consider the group of isometries of Randers metrics.

**Proposition 7.1.** *Let $(M,F)$ be a Randers space with $F$ defined by the Riemannian metric $\alpha$ and the vector field $U$. Then the group of isometries of $(M,F)$ is a closed subgroup of the group of isometries of the Riemannian manifold $(M,\alpha)$.*

*Proof.* Let $\phi$ be an isometry of $(M,F)$. Let $p \in M$ and set $q = \phi(p)$. For every $y \in T_p(M)$ we have

$$F(p,y) = \alpha(p,y) + \langle U|_p, y \rangle = F(q, d\phi_p(y))$$
$$= \alpha(q, d\phi_p(y)) + \langle U|_q, d\phi_p(y) \rangle. \tag{7.2}$$

Substituting $y$ with $-y$ in (7.2), we get

$$\alpha(p,y) - \langle U|_p, y \rangle = \alpha(q, d\phi_p(y)) - \langle U|_q, d\phi_p(y) \rangle. \tag{7.3}$$

Taking the sum of (7.2) and (7.3), we get

$$\alpha(p,y) = \alpha(q, d\phi_p(y)), \quad \langle U|_p, y \rangle = \langle U|_q, d\phi_p(y) \rangle.$$

Thus $\phi$ is an isometry with respect to the underlying Riemannian metric $\alpha$ and for every $p \in M$, we have $d\phi_p(U|_p) = U|_{\phi(p)}$. Therefore $I(M,F)$ is a closed subgroup of $I(M,\alpha)$. $\qquad\square$

Now we turn to homogeneous Randers spaces. A Randers space $(M, F)$ defined by a Riemannian metric $\alpha$ and a 1-form $\beta$ with $\|\beta\| < 1$, or equivalently, a smooth vector field $U$ with $\|U\| < 1$, is called homogeneous if its full group of isometries $I(M, F)$ acts transitively on $M$. By Proposition 7.1, if $(M, F)$ is a homogeneous Randers space, then the Riemannian manifold $(M, \alpha)$ is necessarily homogeneous. Since the isotropy subgroup $H$ of $I(M, F)$ at $x$ is compact, $M$ can be written as a coset space $M = I(M, F)/H$ with compact $H$. Then $I(M, F)/H$ is a reductive homogeneous manifold. So we just need to consider invariant Randers metrics on reductive homogeneous manifolds.

Let $G/H$ be a reductive homogeneous manifold. Let $\mathfrak{g} = \mathrm{Lie}\, G$, $\mathfrak{h} = \mathrm{Lie}\, H$. Fix a reductive decomposition of $\mathfrak{g}$:

$$\mathfrak{g} = \mathfrak{h} + \mathfrak{m} \quad \text{(direct sum of subspace)}, \tag{7.4}$$

where $\mathfrak{m}$ is a subspace of $\mathfrak{g}$ with

$$\mathrm{Ad}(h)\mathfrak{m} \subset \mathfrak{m}, \quad \forall h \in H.$$

By the proof of Proposition 7.1, to construct invariant Randers metrics on $G/H$, we first need to find $G$-invariant vector fields on $G/H$. The following proposition gives a complete description of invariant vector fields.

**Proposition 7.2.** *There exists a bijection between the set of invariant vector fields on $G/H$ and the subspace*

$$V = \{u \in \mathfrak{m} \mid \mathrm{Ad}(h)u = u, \forall h \in H\}.$$

*Proof.* Let $\pi : G \to G/H$ be the natural projection, and $L_g$ and $R_g$ the left and right translations of $G$ by $g$, respectively. The differential $\mathrm{d}\pi$ of the map $\pi$ maps $\mathfrak{g}$ onto the tangent space $T_o(G/H)$ of $G/H$ at the origin $o = \{H\}$. The kernel of $\mathrm{d}\pi$ is $\mathfrak{h}$. The translation $\tau(g) : xH \to gxH$ satisfies

$$\pi \circ Lg = \tau(g) \circ \pi.$$

Moreover, for $h \in H$, we have $\pi \circ R_h = \pi$ and $\mathrm{Ad}(g)u = \mathrm{d}R_{g^{-1}} \circ \mathrm{d}L_g(u)$. Therefore

$$\mathrm{d}\pi \circ \mathrm{Ad}(h)u = \mathrm{d}\tau(h)_o \circ \mathrm{d}\pi(u), \quad u \in \mathfrak{g}.$$

Thus under the isomorphism $\mathfrak{g}/\mathfrak{h} \simeq T_o(G/H)$ the linear transformation $\mathrm{Ad}(h)$ of $\mathfrak{g}/\mathfrak{h}$ corresponds to the linear transformation $\mathrm{d}\tau(h)_o$ of $T_o(G/H)$.

The decomposition $\mathfrak{g} = \mathfrak{h} + \mathfrak{m}$ gives a natural isomorphism

$$\mathfrak{g}/\mathfrak{h} \simeq \mathfrak{m}.$$

Under this isomorphism the linear transformation $\mathrm{d}\tau(h)_o$ of $T_o(G/H)$ corresponds to the linear transformation $\mathrm{Ad}(h)$ of $\mathfrak{m}$.

Now given $u \in V$, let $u_o$ be its image under the isomorphism $\mathfrak{m} \simeq T_o(G/H)$. For $g \in G$, define a tangent vector $U$ at $gH$ by

$$U_{gH} = \mathrm{d}(\tau(g))_o(u_o).$$

If $g_1 H = gH$, then $g^{-1}g_1 \in H$. Since $\mathrm{Ad}(h)u = u, \forall h \in H$, the above argument shows that $\mathrm{d}\tau(g^{-1}g_1)_o u_o = u_o$. Thus $\mathrm{d}\tau(g)_o u_o = \mathrm{d}\tau(g_1)_o u_o$. Therefore $U$ is a well-defined vector field on $G/H$ and it is obviously invariant under the action of $G$. That the correspondence $u \to U$ is a bijection is easy to verify. □

*Remark 7.1.* In the following, we usually denote the invariant vector field generated by $u$ in Proposition 7.2 by $\tilde{u}$.

By Proposition 7.1, the underlying Riemannian metric of an invariant Randers metric on $G/H$ must be invariant. Therefore we first fix an invariant Riemannian metric $\alpha$ on $G/H$ and then consider the invariant Randers metrics on $G/H$ with the underlying Riemannian metric $\alpha$.

The invariant Riemannian metric $\alpha$ induces an inner product $\langle\,,\,\rangle$ on $\mathfrak{g}$ such that

$$\langle \mathrm{Ad}(h)w, \mathrm{Ad}(h)v \rangle = \langle w, v \rangle, \quad w, v \in \mathfrak{g}, \ h \in H. \tag{7.5}$$

The subspace $\mathfrak{m}$ in (7.4) can be taken to be the orthogonal complement of $\mathfrak{h}$ with respect to this inner product.

**Theorem 7.1.** *Let $\alpha$ be an invariant Riemannian metric on $G/H$ and $\mathfrak{m}$ the orthogonal complement of $\mathfrak{h}$ in $\mathfrak{g}$ with respect to the inner product induced on $\mathfrak{g}$ by $\alpha$. Then there exists a bijection between the set of invariant Randers metrics on $G/H$ with the underlying Riemannian metric $\alpha$ and the set*

$$V_1 = \{u \in \mathfrak{m} \mid \mathrm{Ad}(h)u = u, \langle u, u \rangle < 1, \forall h \in H\}.$$

*Proof.* Let $u \in V_1$. By Proposition 7.2, $u$ corresponds to an invariant vector field $\tilde{u}$ on $G/H$. Since $\tilde{u}$ is invariant under the action of $G$, we have

$$\alpha(gH)(\tilde{u}) = \alpha(H)(\tilde{u}) = \langle u, u \rangle < 1.$$

By Lemma 7.1 we can define a Randers metric $F_u$ on $G/H$ by

$$F_u(gH, y) = \alpha(gH)(\tilde{u}) + \langle \tilde{u}, y \rangle, \quad y \in T_{gH}(G/H).$$

Then $F_u$ is obviously invariant under the action of $G$. It is easily seen that the correspondence $u \to F_u$ is a bijection. □

Recall that a Randers metric can also be represented by the navigation data. From (1.11) it follows that if $F$ is a Randers metric on a coset space $G/H$ with navigation data $(h, W)$, then $F$ is invariant under $G$ if and only if both $h$ and $W$ are invariant under $G$.

## 7.2 S-Curvature

In this section we will deduce an explicit formula for the S-curvature of a homogeneous Randers space. In contrast to general formulas in Finsler geometry, which inevitably involve indices with respect to local coordinate systems, we can express the S-curvature of a homogeneous Randers space in terms of the Lie algebra structure and the metric. Such an approach is welcome in Finsler geometry.

We first need to deduce some results concerning the Levi-Civita connection of a homogeneous Riemannian manifold. Let $(G/H, \alpha)$ be a homogeneous Riemannian manifold. Then $G/H$ is a reductive homogeneous manifold, i.e., the Lie algebra of $G$ has a decomposition of the Lie algebra

$$\mathfrak{g} = \mathfrak{h} + \mathfrak{m} \tag{7.6}$$

such that $\mathrm{Ad}\,(h)(\mathfrak{m}) \subset \mathfrak{m}$, $\forall h \in H$. We can identify $\mathfrak{m}$ with the tangent space $T_o(G/H)$. Let $\langle\,,\,\rangle$ be the corresponding inner product on $\mathfrak{m}$.

In the literature, there are several versions of the formula for the connection for Killing vector fields. We will use the formulas in Theorem 2.24. Given $v \in \mathfrak{g}$, we can define a one-parameter transformation group $\varphi_t$, $t \in \mathbb{R}$, of $G/H$ by

$$\phi_t(gH) = (\exp(tv)g)H, \quad g \in G.$$

Then $\varphi_t$ generates a vector field on $G/H$. By the left-invariance of the metric, it is a Killing vector field (this is called the fundamental vector field generated by $v$ in [102]). We denote this vector field by $\hat{v}$ (note that this is different from the notation $\tilde{u}$ in Sect. 7.1). Then for every $v_1, v_2, w \in \mathfrak{m}$, we have

$$\langle \nabla_{\hat{v}_1} \hat{v}_2|_o, w \rangle = \frac{1}{2}(-\langle [v_1, v_2]_\mathfrak{m}, w \rangle + \langle [w, v_1]_\mathfrak{m}, v_2 \rangle + \langle [w, v_2]_\mathfrak{m}, v_1 \rangle), \tag{7.7}$$

where $o = H$ is the origin of the coset space and $[v_1, v_2]_\mathfrak{m}$ denotes the projection of $[v_1, v_2]$ to $\mathfrak{m}$ corresponding to the decomposition (7.6) (caution: for $v_1, v_2 \in \mathfrak{m}$, the Lie bracket $[v_1, v_2]_\mathfrak{m}$ is exactly the negative of the value of the vector field $[\hat{v}_1, \hat{v}_2]$ at the origin).

To apply the formula (7.7) to our study, we need to deduce some formula for the connection in a local coordinate system. Let $u_1, u_2, \ldots, u_n$ be an orthonormal basis of $\mathfrak{m}$ with respect to $\langle\,,\,\rangle$. Then there exists a neighborhood $N$ of $o$ in $G/H$ such that the map

$$(\exp x^1 u_1 \exp x^2 u_2 \cdots \exp x^n u_n)H \mapsto (x^1, x^2, \ldots, x^n) \tag{7.8}$$

defines a local coordinate system on $N$. Now we compute the coordinate vector fields $\frac{\partial}{\partial x^i}$. Let $gH = (x^1, x^2, \ldots, x^n) \in U$. Then

$$\frac{\partial}{\partial x^i}\big|_{gH} = \frac{d}{dt}(\exp x^1 u_1 \cdots \exp x^{i-1} u_{i-1} \exp(t+x^i)u_i \exp x^{i+1} u_{i+1} \cdots \exp x^n u_n \cdot H)\big|_{t=0}$$

$$= \frac{d}{dt}(\exp x^1 u_1 \cdots \exp x^{i-1} u_{i-1} \exp t u_i \exp -x^{i-1} u_{i-1} \cdots \exp -x^1 u_1 \cdot gH)\big|_{t=0}$$

$$= \frac{d}{dt}\exp t(e^{x^1 \mathrm{ad} u_1} \cdots e^{x^{i-1} \mathrm{ad} u_{i-1}}(u_i)) \cdot gH\big|_{t=0}.$$

Set

$$v_i = e^{x^1 \mathrm{ad} u_1} \cdots e^{x^{i-1} \mathrm{ad} u_{i-1}}(u_i).$$

We then have

$$\frac{\partial}{\partial x^i}\big|_{gH} = \hat{v}_i\big|_{gH}. \tag{7.9}$$

Next we compute the Levi-Civita connection of $\alpha$ under the above coordinate system on $N$. Let $u_{n+1}, \ldots, u_m$ be a basis of $\mathfrak{h}$. Then we can write

$$v_i = \sum_{j=1}^m f_i^j u_j,$$

where $f_i^j$, $j = 1, 2, \ldots, n$, are functions of $x^1, \ldots, x^{i-1}$. Hence

$$\frac{\partial}{\partial x^i} = \sum_{j=1}^m f_i^j \hat{u}_j.$$

Therefore we have

$$\nabla_{\frac{\partial}{\partial x^i}} \frac{\partial}{\partial x^j} = \nabla_{\frac{\partial}{\partial x^i}} \left(\sum_{l=1}^m f_j^l \hat{u}_l\right) = \sum_{l=1}^m \left(\frac{\partial f_j^l}{\partial x^i}\right) \hat{u}_l + \sum_{l=1}^m f_j^l \nabla_{\frac{\partial}{\partial x^i}} \hat{u}_l$$

$$= \sum_{l=1}^m \left(\frac{\partial f_j^l}{\partial x^i}\right) \hat{u}_l + \sum_{k,l=1}^m f_i^k f_j^l \nabla_{\hat{u}_k} \hat{u}_l.$$

Since by the symmetry of the Levi-Civita connection we have

$$\nabla_{\frac{\partial}{\partial x^i}} \frac{\partial}{\partial x^j} - \nabla_{\frac{\partial}{\partial x^j}} \frac{\partial}{\partial x^i} = \left[\frac{\partial}{\partial x^i}, \frac{\partial}{\partial x^j}\right] = 0,$$

it suffices to compute $\nabla_{\frac{\partial}{\partial x^i}} \frac{\partial}{\partial x^j}$ for $i \geq j$. Since $f_j^l$ are functions of $x^1, \ldots, x^{j-1}$, we have $\frac{\partial f_j^l}{\partial x^i} = 0$, for $i \geq j$. Thus

$$\nabla_{\frac{\partial}{\partial x^i}} \frac{\partial}{\partial x^j} = \sum_{k,l=1}^m f_i^k f_j^l \nabla_{\hat{u}_k} \hat{u}_l, \quad i \geq j.$$

By the definition of $f_i^j$ we easily see that $f_i^j(0,0,\ldots,0) = \delta_{ij}$. Thus

$$\left(\nabla_{\frac{\partial}{\partial x^i}}\frac{\partial}{\partial x^j}\right)\Big|_o = (\nabla_{\hat{u}_i}\hat{u}_j)\Big|_o, \quad i \geq j.$$

Let $\Gamma_{ij}^k$ be the Christoffel symbols of the connection under the coordinate system, i.e.,

$$\nabla_{\frac{\partial}{\partial x^i}}\frac{\partial}{\partial x^j} = \Gamma_{ij}^k\frac{\partial}{\partial x^k}.$$

Then we have

$$\Gamma_{ij}^k(o)\frac{\partial}{\partial x^k}\Big|_o = (\nabla_{\hat{u}_i}\hat{u}_j)\Big|_o, \quad i \geq j.$$

By (7.9) we see that $\frac{\partial}{\partial x^k}\Big|_o = \hat{v}_k\Big|_o = u_k$. Thus,

$$\Gamma_{ij}^l(o) = \langle\Gamma_{ij}^k(o)u_k, u_l\rangle = \langle\nabla_{\hat{u}_i}\hat{u}_j, \hat{u}_l\rangle|_o, \quad i \geq j. \tag{7.10}$$

By (7.7), we have

$$\Gamma_{ij}^l(o) = \frac{1}{2}(-\langle[u_i, u_j]_\mathfrak{m}, u_l\rangle + \langle[u_l, u_i]_\mathfrak{m}, u_j\rangle + \langle[u_l, u_j]_\mathfrak{m}, u_i\rangle), \quad i \geq j. \tag{7.11}$$

Sometimes it is convenient to express formula (7.11) using the structure constants of the Lie algebra. For $1 \leq i, j \leq m$, let

$$[u_i, u_j] = \sum_{k=1}^m C_{ij}^k u_k.$$

The constants $C_{ij}^k$ are the structure constants of the Lie algebra $\mathfrak{g}$ with respect to the basis $u_1, \ldots, u_n, u_{n+1}, \ldots, u_m$. Using the structure constants, we can rewrite (7.11) as

$$\Gamma_{ij}^l = \Gamma_{ji}^l = \frac{1}{2}(-C_{ij}^l + C_{li}^j + C_{lj}^i), \quad i \geq j. \tag{7.12}$$

Now we give some applications of formulas (7.7)–(7.11) to Randers spaces. By Theorem 7.1, there is a one-to-one correspondence between the invariant Randers metrics on $G/H$ with the underlying Riemannian metric $\alpha$ and the $G$-invariant vector fields on $G/H$ with length $<1$. Furthermore, the $G$-invariant vector fields on $G/H$ are in one-to-one correspondence with the set

$$V = \{u \in \mathfrak{m} \mid \mathrm{Ad}(h)u = u, \forall h \in H\}.$$

Hence the invariant Randers metrics are in one-to-one correspondence with the set

$$V_1 = \{u \in V \mid \langle u, u\rangle < 1\}.$$

Let $u$ be a nonzero element in $V_1$. Select an orthonormal basis $u_1, u_2, \ldots, u_n$ of $\mathfrak{m}$ such that $u_n = u/|u|$ and define the local coordinate system as in (7.8). Since the vector field $\tilde{u}$ generated by $u$ is invariant under the action of $G$, we have

$$\tilde{u}|_{gH} = d\tau_g(u),$$

where $\tau_g$ is the diffeomorphism of $G/H$ defined by $g_1H \mapsto gg_1H$. Thus

$$\tilde{u}|_{gH} = \frac{d}{dt}(\tau_g(\exp(tu)H))|_{t=0}$$

$$= \frac{d}{dt}(\exp x^1 u_1 \exp x^2 u_2 \cdots \exp(x^n + ct)u_n)H|_{t=0} = c\frac{\partial}{\partial x^n}\Big|_{gH},$$

where $c = |u| < 1$.

Our first application is the following.

**Proposition 7.3.** *The Randers metric $F = \alpha + \beta$, generated by $\alpha$ and $u \in V_1$, is a Berwald metric if and only if*

$$\langle [u, v]_{\mathfrak{m}}, v \rangle = 0, \quad \langle [v, w]_{\mathfrak{m}}, u \rangle = 0, \quad \forall v, w \in \mathfrak{m}.$$

*Proof.* By Lemma 7.2, $F$ is of Berwald type if and only if the invariant vector field $\tilde{u}$ is parallel with respect to $\alpha$, i.e., $\Gamma_{ni}^l = \Gamma_{in}^l = 0$, for $i, l = 1, 2, \ldots, n$. By the invariance of $\tilde{u}$, it suffices to check at the origin $o$. By (7.11), this is equivalent to

$$-\langle [u_n, u_i]_{\mathfrak{m}}, u_l \rangle + \langle [u_l, u_i]_{\mathfrak{m}}, u_n \rangle + \langle [u_l, u_n]_{\mathfrak{m}}, u_i \rangle = 0, \quad i, l = 1, 2, \ldots, n. \quad (7.13)$$

Setting $i = l$ in (7.13), we get

$$\langle [u_n, u_i]_{\mathfrak{m}}, u_i \rangle = 0, \quad i = 1, 2, \ldots, n. \quad (7.14)$$

Now it is easily seen that (7.14) is equivalent to

$$\langle [u_n, u_i]_{\mathfrak{m}}, u_l \rangle + \langle [u_n, u_l]_{\mathfrak{m}}, u_i \rangle = 0, \quad i, l = 1, 2, \ldots, n. \quad (7.15)$$

Combining (7.13), (7.14), and (7.15), we complete the proof. $\qquad\qquad\square$

Now we are ready to compute the $S$-curvature of $F$. Since $(G/H, F)$ is homogeneous, we just need to compute at the origin $o = H$. Let $(N, (x^1, x^2, \ldots, x^n))$ be the local coordinate system defined in (7.8). Recall formula (1.10) for the S-curvature in local coordinate systems. If the Finsler metric is of Randers type, it is easy to deduce that (see [44])

$$S = (n+1)\left\{\frac{e_{00}}{2F} - (s_0 + \rho_0)\right\},$$

where $e_{00}, s_0$, and $\rho_0$ are defined in the following way:

I. $e_{00} = e_{ij} y^i y^j$, where $e_{ij} = r_{ij} + b_i s_j + b_j s_i$, $r_{ij} = \frac{1}{2}(b_{i;j} + b_{j;i})$ and the $b_i$ are defined by $\beta = b_i dx^i$. Furthermore, $s_i = b_j s^j{}_i$, and $s^i{}_j$ is defined by $s^i{}_j = a^{ih} s_{hj}$, where $s_{ij} = \frac{1}{2}(\frac{\partial b_i}{\partial x^j} - \frac{\partial b_j}{\partial x^i})$ and $(a^{kl})$ is the inverse matrix of $(a_{ij})$;

II. $s_0 = s_i y^i$;

III. $\rho_0 = \rho_{x^i} y^i$, where $\rho = \ln \sqrt{1 - \|\beta\|}$ and $\|\beta\|$ is the length of the form $\beta$ with respect to $\alpha$.

Now we compute the above quantities for a homogeneous Randers space defined by a Riemannian metric $\alpha$ in $G/H$ and $u \in \mathfrak{m}$ as above. The quantities of type (III) are easy. In fact, $\rho_{x^i} = 0$ for every $i$, since $\beta$, as an invariant form on $G/H$, has constant length. Therefore $\rho_0 = 0$. Next we compute $e_{00}$ and $s_0$.

First, since

$$b_i = \beta\left(\frac{\partial}{\partial x^i}\right) = \left\langle \tilde{u}, \frac{\partial}{\partial x^i} \right\rangle = c\left\langle \frac{\partial}{\partial x^n}, \frac{\partial}{\partial x^i} \right\rangle,$$

we have

$$\frac{\partial b_i}{\partial x^j} = c\frac{\partial}{\partial x^j}\left\langle \frac{\partial}{\partial x^n}, \frac{\partial}{\partial x^i} \right\rangle = c\left(\left\langle \nabla_{\frac{\partial}{\partial x^j}}\frac{\partial}{\partial x^n}, \frac{\partial}{\partial x^i} \right\rangle + \left\langle \frac{\partial}{\partial x^n}, \nabla_{\frac{\partial}{\partial x^j}}\frac{\partial}{\partial x^i} \right\rangle\right).$$

Hence at the origin we have (here we have used the symmetry of the connection: $\nabla_{\frac{\partial}{\partial x^j}}\frac{\partial}{\partial x^i} - \nabla_{\frac{\partial}{\partial x^i}}\frac{\partial}{\partial x^j} = [\frac{\partial}{\partial x^j}, \frac{\partial}{\partial x^i}] = 0$)

$$s_{ij}(o) = \frac{1}{2}c\left(\left\langle \nabla_{\frac{\partial}{\partial x^n}}\frac{\partial}{\partial x^j}, \frac{\partial}{\partial x^i} \right\rangle - \left\langle \nabla_{\frac{\partial}{\partial x^n}}\frac{\partial}{\partial x^i}, \frac{\partial}{\partial x^j} \right\rangle\right)\Big|_o.$$

By (7.7)–(7.11), we have

$$s_{ij}(o) = \frac{1}{2}c\langle [u_i, u_j]_{\mathfrak{m}}, u_n \rangle. \tag{7.16}$$

Since at the origin $(a_{ij}) = I_n$, we have

$$s^i{}_j(o) = a^{ik}(o)s_{kj}(o) = \sum_{k=1}^n \delta_{ik}s_{kj}(o) = s_{ij}(o),$$

therefore

$$s_i(o) = b_l(o)s^l{}_i(o) = cs^n{}_i(o) = cs_{ni}(o).$$

Thus for $y = y^i u_i \in \mathfrak{m}$, we have

$$s_0(y) = y^l s_l(o) = cy^l s_{nl}(o) = \frac{1}{2}c^2 y^l \langle [u_n, u_l]_{\mathfrak{m}}, u_n \rangle$$

$$= \frac{1}{2}\langle [cu_n, y^l u_l]_{\mathfrak{m}}, cu_n \rangle = \frac{1}{2}\langle [u, y]_{\mathfrak{m}}, u \rangle. \tag{7.17}$$

Next we compute $r_{ij}$. First suppose $i \geq j$. In this case we have

$$r_{ij}(o) = \frac{1}{2}(b_{i;j} + b_{j;i})\Big|_o = \frac{1}{2}\left(\frac{\partial b_i}{\partial x^j} - b_l\Gamma^l_{ji} + \frac{\partial b_j}{\partial x^i} - b_l\Gamma^l_{ij}\right)\Big|_o$$

$$= \frac{1}{2}\left(\frac{\partial b_i}{\partial x^j} + \frac{\partial b_j}{\partial x^i}\right)\Big|_o - c\Gamma^n_{ij}(o).$$

By (7.15) and (7.7)–(7.11) we have

$$\frac{1}{2}\left(\frac{\partial b_i}{\partial x^j} + \frac{\partial b_j}{\partial x^i}\right)\Big|_o = -\frac{1}{2}c\langle[u_i,u_j]_m,u_n\rangle, \quad i \geq j. \tag{7.18}$$

Combining (7.11) with (7.18) we get

$$r_{ij}(o) = -\frac{1}{2}c(\langle[u_n,u_i]_m,u_j\rangle + \langle[u_n,u_j]_m,u_i\rangle), \quad i \geq j. \tag{7.19}$$

Note that $r_{ij}$ is symmetric with respect to the indices $i, j$, and the right-hand side of (7.19) is also symmetric with respect to $i, j$. We conclude that (7.19) is also valid for $i \leq j$. On the other hand, a direct computation shows that

$$b_i s_j + b_j s_i|_o = \begin{cases} 0, & \text{for } 0 \leq i,j \leq n-1, \\ \frac{1}{2}c^3\langle[u_n,u_i]_m,u_n\rangle, & \text{for } 1 \leq i \leq n-1, j=n, \\ \frac{1}{2}c^3\langle[u_n,u_j]_m,u_n\rangle, & \text{for } i=n, 1 \leq j \leq n-1, \\ 0, & \text{for } i=j=n. \end{cases}$$

Consequently

$$e_{00}(y) = r_{ij}(o)y^i y^j + (b_i s_j + b_j s_i)|_o y^i y^j$$

$$= -\frac{1}{2}c(\langle[u_n,u_i]_m,u_j\rangle + \langle[u_n,u_j]_m,u_i\rangle)y^i y^j + c^3\langle[u_n,u_j]_m,u_n\rangle y^j y^n$$

$$= -\frac{1}{2}(\langle[cu_n,y^i u_i]_m,y^j u_j\rangle + \langle[cu_n,y^j u_j]_m,y^i u_i\rangle) + \langle[cu_n,u_j y^j]_m,cu_n\rangle cy^n$$

$$= -\langle[u,y]_m,y\rangle + \langle[u,y]_m,u\rangle\langle y,u\rangle$$

$$= \langle[u,y]_m,\langle y,u\rangle u - y\rangle,$$

where we have used the skew-symmetry of the Lie brackets, $[u_n,u_n] = 0$, and the facts that $cu_n = u$, $y^n = \langle$ and $u_n\rangle$. Finally, we obtain the formula for the S-curvature:

$$S(o,y) = (n+1)\left\{\frac{e_{00}(y)}{2F(y)} - (s_o(y) - \rho_0(y))\right\}$$

$$= \frac{n+1}{2}\left\{\frac{\langle[u,y]_m,\langle y,u\rangle u - y\rangle}{F(y)} - \langle[u,y]_m,u\rangle\right\}.$$

We summarize our computation as the following theorem.

**Theorem 7.2.** *Let $G/H$ be a homogeneous manifold and suppose the Lie algebra $\mathfrak{g}$ of $G$ has a reductive decomposition $\mathfrak{g} = \mathfrak{h} + \mathfrak{m}$ with $\mathrm{Ad}(h)\mathfrak{m} \subset \mathfrak{m}$. Let $F$ be an invariant Randers metric on $G/H$ defined by an invariant Riemannian metric $\alpha$ and an $H$-invariant vector $u$ in $\mathfrak{m}$. Then the S-curvature of $F$ is*

$$S(o,y) = \frac{n+1}{2}\left\{\frac{\langle[u,y]_\mathfrak{m},\langle y,u\rangle u - y\rangle}{F(y)} - \langle[u,y]_\mathfrak{m},u\rangle\right\}, \quad y \in \mathfrak{m},$$

*where $o = H$ is the origin of $G/H$, $\langle\,,\rangle$ is the inner product induced by $\alpha$, and we have identified the tangent space $T_o(G/H)$ with $\mathfrak{m}$.*

From the above formula, we can see that $S(u) = S(-u) = 0$. Therefore, for a homogeneous Randers space, the S-curvature has at least two (nonzero) zero points at any tangent space of the manifold.

Next we give some applications of Theorem 7.2. We begin with the following theorem.

**Theorem 7.3.** *Let $(G/H,F)$ be a homogeneous Randers space where $F$ is defined by a $G$-invariant Riemannian metric $\alpha$ and $0 \neq u \in V_1$ as in Theorem 7.2. Then $(M,F)$ has almost isotropic S-curvature if and only if it has vanishing S-curvature.*

*Proof.* We need prove only the "only if" part. Suppose that $F$ has almost isotropic S-curvature. Then there exist a closed 1-form $\eta$ on $G/H$ and a function $c(x)$ on $G/H$ such that

$$S(x,y) = (n+1)(c(x)F(y) + \eta(y)), \quad \forall y \in T(G/H).$$

In particular, at the origin $x = o$ we have

$$\frac{n+1}{2}\left\{\frac{\langle[u,y]_\mathfrak{m},\langle y,u\rangle u - y\rangle}{F(y)} - \langle[u,y]_\mathfrak{m},u\rangle\right\} = (n+1)(c(o)F(y) + \eta(y)), \quad (7.20)$$

for every $y \in \mathfrak{m}$. Considering the values at $y = u$ and $y = -u$ we get

$$c(o)F(u) + \eta(u) = 0, \quad c(o)F(-u) - \eta(u) = 0.$$

Thus $c(o)(F(u) + F(-u)) = 0$. Therefore we have $c(o) = 0$. Hence

$$\left\{\frac{\langle[u,y]_\mathfrak{m},\langle y,u\rangle u - y\rangle}{F(y)} - \langle[u,y]_\mathfrak{m},u\rangle\right\} = 2\eta(y). \quad (7.21)$$

Writing $y = y^i u_i$, where $u_i$ is the orthonormal basis of $\mathfrak{m}$ as above, we can rewrite (7.21) as

$$(2\eta(y) + \langle[u,y]_\mathfrak{m},u\rangle)\sqrt{\sum_{i=1}^{n}(y^i)^2}$$

$$= \langle[u,y]_\mathfrak{m},\langle y,u\rangle u - y\rangle - (2\eta(y) + \langle[u,y]_\mathfrak{m},u\rangle) \times \langle u,y\rangle.$$

Note that the right-hand side of the above equation is a polynomial in $y^i$. This can hold only when

$$2\eta(y) + \langle [u,y]_{\mathrm{m}}, u \rangle = 0.$$

Therefore we have

$$\langle [u,y]_{\mathrm{m}}, \langle y,u \rangle u - y \rangle = 0, \quad \forall y \in \mathrm{m}. \tag{7.22}$$

Now the subspace $\mathrm{m}$ has a decomposition

$$\mathrm{m} = L(u) + L(u)^{\perp},$$

where $L(u)$ is the span of $u$. By (7.22), for every $y_2 \in L(u)^{\perp}$ we have

$$\langle [u,y_2], y_2 \rangle = 0.$$

Hence for every $y = y_1 + y_2$, $y_1 \in L(u)$, $y_2 \in L(u)^{\perp}$ we have

$$\begin{aligned}
0 &= \langle [u,y]_{\mathrm{m}}, \langle y,u \rangle u - y \rangle = \langle [u,y_2]_{\mathrm{m}}, \langle y_1,u \rangle u - y_1 \rangle \\
&= \langle [u,y_2]_{\mathrm{m}}, \langle y_1,u \rangle u - \frac{\langle y_1,u \rangle}{\langle u,u \rangle} u \rangle = \langle y_1,u \rangle \times \left( 1 - \frac{1}{\langle u,u \rangle} \right) \times \langle [u,y_2]_{\mathrm{m}}, u \rangle.
\end{aligned}$$

Since $\langle u,u \rangle < 1$, the above equality implies that

$$\langle [u,y_2]_{\mathrm{m}}, u \rangle = 0, \quad \forall y_2 \in L(u)^{\perp},$$

or equivalently

$$\langle [u,y]_{\mathrm{m}}, u \rangle = 0, \quad \forall y \in \mathrm{m}. \tag{7.23}$$

Combining (7.22) and (7.23), we conclude that the S-curvature of $F$ vanishes at $o$. Since $(G/H, F)$ is homogeneous, the S-curvature vanishes everywhere. $\square$

For $y \in \mathrm{m}$, we define a linear transformation $\mathrm{ad}_{\mathrm{m}}(y)$ as

$$\mathrm{ad}_{\mathrm{m}}(y)(v) = [y,v]_{\mathrm{m}}.$$

The proof of Theorem 7.3 has the following corollary.

**Corollary 7.1.** *The homogeneous Randers space in Theorem 7.2 has almost isotropic S-curvature if and only if* $\mathrm{ad}_{\mathrm{m}}(u)$ *is skew-symmetric with respect to the inner product* $\langle, \rangle$. *In particular, it has vanishing S-curvature if and only if* $\mathrm{ad}_{\mathrm{m}}(u)$ *is skew-symmetric with respect to the inner product* $\langle, \rangle$.

*Proof.* If $\mathrm{ad}_{\mathrm{m}}(u)$ is skew-symmetric with respect to $\langle, \rangle$, then for every $y \in \mathrm{m}$, we have

$$\langle [u,y]_{\mathrm{m}}, y \rangle = 0,$$

and

$$\langle [u,y]_{\mathfrak{m}}, u \rangle = -\langle y, [u,u]_{\mathfrak{m}} \rangle = 0.$$

By Theorem 7.2, $F$ has vanishing S-curvature. On the other hand, if $F$ has almost isotropic S-curvature, then (7.22) and (7.23) hold. Thus

$$\langle [u,y]_{\mathfrak{m}}, y \rangle = 0,$$

for every $y \in \mathfrak{m}$. Hence $\mathrm{ad}_{\mathfrak{m}}(u)$ is skew-symmetric with respect to $\langle , \rangle$.  $\square$

As another application of the above formulas, we have the following result.

**Proposition 7.4.** *Let $(G/H, F)$ be a homogeneous Randers space defined by $\alpha$ and $u \neq 0$ as in Theorem 7.2. Then $F$ is of Douglas type if and only if*

$$\langle [v_1, v_2]_{\mathfrak{m}}, u \rangle = 0, \quad \forall v_1, v_2 \in \mathfrak{m}.$$

*Furthermore, if $F$ is of Douglas type and has almost isotropic S-curvature, then $F$ is of Berwald type.*

*Proof.* According to [16], $F$ is of Douglas type if and only if the corresponding 1-form $\beta$ is closed, i.e., $d\beta = 0$. Since $\beta$ is $G$-invariant, it suffices to prove that $d\beta|_o = 0$. Using the local coordinate of (7.8), we see that this is equivalent to

$$s_{ij}(0) = \frac{1}{2} \left( \frac{\partial b_i}{\partial x^j} - \frac{\partial b_j}{\partial x^i} \right) = 0, \quad \forall i, j.$$

By (7.16), this is equivalent to

$$\frac{1}{2} c \langle [u_i, u_j]_{\mathfrak{m}}, u_n \rangle = 0, \quad \forall i, j,$$

that is,

$$\langle [v_1, v_2]_{\mathfrak{m}}, u \rangle = 0, \quad \forall v_1, v_2 \in \mathfrak{m}.$$

This proves the first assertion. If $F$ has almost isotropic S-curvature, then by Corollary 7.1, we have
$$\langle [u,v]_{\mathfrak{m}}, v \rangle = 0.$$
By Proposition 7.3, if $F$ is in addition of Douglas type, then $F$ must be of Berwald type.  $\square$

Finally we consider the special case of left-invariant Randers metrics on a Lie group. Let $F = \alpha + \beta$ be a left-invariant Randers metric on a connected Lie group $G$. Then for every $g \in G$ and $y \in \mathfrak{g} = T_e(G)$, we have $F(dL_g(y)) = F(y)$, where $L_g$ is the left translation defined by $g$. Hence

$$\alpha(dL_g(y)) + \beta(dL_g(y)) = \alpha(y) + \beta(y). \tag{7.24}$$

Substituting $y$ with $-y$ in (7.24) we get

$$\alpha(\mathrm{d}L_g(y)) - \beta(\mathrm{d}L_g(y)) = \alpha(y) - \beta(y). \tag{7.25}$$

Combining (7.24) and (7.25) we have

$$\alpha(\mathrm{d}L_g(y)) = \alpha(y), \quad \beta(\mathrm{d}L_g(y)) = \beta(y).$$

This means that both the Riemannian metric $\alpha$ and the 1-form $\beta$ are invariant under the left translations of $G$. The left-invariant Riemannian metric corresponds to an inner product $\langle\,,\,\rangle$ on $\mathfrak{g}$ and the 1-from $\beta$ corresponds to a left-invariant vector field on $G$, i.e., an element of $\mathfrak{g}$, with length less than 1. On the other hand, every inner product on $\mathfrak{g}$ defines a left-invariant Riemannian metric on $G$, and this metric together with any element of $\mathfrak{g}$ of length less than 1 defines a left-invariant Randers metric on $\mathfrak{g}$. By Theorem 7.2 we have the following.

**Proposition 7.5.** *Let $G$ be an $n$-dimensional connected Lie group with Lie algebra $\mathfrak{g}$. Let $\langle\,,\,\rangle$ be an inner product on $\mathfrak{g}$ and $u \in \mathfrak{g}$ with $\langle u,u\rangle < 1$. Then the left-invariant Randers metric $F$ on $G$ defined by $\langle\,,\,\rangle$ and $u$ has S-curvature*

$$S(e,y) = \frac{n+1}{2}\left\{ \frac{\langle[u,y],\langle y,u\rangle u - y\rangle}{F(y)} - \langle[u,y],u\rangle \right\}.$$

*Here $F$ has almost isotropic S-curvature if and only if $F$ has vanishing S-curvature, if and only if the linear endomorphism $\mathrm{ad}(u)$ of $\mathfrak{g}$ is skew-symmetric with respect to the inner product $\langle\,,\,\rangle$.*

As an application, we classify left-invariant Randers metrics with almost isotropic S-curvature on a connected nilpotent Lie group.

**Proposition 7.6.** *Let $G$ be a connected nilpotent Lie group with Lie algebra $\mathfrak{g}$ and $\langle\,,\,\rangle$ an inner product on $\mathfrak{g}$. Suppose $u$ is an element in $\mathfrak{g}$ with $\langle u,u\rangle < 1$. Then the Randers metric constructed by $\langle\,,\,\rangle$ and $u$ has almost isotropic S-curvature (and hence vanishing S-curvature) if and only if $u \in C(\mathfrak{g})$, where $C(\mathfrak{g})$ denotes the center of $\mathfrak{g}$.*

*Proof.* If $u$ lies in the center of $\mathfrak{g}$, then $\mathrm{ad}(u) = 0$. Hence it is skew-symmetric with respect to $\langle\,,\,\rangle$. By Proposition 7.5, the corresponding Randers metric $F$ has vanishing S-curvature. On the other hand, if $F$ has vanishing S-curvature, then $\mathrm{ad}(u)$ is skew-symmetric with respect to $\langle\,,\,\rangle$. Thus $\mathrm{ad}(u)$ is a (complex) semisimple endomorphism of $\mathfrak{g}$. On the other hand, since the Lie algebra $\mathfrak{g}$ is nilpotent, Engel's theorem implies that $\mathrm{ad}(u)$ is a nilpotent endomorphism of $\mathfrak{g}$. This forces $\mathrm{ad}(u) = 0$. Hence $u \in C(\mathfrak{g})$. $\qquad\square$

Since a nonzero nilpotent Lie algebra has nonzero center, Proposition 7.6 implies that for every inner product on the Lie algebra $\mathfrak{g}$, there exists $u$ in $\mathfrak{g}$ such that $F$ has vanishing S-curvature. However, such a Randers metric may not be of Berwald type, as indicated by the next examples.

*Example 7.1.* A real Lie algebra $\mathfrak{n}$ is called two-step nilpotent if $[\mathfrak{n},\mathfrak{n}] \neq 0$ and $[[\mathfrak{n},\mathfrak{n}],\mathfrak{n}] = 0$. A two-step nilpotent Lie algebra is called nonsingular if for every $x \in \mathfrak{n}\backslash\mathfrak{z}(\mathfrak{n})$, where $\mathfrak{z}(\mathfrak{n})$ denotes the center of $\mathfrak{n}$, the linear map $\mathrm{ad}(x) : \mathfrak{n} \to \mathfrak{z}(\mathfrak{n})$ defined by $\mathrm{ad}(x)(y) = [x,y]$ is surjective. Suppose $N$ is a Lie group with Lie algebra $\mathfrak{n}$ that is nonsingular two-step nilpotent. Let $\langle\,,\,\rangle$ be an inner product on $\mathfrak{n}$ and $u \in \mathfrak{n}$ such that $\langle u,u \rangle < 1$. According to Proposition 7.6, the corresponding left-invariant Randers metric $F$ has almost isotropic (hence vanishing) S-curvature if and only if $u \in \mathfrak{z}(\mathfrak{n})$. Since $\mathfrak{n}$ is nonsingular, we have $[\mathfrak{n},\mathfrak{n}] = \mathfrak{z}(\mathfrak{n})$. Now suppose $F$ is in addition of Berwald type. Then by Proposition 7.3, we have $u \perp [\mathfrak{n},\mathfrak{n}] = \mathfrak{z}(\mathfrak{n})$. Hence $u = 0$. Thus on a nonsingular two-step nilpotent Lie group, there exists no left-invariant Randers metric that is non-Riemannian and of Berwald type. But there always exist non-Riemannian ones with vanishing S-curvature.

The most important nonsingular two-step nilpotent Lie algebras are the Heisenberg Lie algebras. Let $\mathfrak{n}$ be a real vector space with basis $x_1,\ldots,x_n$, $y_1,\ldots,y_n$, $z$, $n \geq 1$. Define brackets as follows:

$$[x_i,x_j] = [y_i,y_j] = 0, \quad [x_i,y_j] = \delta_{ij}z, \quad [x_i,z] = [y_j,z] = 0.$$

Then it is easily seen that $\mathfrak{n}$ is a two-step nilpotent Lie algebra with center spanned by $z$. This Lie algebra is called the Heisenberg Lie algebra. It is also easily seen that $\mathfrak{n}$ is nonsingular. Hence on a Lie group with Heisenberg Lie algebra (i.e., Heiserberg Lie group) there exists no non-Riemannian left-invariant Randers metric of Berwald type. However, given any inner product $\langle\,,\,\rangle$ on $\mathfrak{n}$, if we set

$$u = c\frac{z}{\sqrt{\langle z,z \rangle}}, \quad |c| < 1,$$

then the corresponding Randers metric has vanishing S-curvature.

*Example 7.2.* Let $G$ be a connected compact Lie group with Lie algebra $\mathfrak{g}$. Fix an $\mathrm{Ad}(G)$-invariant inner product $\langle\,,\,\rangle$ on $\mathfrak{g}$. The corresponding left-invariant Riemannian metric $Q$ is then bi-invariant under the action of $G$. Hence $(G,Q)$ is a globally symmetric Riemannian space. Since $\langle\,,\,\rangle$ is $\mathrm{Ad}(G)$-invariant, for every $u \in \mathfrak{g}$, $\mathrm{ad}(u)$ is skew-symmetric with respect to $\langle\,,\,\rangle$. Therefore, for every $u \in \mathfrak{g}$ with $\langle u,u \rangle < 1$, the left-invariant Randers metric $F$ constructed by $\langle\,,\,\rangle$ and $u$ has vanishing S-curvature. Therefore $F$ is of Berwald type if and only if $u \perp [\mathfrak{g},\mathfrak{g}]$. In particular, if $G$ is semisimple, then no such Randers metric can be non-Riemannian and of Berwald type. However, if $G$ is not semisimple, then $\mathfrak{g} \neq [\mathfrak{g},\mathfrak{g}]$. For every $u \in [\mathfrak{g},\mathfrak{g}]^{\perp}$, the corresponding Randers metric must be of Berwald type. This method provides a convenient way to construct globally defined Berwald spaces that are neither Riemannian nor locally Minkowskian (see [16] for details).

## 7.3 Homogeneous Einstein–Randers Spaces

We now turn to the study of homogeneous Einstein–Randers spaces. Recall that a Finsler space $(M, F)$ is called Einstein if its Ricci scalar is a scalar function $K(x)$ on $M$. In the case that $(M, F)$ is homogeneous, it is obvious that the function $K(x)$ is necessarily constant.

The main result of this section is the following:

**Theorem 7.4.** *A homogeneous Einstein–Randers space with negative Ricci curvature is Riemannian.*

Note that Bao and Robles have proved in [19] that a connected compact Einstein–Randers space with negative Ricci scalar must be Riemannian. However, the above theorem does not overlap this result, since by Theorem 2.27, a homogeneous Einstein–Riemannian manifold with negative Ricci curvature must be noncompact, and by Theorem 7.5 below, the underlying homogeneous Riemannian manifold (which is Einstein) of a homogeneous Einstein–Randers space must have negative Ricci curvature.

The remainder of this section is devoted to proving Theorem 7.4. Suppose $G/H$ is a reductive homogeneous manifold with an invariant Randers metric $F$ with navigation data $(h, W)$. As before, there is a decomposition of the Lie algebra:

$$\mathfrak{g} = \mathfrak{h} + \mathfrak{m} \quad \text{(direct sum of subspaces)},$$

where $\mathfrak{g}$, resp. $\mathfrak{h}$, is the Lie algebra of $G$, resp. $H$, and $\mathfrak{m}$ is a subspace of $\mathfrak{g}$ satisfying $\mathrm{Ad}(h)(\mathfrak{m}) \subset \mathfrak{m}$. We identify the tangent space $T_{eH}(G/H)$ of $G/H$ at the origin $eH$ with $\mathfrak{m}$, through the map $X \mapsto \frac{d}{dt}(\exp(tX)H)|_{t=0}$, $X \in \mathfrak{m}$. Under this identification, the isotropy representation of $H$ at $T_{eH}(G/H)$ corresponds to the adjoint representation of $H$ at $\mathfrak{m}$. Then $W$ corresponds to an $H$-fixed vector $w$ in $\mathfrak{m}$ with $h(w, w) < 1$.

Now we study the Killing vector fields of the invariant Riemannain metric on $G/H$. If $z \in \mathfrak{m}$ is an $H$-fixed vector in $\mathfrak{m}$ and $Z$ is the corresponding $G$-invariant vector field, then $Z$ is a Killing vector field if and only if the one-parameter transformation group

$$\varphi_t : G/H \to G/H, \quad gH \mapsto g\exp(tz)H, \quad t \in \mathbb{R},$$

consists of isometries of $h$. In particular, for every $z_1, z_2 \in \mathfrak{m}$, we have

$$h(z_1, z_2) = h(d\varphi_t(z_1), d\varphi_t(z_2)). \tag{7.26}$$

Now we compute $d(\varphi_t)(z_i)$. Since $z_i$ is the initial vector of the curve $(\exp(sz_i))H$ and

$$\varphi_t(\exp(sz_i)H) = \exp(sz_i)\exp(tX)H,$$

we have

$$d(\varphi_t)(z_i) = \frac{d}{ds}(\exp(sz_i)\exp(tX)H)|_{s=0}.$$

Now

$$\exp(sZ_i)\exp(tX)H = \exp(tX)\exp(-tX)\exp(sz_i)\exp(tX)H. \tag{7.27}$$

Note that

$$\exp(-tz)\exp(sz_i)\exp(tz) = \exp(\mathrm{Ad}(\exp(tz))(sz_i)) = \exp(se^{\mathrm{ad}(tz)}(z_i)).$$

Taking the derivative with respect to $s$ in (7.27), we get

$$d(\varphi_t)(z_i) = \frac{d}{ds}\left[\exp(tz)\exp(se^{\mathrm{ad}(tz)}(z_i))H\right]|_{s=0} = dL_{\exp tz}\left[e^{\mathrm{ad}(tz)}(z_i)\right]_{\mathfrak{m}}, \tag{7.28}$$

where $L_{\exp(tz)}$ is the transformation of $G/H$ defined by $gH \to \exp(tz)gH$. Since $h$ is $G$-invariant, $L_{\exp(tz)}$ are all isometries. Therefore (7.26) and (7.28) imply that

$$h(z_1, z_2) = h([e^{\mathrm{ad}(tz)}(z_1))]_{\mathfrak{m}}, [e^{\mathrm{ad}(tz)}(z_2))]_{\mathfrak{m}}), \quad \forall t \in \mathbb{R}.$$

Taking the derivative with respect to $t$ and considering the value at $t = 0$, we get

$$h([z,z_1]_{\mathfrak{m}}, z_2) + h(z_1, [z,z_2]_{\mathfrak{m}}) = 0, \quad \forall z_1, z_2 \in \mathfrak{m}. \tag{7.29}$$

Conversely, if (7.29) holds, then a backward argument implies that $Z$ is a Killing vector field of $h$.

Summarizing, we have proved the following result.

**Proposition 7.7.** *Suppose $G$ is a connected Lie group and $H$ is a closed subgroup such that $G/H$ is a reductive homogeneous space with a reductive decomposition $\mathfrak{g} = \mathfrak{h} + \mathfrak{m}$. Let $h$ be a $G$-invariant Riemannian metric on $G/H$ and suppose $z \in \mathfrak{m}$ is an $H$-fixed vector. Then the corresponding invariant vector field $Z$ on $G/H$ is a Killing vector field with respect to $h$ if and only if $z$ satisfies (7.29).*

Combining Theorem 1.16 with Proposition 7.7, we get a characterization of homogeneous Einstein–Randers spaces.

**Theorem 7.5.** *Let $G$ be a connected Lie group and $H$ a closed subgroup of $G$ such that $G/H$ is a reductive homogeneous space with a reductive decomposition $\mathfrak{g} = \mathfrak{h} + \mathfrak{m}$. Suppose $h$ is a $G$-invariant Riemannian metric on $G/H$ and $w \in \mathfrak{m}$ is an $H$-fixed vector with $h(w,w) < 1$. Let $W$ be the corresponding vector field on $G/H$. Then the Randers metric $F$ with navigation data $(h,W)$ is Einstein with Ricci constant $K$ if and only if $h$ is Einstein with Ricci constant $K$ and $w$ satisfies (7.29).*

*Proof.* The "if" part is obvious. Now we prove the "only if" part. Assume that $F$ is Einstein with Ricci constant $K$. If $K \geq 0$, then by Theorem 1.16, we see that $h$ must be Einstein with Ricci constant $K$ and the constant $\sigma$ there has to be 0. Hence $W$ must be a Killing vector field. If $K < 0$, then we have the following two cases:

1. $h$ is Ricci flat. If $W$ is not a Killing vector field, then $\sigma \neq 0$. Hence the Ricci constant of $F$ is $-\frac{1}{16}\sigma^2 < 0$. Now Theorem 2.26 asserts that a Ricci-flat homogeneous Riemannian manifold must be locally Euclidean. Thus $h$ is of constant sectional curvature 0. But then by Theorem 1.15, $F$ must be of constant flag curvature $-\frac{1}{16}\sigma^2$. Since a homogeneous Finsler space must be complete and its Cartan tensor is invariant, the norm of the tensors $A$ and $L$ must be bounded. Then the Akbar–Zadeh theorem (Theorem 1.14) implies that $F$ must be Riemannian. Hence $W = 0$. This is a contradiction. Hence $W$ must be a Killing vector field.

2. $h$ is not Ricci flat. Then Theorem 1.16 ensures that $\sigma = 0$. Hence $W$ must be a Killing vector field.

This completes the proof of the theorem.                                    □

Now we can give the proof of the main result of this section.

*Proof of Theorem 7.4.* Assume that $(M,F)$ is a homogeneous Einstein–Randers metric with negative Ricci curvature $K$. Then we can write $M$ as a coset space $G/H$ of a Lie group $G$ with $G/H$ reductive. Let $\mathfrak{g}, \mathfrak{h}$, $\mathfrak{m}$, $h$, $W$, $w$ be as above. Then by Theorem 1.16, $h$ is Einstein with Ricci constant $K$ and $w$ satisfies (7.29). Now we use the formula for Ricci curvature of homogeneous Riemannian manifolds. Let $x_1, x_2, \dots, x_n$ be an orthonormal basis of $\mathfrak{m}$ with respect to $h$ and suppose $U$ is the bilinear map from $\mathfrak{m} \times \mathfrak{m}$ to $\mathfrak{m}$ defined by

$$2h(U(z_1,z_2),z_3) = h([z_3,z_1]_{\mathfrak{m}},z_2) + h(z_1,[z_3,z_2]_{\mathfrak{m}}), \quad z_1,z_2,z_3 \in \mathfrak{m}.$$

Set

$$z = \sum_{i=1}^{n} U(z_i,z_i).$$

Then by Corollary 2.1, the Ricci curvature of $h$ can be expressed as

$$r(y,y) = -\frac{1}{2}\sum_{i=1}^{n} |[y,x_i]_{\mathfrak{m}}|^2 - \frac{1}{2}\sum_{i=1}^{n} h([y,[y,x_i]_{\mathfrak{m}}]_{\mathfrak{m}},x_i)$$

$$-\sum_{i=1}^{n} h([y,[y,x_i]_{\mathfrak{h}}]_{\mathfrak{m}},x_i) + \frac{1}{4}\sum_{i,j=1}^{n} h([x_i,x_j]_{\mathfrak{m}},y)^2$$

$$-h([z,y]_{\mathfrak{m}},y), \quad y \in \mathfrak{m}.$$

Now we consider the value at $w$. By (7.29) we have

$$h([w,[w,x_i]_{\mathfrak{m}}]_{\mathfrak{m}},x_i) = -h([w,x_i]_{\mathfrak{m}},[w,x_i]_{\mathfrak{m}}) = -|[w,x_i]_{\mathfrak{m}}|^2.$$

Therefore the first term and the second term of $r(w,w)$ sum to 0. Since $w$ commutes with the subalgebra $\mathfrak{h}$, the third term is equal to 0. Now by (7.29),

$$h([z,w]_{\mathfrak{m}},w) = -h(z,[w,w]_{\mathfrak{m}}) = 0.$$

Therefore we have

$$r(w,w) = \frac{1}{4} \sum_{i,j=1}^{n} h([x_i,x_j]_{\mathrm{m}},w)^2.$$

The Einstein condition thus implies that

$$\frac{1}{4} \sum_{i,j=1}^{n} h([x_i,x_j]_{\mathrm{m}},w)^2 = Kh(w,w).$$

Therefore, by the assumption $K < 0$, we have $w = 0$. Thus $F$ must be Riemannian and the theorem is proved. □

## 7.4 Homogeneous Einstein–Randers Metrics on Spheres

In this section we will describe homogeneous Einstein–Randers metrics on spheres and give the complete classification of them under isometries. The study of homogeneous Einstein–Riemannian manifolds is an important subject in differential geometry. However, up to now there have appeared only very few useful existence (or nonexistence) theorems in the literature. In the noncompact case, all the known examples of nonflat homogeneous Einstein–Riemannian manifolds can be viewed as left-invariant metrics on a solvable Lie group. However, up to now the conjecture of Alekseevskii has not been proved. So we still don't know whether there are examples of homogeneous Einstein–Riemannian metrics that cannot be realized as left-invariant Riemannian metrics on a solvable Lie group. In the compact case, the problem is also far from being fully understood; see [28] for a survey on this topic. In the generalized Finsler case, the problem is more complicated. A complete treatment seems to be unreachable. Our first goal is to get a classification of homogeneous Einstein–Randers spaces. In this section, we will perform this for Randers metrics on spheres.

We first set up the strategy of the classification. By Theorem 7.5, the classification of homogeneous Einstein–Randers metrics on a sphere $S^n$ can be reduced to the following two steps:

1. A classification of Einstein–Riemannian metrics on $S^n$ that are invariant under a transitive subgroup $G$ of transformations.
2. A complete description of $G$-invariant Killing vector fields on $S^n = G/H$, where $H$ is the isotropy subgroup of $G$ at a point.

The classification of Einstein–Riemannian metrics on $S^n$ was established by Ziller in [182]. Before stating Ziller's classification, we first study invariant vector fields on spheres when we write the spheres as coset spaces.

The compact Lie groups that admit an effective transitive action on the $n$-dimensional sphere $S^n$ have been classified by Montgomery–Samelson and Borel [31, 121]. The results are summarized in Table 7.1.

**Table 7.1**   Compact Lie groups acting transitively on spheres

| Spheres | $G$ | $H$ | Isotropy representation |
|---|---|---|---|
| $S^n$ | $SO(n+1)$ | $SO(n)$ | Irreducible |
| $S^{2n+1}$ | $SU(n+1)$ | $SU(n)$ | $\mathfrak{m} = \mathfrak{m}_0 \oplus \mathfrak{m}_1$ |
| $S^{2n+1}$ | $U(n+1)$ | $U(n)$ | $\mathfrak{m} = \mathfrak{m}_0 \oplus \mathfrak{m}_1$ |
| $S^{4n+3}$ | $Sp(n+1)$ | $Sp(n)$ | $\mathfrak{m} = \mathfrak{m}_0 \oplus \mathfrak{m}_1$ |
| $S^{4n+3}$ | $Sp(n+1)Sp(1)$ | $Sp(n)Sp(1)$ | $\mathfrak{m} = \mathfrak{m}_1 \oplus \mathfrak{m}_2$ |
| $S^{4n+3}$ | $Sp(n+1)U(1)$ | $Sp(n)U(1)$ | $\mathfrak{m} = \mathfrak{m}_0 \oplus \mathfrak{m}_1 \oplus \mathfrak{m}_2$ |
| $S^{15}$ | $Spin(9)$ | $Spin(7)$ | $\mathfrak{m} = \mathfrak{m}_1 \oplus \mathfrak{m}_2$ |
| $S^7$ | $Spin(7)$ | $G_2$ | Irreducible |
| $S^6$ | $G_2$ | $SU(3)$ | Irreducible |

Each transitive group yields a coset space diffeomorphic to $S^n$. To determine all the invariant vector fields on the spheres, we have to consider all such coset spaces. We will consider only the following two cases: $S^{2n+1} = SU(n+1)/SU(n)$; $S^{4n+3} = Sp(n+1)/Sp(n)$. Other cases can be treated similarly. Now we determine invariant vector fields in these cases.

*Case 1.* $S^{2n+1} = SU(n+1)/SU(n)$. It is a reductive homogeneous space and the isotropy subgroup is connected. Set $G_1 = SU(n+1)$, $H_1 = SU(n)$, $\mathfrak{g}_1 = \mathrm{Lie}\,G_1 = \mathfrak{su}(n+1)$, $\mathfrak{h}_1 = \mathrm{Lie}\,H_1 = \mathfrak{su}(n)$. The reductive decomposition can be taken as $\mathfrak{g}_1 = \mathfrak{h}_1 + \mathfrak{m}_1$, where $\mathfrak{m}_1$ is the direct sum of the following subspaces of $\mathfrak{g}$:

$$\mathfrak{m}_1' = \left\{ \begin{pmatrix} 0 & \alpha \\ -\bar{\alpha}' & 0 \end{pmatrix} \right\}, \mathfrak{m}_1'' = \left\{ \begin{pmatrix} -\frac{c\sqrt{-1}}{n}I_n & 0 \\ 0 & c\sqrt{-1} \end{pmatrix} \right\}, \qquad (7.30)$$

where $\alpha' = (x_1, x_2, \ldots, x_n) \in \mathbb{C}^n$, $c \in \mathbb{R}$, and $I_n$ is the identity matrix of order $n$. The isotropy representation keeps the above two subspaces invariant. The isotropy action of $SU(n)$ on $\mathfrak{m}_1''$ is trivial, and the action on $\mathfrak{m}_1'$ can be described as

$$\rho(A)(X) = Y,$$

where

$$X = \begin{pmatrix} 0 & \alpha \\ -\bar{\alpha}' & 0 \end{pmatrix} \in \mathfrak{m}_1', \quad Y = \begin{pmatrix} 0 & A\alpha \\ -(\overline{A\alpha})' & 0 \end{pmatrix}.$$

From this we conclude that the subspace $V$ of the invariant vectors of the $H_1$-action on $\mathfrak{m}_1$ is just $\mathfrak{m}_1''$.

*Case 2.* Let us we consider the reductive homogeneous spaces $S^{4n+3} = Sp(n+1)/Sp(n)$. The isotropy subgroup is also connected. Recall that the Lie algebra $\mathfrak{sp}(n)$ of $Sp(n)$ consists of the skew-Hermitian matrices in $\mathrm{End}(\mathbb{C}^{2n})$ of the form

$$\begin{pmatrix} A & -\bar{B} \\ B & \bar{A} \end{pmatrix}, \quad \bar{A}' = -A, \quad B' = B,$$

where $A$ and $B$ are $n \times n$ complex matrices. A reductive decomposition of the Lie algebra can be taken as $\mathfrak{sp}(n+1) = \mathfrak{sp}(n) + \mathfrak{m}_2$, where the subspace $\mathfrak{m}_2$ is defined by

$$\mathfrak{m}_2 = \begin{pmatrix} 0 & \alpha & 0 & -\overline{\beta} \\ -\overline{\alpha}' & c\sqrt{-1} & -\overline{\beta}' & -a+b\sqrt{-1} \\ 0 & \beta & 0 & \overline{\alpha} \\ \beta' & a+b\sqrt{-1} & -\alpha' & -c\sqrt{-1} \end{pmatrix}, \tag{7.31}$$

where $\alpha, \beta \in \mathbb{C}^n$, and $a, b, c$ are real numbers. Now we determine the subspace $V$ of $\mathrm{Sp}(n)$-invariant vectors in $\mathfrak{m}_2$. Let

$$P_1 = \begin{pmatrix} 0 & \alpha & 0 & -\overline{\beta} \\ -\overline{\alpha}' & c\sqrt{-1} & -\overline{\beta}' & -a+b\sqrt{-1} \\ 0 & \beta & 0 & \overline{\alpha} \\ \beta' & a+b\sqrt{-1} & -\alpha' & -c\sqrt{-1} \end{pmatrix} \in \mathfrak{m}_2$$

and

$$P_2 = \begin{pmatrix} A & 0 & -\overline{B} & 0 \\ 0 & 0 & 0 & 0 \\ B & 0 & \overline{A} & 0 \\ 0 & 0 & 0 & 0 \end{pmatrix} \in \mathfrak{sp}(n).$$

Then $P_1 \in V$ if and only if $P_1 P_2 = P_2 P_1$ for every skew-Hermitian matrix $A$ and symmetric matrix $B$. By a direct computation, it is easily seen that this is equivalent to the equations

$$A\alpha - \overline{B}\beta = 0, \quad \overline{A}\beta + B\alpha = 0.$$

Obviously, this is equivalent to the condition $\alpha = \beta = 0$. Therefore, for the reductive homogeneous spaces $S^{4n+3} = \mathrm{Sp}(n+1)/\mathrm{Sp}(n)$, the subspace $V$ is

$$\mathfrak{m}_2' = \left\{ \begin{pmatrix} 0 & 0 & 0 & 0 \\ 0 & c\sqrt{-1} & 0 & -a+b\sqrt{-1} \\ 0 & 0 & 0 & 0 \\ 0 & a+b\sqrt{-1} & 0 & -c\sqrt{-1} \end{pmatrix} \middle| a, b, c \in \mathbb{R} \right\}. \tag{7.32}$$

For convenience, we denote the orthogonal complement of $\mathfrak{m}_2'$ in $\mathfrak{m}_2$ (with respect to $-B$, where $B$ is the Killing form of $\mathfrak{sp}(n+1)$) by $\mathfrak{m}_2''$. A direct computation then shows that

$$\mathfrak{m}_2'' = \left\{ \begin{pmatrix} A & 0 & -\overline{B} & 0 \\ 0 & 0 & 0 & 0 \\ B & 0 & \overline{A} & 0 \\ 0 & 0 & 0 & 0 \end{pmatrix} \middle| A, B \in \mathbb{C}^n, \overline{A}' = -A, B' = B \right\}. \tag{7.33}$$

To summarize, we have proved the following theorem:

**Theorem 7.6.** *The set of* $SU(n+1)$*-invariant vector fields on the coset space* $S^{2n+1} = SU(n+1)/SU(n)$ *is a vector space of dimension 1. The set of* $Sp(n+1)$*-invariant vector fields on the coset space* $S^{4n+3} = Sp(n+1)/Sp(n)$ *is a vector space of dimension 3.*

*Remark 7.2.* A similar method can be used to study the isotropy representation of other coset spaces. The results are indicated in the last column of Table 7.1, where $\mathfrak{m}_0$ denotes the set of $H$-fixed vectors in $\mathfrak{m}$, and $\mathfrak{m}_i$, $i = 1, 2, \ldots$, denote the irreducible $H$-invariant subspaces, where the action of $H$ is nontrivial. This result will be useful in our study of homogeneous Randers spaces of positive flag curvature in Sect. 7.6.

Now we can state Ziller's classification of homogeneous Einstein–Riemannian metrics on spheres. This classification also presented some examples of homogeneous Einstein–Riemannian metrics that were new before the publication of [182].

**Lemma 7.3 (Ziller [182]).** *On $S^{2n}$, every homogeneous Einstein–Riemannian metric must be one of the standard ones. On $S^{2n+1}$ with n odd, every homogeneous Einstein–Riemannian metric must be one of the standard ones. If we write $S^{2n+1}$ as* $SU(n+1)/SU(n)$*, then the standard Riemannian metric with constant sectional curvature 1 can be written as*

$$h(X_1, X_2) = c_1 c_2 + \operatorname{Re} \alpha_1' \overline{\alpha}_2, \qquad (7.34)$$

*where*

$$X_i = \begin{pmatrix} -\dfrac{c_i\sqrt{-1}}{n}E & \alpha_i \\ -\overline{\alpha}_i' & c_i\sqrt{-1} \end{pmatrix} \in \mathfrak{m}_1.$$

*On $S^{4n+3}$ there are two homogeneous $Sp(n+1)$-invariant Einstein–Riemannian metrics:*

$$h_1 = -2B\big|_{\mathfrak{m}_2' \times \mathfrak{m}_2'} - B\big|_{\mathfrak{m}_2'' \times \mathfrak{m}_2''} \qquad (7.35)$$

*and*

$$h_2 = -B\big|_{\mathfrak{m}_2'' \times \mathfrak{m}_2''} - \frac{2}{2n+3}B\big|_{\mathfrak{m}_2' \times \mathfrak{m}_2'}, \qquad (7.36)$$

*where $\mathfrak{m}_2'$, $\mathfrak{m}_2''$ are defined in (7.32) and (7.33), and B is the Killing form of $\mathfrak{g}$. The metric $h_1$ is the standard metric of constant sectional curvature and $h_2$ is an Einstein metric of nonconstant sectional curvature.*

*Moreover, On $S^{15}$ there is a $\operatorname{spin}(15)$-invariant Einstein Riemannian metric, when we write $S^{15}$ as the coset space $\operatorname{spin}(9)/\operatorname{spin}(7)$.*

*Up to isometries and homotheties, the above are all the homogeneous Einstein–Riemannian metrics on spheres.*

*Remark 7.3.* The $\operatorname{spin}(9)$-invariant Einstein Riemannian metric was found by Bourguinon and Karcher, see [182]. However, from Table 7.1, one sees that the isotropy representation of the coset space $\mathrm{spin}(9)/\mathrm{spin}(7)$ admits no nonzero fixed vector. Therefore, this Einstein metric will not produce new Einstein-Randers metrics. Therefore we need not consider this metric in the following.

Now we describe homogeneous Einstein–Randers metrics on spheres. We first study the reductive homogeneous space $\mathrm{Sp}(n+1)/\mathrm{Sp}(n)$. The $\mathrm{Sp}(n+1)$-invariant vector fields have already been described in Theorem 7.6. Therefore we need only find out the homotheties among these vector fields. Since Ricci-flat homogeneous Riemannian manifolds must be flat, $h_2$ cannot be Ricci-flat. Therefore, homotheties of $h_2$ are just the Killing vector fields. So it suffices to determine the Killing vector fields among the invariant vector fields. By a direct computation, we find that all the invariant vector fields $W$ satisfy (7.29) for the two metrics above. Thus, by Theorem 7.5, $W$ is a Killing vector field. Using the Riemannian metric $h_1$ in (7.35), we obtain a family of $\mathrm{Sp}(n+1)$-invariant homogeneous Einstein–Randers metrics $F_1$ on $S^{4n+3}$:

$$F_1(y) = \frac{\sqrt{[h(W,y)]^2 + h(y,y)\lambda}}{\lambda} - \frac{h(W,y)}{\lambda}$$

$$= \frac{\sqrt{64(n+2)^2(aa_1 + bb_1)^2 + 8\lambda(n+2)(|\alpha|^2 + |\beta|^2 + a_1^2 + b_1^2)}}{\lambda}$$

$$- \frac{8(n+2)(aa_1 + bb_1)}{\lambda}, \tag{7.37}$$

where

$$y = \begin{pmatrix} 0 & \alpha & 0 & -\overline{\beta} \\ -\overline{\alpha}' & c_1\sqrt{-1} & -\overline{\beta}' & -a_1 + b_1\sqrt{-1} \\ 0 & \beta & 0 & \overline{\alpha} \\ \beta' & a_1 + b_1\sqrt{-1} & -\alpha' & -c_1\sqrt{-1} \end{pmatrix} \in \mathfrak{m}_2$$

and

$$W = \begin{pmatrix} 0 & 0 & 0 & 0 \\ 0 & c\sqrt{-1} & 0 & -a + b\sqrt{-1} \\ 0 & 0 & 0 & 0 \\ 0 & a + b\sqrt{-1} & 0 & -c\sqrt{-1} \end{pmatrix} \in \mathfrak{m}_2',$$

$\lambda = 1 - h_1(W,W) = 1 - 16(n+1)(a^2 + b^2 + c^2)$, $|\cdot|$ denotes the standard norm of $\mathbb{C}^n$, and $a^2 + b^2 + c^2 < \frac{1}{16(n+1)}$. This family of metrics are non-Riemannian if and only if $a^2 + b^2 + c^2 \neq 0$. Similarly, using the metric $h_2$ in (7.36), we obtain a family of $\mathrm{Sp}(n+1)$-invariant homogeneous Einstein–Randers metrics $F_2$ on $S^{4n+3}$ too:

$$F_2(y) = \frac{\sqrt{[h(W,y)]^2 + h(y,y)\lambda}}{\lambda} - \frac{h(W,y)}{\lambda}$$

$$= \frac{\sqrt{\frac{64(n+2)^2(aa_1+bb_1)^2}{(2n+3)^2} + 8\lambda(n+2)(|\alpha|^2+|\beta|^2) + \frac{8\lambda(n+2)(a_1^2+b_1^2)}{2n+3}}}{\lambda}$$

$$- \frac{8(n+2)(aa_1+bb_1)}{\lambda(2n+3)}, \tag{7.38}$$

where $y$ and $W$ are as above, $\lambda = 1 - h_2(W,W) = 1 - \frac{16(n+1)}{2n+3}(a^2+b^2+c^2)$, $|\cdot|$ denotes the standard norm of $\mathbb{C}^n$, and $a^2+b^2+c^2 < \frac{2n+3}{16(n+1)}$. This family of metrics are non-Riemannian if and only if $a^2+b^2+c^2 \neq 0$.

Next we consider the homogeneous manifold $S^{2n+1} = SU(n+1)/SU(n)$. It is known that the $SU(n+1)$-invariant Riemannian–Einstein metric is a metric of constant sectional curvature, and the metric $h$ can be written as (7.34). A direct computation (although somewhat tedious) shows that all the $SU(n+1)$-invariant vector fields corresponding to an element $W \in \mathfrak{m}_1''$ in (7.30) satisfy (7.29). Thus by Theorem 7.5, all the $SU(n+1)$-invariant vector fields are Killing vector fields. Therefore, we obtain a family of $SU(n+1)$-invariant homogeneous Einstein–Randers metrics $F$ on $S^{2n+1}$:

$$F(y) = \frac{\sqrt{[h(W,y)]^2+h(y,y)\lambda}}{\lambda} - \frac{h(W,y)}{\lambda} = \frac{\sqrt{c^2c_1^2+\lambda(c_1^2+|\alpha|^2)}}{\lambda} - \frac{cc_1}{\lambda}, \tag{7.39}$$

where $h$ is given by (7.34),

$$y = \begin{pmatrix} -\frac{c_1\sqrt{-1}}{n}E & \alpha \\ -\bar{\alpha}' & c_1\sqrt{-1} \end{pmatrix} \in \mathfrak{m}_1, \quad W = \begin{pmatrix} -\frac{c\sqrt{-1}}{n}E & 0 \\ 0 & c\sqrt{-1} \end{pmatrix} \in \mathfrak{m}_1'',$$

and $\lambda = 1 - h(W,W) = 1 - c^2$ with $c^2 < 1$. This family of metrics are non-Riemannian if and only if $c \neq 0$.

Now we can state our main theorem of this section:

**Theorem 7.7.** *Every homogeneous Einstein–Randers metric on the even-dimensional sphere $S^{2n}$ must be the standard Riemannian metric (up to a homothety). On $S^{2n+1}$ with $n$ even, up to homotheties there is one family of non-Riemannian homogeneous Einstein–Randers metrics. If we write $S^{2n+1}$ as $SU(n+1)/SU(n)$, then the metrics can be written as in (7.39). This family of metrics have constant flag curvature. On $S^{4n+3}$, up to homothety there are two families of homogeneous Einstein–Randers metrics. If we write $S^{4n+3}$ as $Sp(n+1)/Sp(n)$, then the first family can be written as in (7.37) and the second family can be written as in (7.38). Metrics in (7.37) have constant flag curvature and metrics in (7.38) have nonconstant flag curvature.*

*Proof.* Table 7.1 shows that if $n \neq 3$, then the only Lie group that has an effective transitive action on the even-dimensional sphere $S^{2n}$ is $SO(2n)$. The isotropy subgroup of SO(2n) at $p = (1,0,\ldots,0)$ is SO(2n − 1), and the pair $(SO(2n), SO(2n − 1))$ is an irreducible Riemannian symmetric pair of compact type. The isotropy representation has no nonzero fixed points. Therefore every SO(2n)-invariant Randers metric on $S^{2n}$ must be Riemannian. On the other hand, the sphere $S^6$ can also be written as $G_2/SU(3)$. But it is easily seen that the isotropy representation of this coset space is transitive on the unit sphere with respect to a certain $G_2$-invariant Riemannian metric. In particular, the isotropy representation must also be irreducible, and hence it has no fixed points. This proves the first assertion.

Next we consider $S^{2n+1}$ with $n$ even. Since the only SU(n + 1)-invariant Einstein–Riemannian metric on $S^{2n+1}$ is the standard one (up to a homothety), and the SU (n + 1)-invariant vector fields have been completely determined in Theorem 7.6, we get a family of homogeneous non-Riemannian Einstein–Randers metrics as described in (7.39). Since the underlying Riemannian metric has constant sectional curvature, these metrics have constant flag curvature. Now we prove that up to homotheties these are all the homogeneous non-Riemannian Einstein–Randers metrics on $S^{2n+1}$. Suppose $F$ is an arbitrary Randers metric with the above properties. Then by Lemma 7.3 the underlying Riemannian metric must be the standard one. Thus $F$ must be also of constant flag curvature. Note that in a homogeneous Randers metric the invariant vector field has constant length with respect to the Riemannian metric. By Bao–Robles–Shen's result in [21], up to isometry there is only one family of Randers metrics with constant flag curvature on $S^{2n+1}$ where the external influence vector field $W$ has constant length. This proves the second assertion. A similar argument can be used to prove the third assertion. Finally, $S^{15}$ can also be written as Spin(9)/Spin(7). However, as noted by Ziller [183], the isotropy representation of the reductive homogeneous space Spin(9)/Spin(7) has no nonzero fixed points. Therefore this coset space cannot produce any new non-Riemannian homogeneous Randers metrics. This completes the proof of the theorem.                                                                                    □

*Remark 7.4.* From the proof we see that the metrics in (7.37) and the metrics in (7.39) are isometric if we write $S^{4n+1} = SU(2n+1)/SU(2n) = Sp(n+1)/Sp(n)$. This family of metrics has also been described in a different way in Bao–Robles–Shen's article [21]. However, the merit of this new description is that we find that these metrics are homogeneous. The metrics in (7.38) are new and they are the only homogeneous Einstein–Randers metrics on spheres with nonconstant flag curvature.

Finally, we give a classification of the above-described Einstein–Randers metrics under isometries.

By Remark 7.4, the only metrics we need to study are the metrics $F_2$ on $S^{4n+1}$ when it is written as Sp(n + 1)/Sp(n). The problem here is thus to find the conditions such that the Randers metrics $F_2$ and $F_2'$, which solve Zermelo's navigation problem for the Riemannian metric $h_2$ under the external influences $W$ and $W'$, respectively, are isometric. Let $o$ denote the point $(1,0,\ldots,0) \in \mathbb{H}^{n+1}$. Then $T_o S^{4n+3} = \{(\alpha,\xi)' | \alpha + \overline{\alpha} = 0, \xi \in \mathbb{H}^n\}$. We can identify the isotropy subgroup Sp(n) at $o$ with a subgroup of Sp(n + 1) through the map

$$A \hookrightarrow \begin{pmatrix} 1 & 0 \\ 0 & A \end{pmatrix}, \quad A \in \mathrm{Sp}(n).$$

Suppose $(\alpha, \xi)'$ is the restriction of an $\mathrm{Sp}(n+1)$-invariant vector field at $o$. Let $A, B \in \mathrm{Sp}(n+1)$. If $Ao = Bo$, then $B^{-1}A \in \mathrm{Sp}(n)$. Since $(\alpha, \xi)'$ is invariant under $\mathrm{Sp}(n+1)$, we have $A_*(\alpha, \xi)' = A(\alpha, \xi)' = B(\alpha, \xi)' = B_*(\alpha, \xi)'$. This implies that $B^{-1}A(\alpha, \xi)' = (\alpha, \xi)'$. Hence $(\alpha, \xi)'$ is the restriction of an $\mathrm{Sp}(n+1)$-invariant vector field at $o$ if and only if $(\alpha, \xi)'$ is a fixed point of the linear isotropy representation of $\mathrm{Sp}(n)$, namely

$$\begin{pmatrix} 1 & 0 \\ 0 & A \end{pmatrix} \begin{pmatrix} \alpha \\ \xi \end{pmatrix} = \begin{pmatrix} \alpha \\ A\xi \end{pmatrix} = \begin{pmatrix} \alpha \\ \xi \end{pmatrix}, \quad \forall A \in \mathrm{Sp}(n).$$

So, $\xi = 0$ if and only if $(\alpha, \xi)'$ is the restriction of an $\mathrm{Sp}(n+1)$-invariant vector field at $o$.

Let $W$ and $W'$ be two $\mathrm{Sp}(n+1)$-invariant vector fields and

$$W|_o = (\alpha, 0)', \quad W'|_o = (\beta, 0)'.$$

Recall that the full isometric group of $\mathrm{Sp}(n+1)$-invariant metrics of $(S^{4n+3}, h_2)$ is $\mathrm{Sp}(n+1) \times \mathrm{Sp}(1)$, with $\mathrm{Sp}(1)$ acting by right multiplication (see [182]). Since $W$ and $W'$ are $\mathrm{Sp}(n+1)$-invariant, the only maps we can choose are elements from $\mathrm{Sp}(1)$. Let $\eta_0 \in \mathrm{Sp}(n+1)$, and let $L(\eta_0)$, $R(\eta)$ denote the left multiplication of $\eta_0$ and right multiplication of $\eta$, respectively. It is obvious that $L(\eta_0)R(\eta)x = R(\eta)L(\eta_0)x, \forall x \in S^{4n+3}$. Thus, $L(\eta_0)_*R(\eta)_*W = R(\eta)_*L(\eta_0)_*W = R(\eta)_*W$. This implies that $R(\eta)_*W$ is also an $\mathrm{Sp}(n+1)$-invariant vector field. Thus $R(\eta)_*W = W'$ if and only if $R(\eta)_*W|_{o\eta} = W'|_{o\eta}$. But $\overline{\eta} \oplus \mathrm{id}_n = \eta^{-1} \oplus \mathrm{id}_n \in \mathrm{Sp}(n+1)$ maps $o\eta$ to $o$ ($\mathrm{id}_n$ is the identity map of $\mathbb{H}^n$). Hence $R(\eta)_*W|_{o\eta} = W'|_{o\eta}$ if and only if $(L(\overline{\eta})_*R(\eta)_*W)|_o = W'|_o$, since $W$ and $W'$ are $\mathrm{Sp}(n+1)$-invariant. In other words, $\overline{\eta}\alpha\eta = \beta$ or $\alpha\eta = \eta\beta$. Let

$$\eta = x_0 + x_1\mathbf{i} + x_2\mathbf{j} + x_3\mathbf{k}, \quad \alpha = a_1\mathbf{i} + a_2\mathbf{j} + a_3\mathbf{k}, \quad \beta = b_1\mathbf{i} + b_2\mathbf{j} + b_3\mathbf{k},$$

where $\mathbf{i}, \mathbf{j}, \mathbf{k}$ are the standard imaginary quaternion units. Then the equality $\alpha\eta = \eta\beta$ is equivalent to the following system of linear equations:

$$\begin{cases} (a_1 - b_1)x_1 + (a_2 - b_2)x_2 + (a_3 - b_3)x_3 = 0, \\ -(a_1 - b_1)x_0 - (a_3 + b_3)x_2 + (a_2 + b_2)x_3 = 0, \\ -(a_2 - b_2)x_0 + (a_3 + b_3)x_1 - (a_1 + b_1)x_3 = 0, \\ -(a_3 - b_3)x_0 - (a_2 + b_2)x_1 + (a_1 + b_1)x_2 = 0. \end{cases}$$

Set

$$P = \begin{pmatrix} 0 & a_1 - b_1 & a_2 - b_2 & a_3 - b_3 \\ -(a_1 - b_1) & 0 & -(a_3 + b_3) & a_2 + b_2 \\ -(a_2 - b_2) & a_3 + b_3 & 0 & -(a_1 + b_1) \\ -(a_3 - b_3) & -(a_2 + b_2) & a_1 + b_1 & 0 \end{pmatrix}.$$

Then there exists an isometric map $\eta \in \mathrm{Sp}(1)$ satisfying $\alpha\eta = \eta\beta$ if and only if $\det P = 0$. By a direct computation, we obtain $\det P = (|\alpha|^2 - |\beta|^2)^2$, where $|\alpha|^2 = a_1^2 + a_2^2 + a_3^2$ and $|\beta|^2 = b_1^2 + b_2^2 + b_3^2$. So the above system of linear equations has nonzero solutions if and only if $|\alpha|^2 = |\beta|^2$. From (7.29), one easily sees that $h_2(W, W) = h_2(W', W')$ if and only if $|\alpha|^2 = |\beta|^2$. Suppose $\eta_0$ is such a nonzero solution and $|\eta_0| = 1$ (otherwise we can replace $\eta_0$ by $\frac{\eta_0}{|\eta_0|}$), i.e., $\eta_0 \in \mathrm{Sp}(1)$. Then $(\eta_0)_* W = W'$. Therefore, the Randers metrics $F_2$ and $F_2'$ that solve Zermelo's problem of navigation on the Riemannian manifold $(S^{4n+3}, h_2)$ under the external influences $W$ and $W'$, respectively, are isometric if and only if $h_2(W, W) = h_2(W', W')$.

This method can also be used to treat the the metrics in (7.37) and (7.39). However, since the metrics in (7.37) and (7.39) have constant flag curvature, the problem of classification for this metrics under isometries has already been settled in Bao–Robles–Shen's article [21]. The conclusion is that two metrics in (7.37) (resp. (7.39)) are isometric if and only if the corresponding invariant vector fields have the same length.

Let us summarize the above as the following theorem.

**Theorem 7.8.** *On $S^{4n+3} = \mathrm{Sp}(n+1)/\mathrm{Sp}(n)$, there are two families of homogeneous $\mathrm{Sp}(n+1)$-invariant Einstein–Randers metrics $F_1$ and $F_2$ given by (7.37) and (7.38) respectively that solve Zermelo's problem of navigation on the Riemannian manifold $(S^{4n+3}, h_i)$ under the external influence $W$. The metrics in the family $F_1$ are of constant flag curvature and those in the family $F_2$ are of nonconstant flag curvature. If $(S^{4n+3}, F_i)$ and $(S^{4n+3}, F_i')$ solve Zermelo's problem of navigation on the Riemannian manifold $(S^{4n+3}, h_i)$ under the external influences $W$ and $W'$, then the metrics are isometric if and only if $h_i(W, W) = h_i(W', W') (i = 1, 2)$. On $S^{2n+1} = \mathrm{SU}(n+1)/\mathrm{SU}(n)$ with $n$ even, there is only one family of homogeneous $\mathrm{SU}(n+1)$-invariant Einstein–Randers metrics $F$ given by (7.39) that solve Zermelo's problem of navigation on the Riemannian manifold $(S^{2n+1}, h)$ under the external influence $W$. These metrics are of constant flag curvature. Similarly, if $(S^{2n+1}, F)$ and $(S^{2n+1}, F')$ solve Zermelo's problem of navigation on the Riemannian manifold $(S^{2n+1}, h)$ under the external influences $W$ and $W'$, respectively, then they are isometric if and only if $h(W, W) = h(W', W')$.*

## 7.5  Homogeneous Ricci-Quadratic Randers Spaces

In this section we prove an interesting result on homogeneous Ricci-quadratic Randers spaces. A Finsler metric is said to be R-quadratic if its Riemann curvature is quadratic [44]. R-quadratic metrics were first introduced by Basco and Matsumoto [11]. They form a rich class of Finsler spaces. For example, all Berwald metrics are R-quadratic, and some non-Berwald R-quadratic Finsler metrics have been constructed in [21, 113]. There are many interesting publications related to this subject (see, for example, [139]).

A Finsler metric is called Ricci-quadratic if $\mathrm{Ric}(x, y)F^2(y)$ is quadratic in $y$. It is obvious that the notion of Ricci-quadratic metrics is weaker than that of R-quadratic metrics. Hence any R-quadratic Finsler space must be Ricci-quadratic. In particular, every Berwald space must be Ricci-quadratic. It has been shown that there are many non-Berwald spaces that are Ricci-quadratic. In general, it is quite difficult to characterize Ricci-quadratic metrics. Li and Shen considered the case of Randers metrics in [113] and obtained a characterization of Ricci-quadratic properties of such spaces, using some complicated calculations in local coordinate systems. Their results are rather complicated (see Theorem 7.10 below). In this section we shall prove the following result.

**Theorem 7.9.** *A homogeneous Randers space is Ricci-quadratic if and only if it is of Berwald type.*

We first use Killing vector fields to present some formulas about the Levi-Civita connection of homogeneous Riemannian manifolds. This is a continuation of the study of Sect. 7.2. We will follow the method and notation used there.

Let $(G/H, \alpha)$ be a homogeneous Riemannian manifold with a reductive decomposition $\mathfrak{g} = \mathfrak{h} + \mathfrak{m}$. As before, we identify $\mathfrak{m}$ with the tangent space $T_o(G/H)$ of the origin $o = H$. We shall use the notation $\langle\,,\rangle$ to denote the Riemannian metric on the manifold as well as its restriction to $\mathfrak{m}$. Note that it is an $\mathrm{Ad}(H)$-invariant inner product on $\mathfrak{m}$. Hence we have

$$\langle [x,u],v \rangle + \langle [x,v],u \rangle = 0, \quad \forall x \in \mathfrak{h}, \quad \forall u,v \in \mathfrak{m},$$

which is equivalent to

$$\langle [x,u],u \rangle = 0, \quad \forall x \in \mathfrak{h}, \quad \forall u \in \mathfrak{m}.$$

Given $v \in \mathfrak{g}$, the fundamental vector field $\hat{v}$ generated by $v$, i.e.,

$$\hat{v}_{gH} = \frac{d}{dt} \exp(tv)gH \Big|_{t=0}, \quad \forall g \in G,$$

is a Killing vector field.

Let $\hat{X}, \hat{Y}, \hat{Z}$ be Killing vector fields on $G/H$ and let $U,V$ be arbitrary smooth vector fields on $G/H$. Then we have (see [28] for the notation, in particular, pages 40, 182, 183)

$$[\hat{X},\hat{Y}] = -[X,Y]\hat{},\tag{7.40}$$

$$\hat{X}\langle U,V \rangle = \langle [\hat{X},U],V \rangle + \langle [\hat{X},V],U \rangle,\tag{7.41}$$

$$\langle \nabla_{\hat{X}}\hat{Y},\hat{Z} \rangle = -\frac{1}{2}\left( \langle [X,Y]\hat{},\hat{Z} \rangle + \langle [X,Z]\hat{},\hat{Y} \rangle + \langle [Y,Z]\hat{},\hat{X} \rangle \right).\tag{7.42}$$

We shall prove only (7.41). In fact, by (c) of Theorem 1.81 of [28], we have

$$\hat{X}\langle U,V \rangle = \langle \nabla_{\hat{X}}U,V \rangle + \langle U,\nabla_{\hat{X}}V \rangle$$
$$= \langle \nabla_U\hat{X},V \rangle + \langle [\hat{X},U],V \rangle + \langle U,\nabla_V\hat{X} \rangle + \langle U,[\hat{X},V] \rangle$$
$$= \langle [\hat{X},U],V \rangle + \langle [\hat{X},V],U \rangle.$$

Let $u_1, u_2, \ldots, u_n$ be an orthonormal basis of $\mathfrak{m}$ with respect to $\langle\,,\,\rangle$. We extend it to a basis $u_1, u_2, \ldots, u_m$ of $\mathfrak{g}$. As in (7.8), there is a local coordinate system on a neighborhood $U$ of $o$ defined by the map

$$(\exp(x^1 u_1)\exp(x^2 u_2)\cdots\exp(x^n u_n))H \to (x^1, x^2, \ldots, x^n).$$

Let $gH = (x^1, x^2, \ldots, x^n) \in U$. Then as in Sect. 7.2, we have

$$\left.\frac{\partial}{\partial x^i}\right|_{gH} = \frac{d}{dt}(\exp t e^{x^1 \mathrm{ad} u_1}\cdots e^{x^{i-1}\mathrm{ad} u_{i-1}}(u_i)\cdot gH)|_{t=0}.$$

Set

$$e^{x^1 \mathrm{ad} u_1}\cdots e^{x^{i-1}\mathrm{ad} u_{i-1}}(u_i) = f_i^a u_a. \tag{7.43}$$

We have

$$\left.\frac{\partial}{\partial x^i}\right|_{gH} = f_i^a \hat{u}_a|_{gH}.$$

*Remark 7.5.* In the following, the range of the indices are governed as follows:

- The indices $a, b, c, \ldots$ range from 1 to $m$.
- The indices $i, j, k, \ldots$ range from 1 to $n$.
- The indices $\lambda, \mu, \ldots$ range from $n+1$ to $m$.

Let $\Gamma_{ij}^l$ be the Christoffel symbols under the coordinate system, i.e.,

$$\nabla_{\frac{\partial}{\partial x^i}}\frac{\partial}{\partial x^j} = \Gamma_{ij}^k\frac{\partial}{\partial x^k}.$$

Then

$$\Gamma_{ij}^l\frac{\partial}{\partial x^l} = \nabla_{\frac{\partial}{\partial x^i}}\frac{\partial}{\partial x^j} = \frac{\partial f_j^a}{\partial x^i}\hat{u}_a + f_i^b f_j^a \nabla_{\hat{u}_b}\hat{u}_a. \tag{7.44}$$

From (7.43), we see that $f_i^a$ are functions of $x^1, \ldots, x^{i-1}$. Thus

$$\frac{\partial f_j^a}{\partial x^i} = 0, \quad i \geq j.$$

Therefore (7.44) becomes

$$\Gamma_{ij}^l\frac{\partial}{\partial x^l} = f_i^b f_j^a \nabla_{\hat{u}_b}\hat{u}_a, \quad i \geq j.$$

Differentiating the above equation with respect to $x_k$, we get

$$\frac{\partial \Gamma_{ij}^l}{\partial x^k}\frac{\partial}{\partial x^l} + \Gamma_{ij}^s \Gamma_{ks}^l\frac{\partial}{\partial x^l} = \frac{\partial f_i^b f_j^a}{\partial x^k}\nabla_{\hat{u}_b}\hat{u}_a + f_i^b f_j^a f_k^c \nabla_{\hat{u}_c}\nabla_{\hat{u}_b}\hat{u}_a, \quad i \geq j. \tag{7.45}$$

Differentiating (7.43) with respect to $x_k$ and letting $(x^1, \ldots, x^n) \to 0$, we obtain

$$\frac{\partial f_i^a}{\partial x^k}(0) = f(k,i)C_{ki}^a,$$

where $C_{ab}^c$ are the structure constants of $\mathfrak{g}$ and $f(k,l)$ is defined by

$$f(k,i) := \begin{cases} 1, & k < i, \\ 0, & k \geq i. \end{cases}$$

Considering the value at the origin $o$, we get the following result.

**Lemma 7.4.** *The following equalities hold:*

$$\Gamma_{ij}^l(o) = f(i,j)C_{ij}^l + \langle \nabla_{\hat{u}_i}\hat{u}_j, \hat{u}_l \rangle, \tag{7.46}$$

$$\left.\frac{\partial \Gamma_{ij}^l}{\partial x^k}\right|_o = -\Gamma_{ij}^s(\Gamma_{ks}^l + \langle \nabla_{\hat{u}_k}\hat{u}_l, \hat{u}_s \rangle) + f(k,j)C_{kj}^a \langle \nabla_{\hat{u}_i}\hat{u}_a, \hat{u}_l \rangle$$

$$+ f(k,i)C_{ki}^s \langle \nabla_{\hat{u}_s}\hat{u}_j, \hat{u}_l \rangle + \hat{u}_k \langle \nabla_{\hat{u}_i}\hat{u}_j, \hat{u}_l \rangle, \quad i \geq j. \tag{7.47}$$

*Proof.* From (7.43) we know that $f_i^a(0) = \delta_i^a$ and $\frac{\partial}{\partial x^k}|_o = \hat{u}_k|_o = u_k$. Thus, by (7.44),

$$\Gamma_{ij}^l(o) = \left\langle \Gamma_{ij}^s \frac{\partial}{\partial x^s}, \hat{u}_l \right\rangle \bigg|_o = \left\langle \frac{\partial f_j^a}{\partial x^i}\hat{u}_a + f_i^b f_j^a \nabla_{\hat{u}_b}\hat{u}_a, \hat{u}_l \right\rangle \bigg|_o.$$

From this (7.46) follows. On the other hand, by (7.45) we have

$$\left.\frac{\partial \Gamma_{ij}^l}{\partial x^k}\right|_o = -\Gamma_{ij}^s \Gamma_{ks}^l + f(k,j)C_{kj}^a \langle \nabla_{\hat{u}_i}\hat{u}_a, \hat{u}_l \rangle$$

$$+ f(k,i)C_{ki}^a \langle \nabla_{\hat{u}_a}\hat{u}_j, \hat{u}_l \rangle + \langle \nabla_{\hat{u}_k}\nabla_{\hat{u}_i}\hat{u}_j, \hat{u}_l \rangle, \quad i \geq j.$$

Considering the values at the origin, we easily deduce that

$$\langle \nabla_{\hat{u}_k}\nabla_{\hat{u}_i}\hat{u}_j, \hat{u}_l \rangle = \hat{u}_k \langle \nabla_{\hat{u}_i}\hat{u}_j, \hat{u}_l \rangle - \langle \nabla_{\hat{u}_i}\hat{u}_j, \nabla_{\hat{u}_k}\hat{u}_l \rangle,$$

that

$$f(k,i)C_{ki}^a \langle \nabla_{\hat{u}_a}\hat{u}_j, \hat{u}_l \rangle = f(k,i)C_{ki}^s \langle \nabla_{\hat{u}_s}\hat{u}_j, \hat{u}_l \rangle,$$

and that

$$\langle \nabla_{\hat{u}_i}\hat{u}_j, \nabla_{\hat{u}_k}\hat{u}_l \rangle = \Gamma_{ij}^s \langle \nabla_{\hat{u}_k}\hat{u}_l, \hat{u}_s \rangle, \quad i \geq j.$$

From the above four equations (7.47) follows. $\square$

To prove Theorem 7.9, we also need the following lemma.

**Lemma 7.5.** *For* $u_i, u_j, u_k, u_l \in \mathfrak{m}, u_\lambda \in \mathfrak{h}$, *we have*

$$\langle \nabla_{\hat{u}_i} \hat{u}_j, \hat{u}_l \rangle |_o = -\frac{1}{2}(C_{ij}^l + C_{il}^j + C_{jl}^i), \tag{7.48}$$

$$\langle \nabla_{\hat{u}_i} \hat{u}_\lambda, \hat{u}_j \rangle |_o = \langle [u_j, u_\lambda]_{\mathfrak{m}}, u_i \rangle = C_{j\lambda}^i, \tag{7.49}$$

$$\hat{u}_k \langle \nabla_{\hat{u}_i} \hat{u}_j, \hat{u}_l \rangle |_o = \frac{1}{2} \left( C_{ka}^l C_{ij}^a + C_{ka}^j C_{il}^a + C_{ka}^i C_{jl}^a + C_{ij}^s C_{kl}^t \delta_{st} + C_{il}^s C_{kj}^t \delta_{st} + C_{jl}^s C_{ki}^t \delta_{st} \right), \tag{7.50}$$

*where* $[u_j, u_\lambda]_{\mathfrak{m}}$ *denotes the projection of* $[u_j, u_\lambda]$ *to* $\mathfrak{m}$.

*Proof.* First, (7.48) is an alternative formulation of (7.42) at the origin in terms of the structure constants. Using the invariance of $\mathrm{ad}(u)_\lambda$, (7.49) can also be deduced from (7.42). Finally, by (7.42) we have

$$\hat{u}_k \langle \nabla_{\hat{u}_i} \hat{u}_j, \hat{u}_l \rangle = -\frac{1}{2} \hat{u}_k \left( \langle \widehat{[u_i, u_j]}, \hat{u}_l \rangle + \langle \widehat{[u_i, u_l]}, \hat{u}_j \rangle + \langle \widehat{[u_j, u_l]}, \hat{u}_i \rangle \right).$$

Considering the value at the origin $o$ and taking into account (7.40) and (7.41), we can deduce from the above equation that

$$\begin{aligned}
\hat{u}_k \langle \nabla_{\hat{u}_i} \hat{u}_j, \hat{u}_l \rangle |_o = \frac{1}{2} (&\langle \left[ u_k, [u_i, u_j] \right]_{\mathfrak{m}}, u_l \rangle + \langle \left[ u_k, [u_i, u_l] \right]_{\mathfrak{m}}, u_j \rangle \\
&+ \langle \left[ u_k, [u_j, u_l] \right]_{\mathfrak{m}}, u_i \rangle + \langle [u_i, u_j]_{\mathfrak{m}}, [u_k, u_l]_{\mathfrak{m}} \rangle \\
&+ \langle [u_i, u_l]_{\mathfrak{m}}, [u_k, u_j]_{\mathfrak{m}} \rangle + \langle [u_j, u_l]_{\mathfrak{m}}, [u_k, u_i]_{\mathfrak{m}} \rangle),
\end{aligned}$$

from which (7.50) follows.                                                                 $\square$

By the two lemmas above, at the origin $o$ we have

$$\Gamma_{ni}^j - \Gamma_{nj}^i = \langle \nabla_{\hat{u}_n} \hat{u}_i, \hat{u}_j \rangle - \langle \nabla_{\hat{u}_n} \hat{u}_j, \hat{u}_i \rangle = C_{ji}^n.$$

Now we recall some notation about Randers spaces used by Li and Shen in the study of Ricci-quadratic Randers spaces. Let

$$F = \alpha + \beta = \sqrt{a_{ij}(x) y^i y^j} + b_i(x) y^i$$

be a Randers metric. Fix a local coordinate system. Let $\nabla \beta = b_{i|j} y^i dx^j$ denote the covariant derivative of $\beta$ with respect to $\alpha$. Define

$$r_{ij} := \frac{1}{2}(b_{i|j} + b_{j|i}), \qquad s_{ij} := \frac{1}{2}(b_{i|j} - b_{j|i}), \qquad s_j := b^i s_{ij}, \qquad t_j := s_m s^m{}_j.$$

We use $a_{ij}$ to raise and lower the indices of tensors defined by $b_i$ and $b_{i|j}$. The index "0" means the contraction with $y^i$. For example, $s_0 = s_i y^i$ and $r_{00} = r_{ij} y^i y^j$, etc. For Ricci-quadratic metrics on Randers spaces, we have the following theorem

**Theorem 7.10 (Li–Shen [113]).** *Let* $F = \alpha + \beta$ *be a Randers metric on an n-dimensional manifold. Then it is Ricci-quadratic if and only if*

$$r_{00} + 2s_0\beta = 2\widetilde{c}(\alpha^2 - \beta^2), \tag{7.51}$$

$$s^k{}_{0|k} = (n-1)A_0, \tag{7.52}$$

*where* $\widetilde{c} = \widetilde{c}(x)$ *is a scalar function and* $A_k := 2\widetilde{c}s_k + \widetilde{c}^2 b_k + t_k + \frac{1}{2}\widetilde{c}_k$ *with* $\widetilde{c}_k = \frac{\partial \widetilde{c}}{\partial x^k}$.

Now we consider homogeneous Randers spaces. Let $(G/H, F)$ be a homogeneous Randers space and let $(U, (x^1, \ldots, x^n))$ be a local coordinate system as in Sect. 7.2. Suppose the Randers metric is generated by the pair $(\langle\,,\,\rangle, u)$ and denote by $\widetilde{u}$ the vector field on $G/H$ generated by $u$. Moreover, we can suppose $u = cu_n (c < 1)$. Then by (7.9), we have

$$\widetilde{u}\big|_{gH} = c\frac{\partial}{\partial x^n}\Big|_{gH}.$$

The quantities $b_i$, $\frac{\partial b_i}{\partial x^j}$, $b_{i|j}$, $r_{ij}$, $s_{ij}$, and $s_j$ were computed in Sect. 7.2. Now we prove the following lemma.

**Lemma 7.6.** *Let* $(G/H, F)$ *be a homogeneous Randers space as above. Then* (7.51) *implies that*

$$\langle [y, u]_{\mathrm{m}}, y \rangle = 0, \quad \forall y \in \mathrm{m}. \tag{7.53}$$

*Proof.* Considering the value at $o$, and using (7.46), (7.48), and the formulas in Sect. 7.2, we get

$$(r_{00} + 2s_0\beta)|_o = c\Gamma^0_{n0} + 2c\frac{c}{2}(\Gamma^n_{n0} - \Gamma^0_{nn})\langle u, y \rangle$$

$$= cC^0_{0n} + c^2 C^n_{n0}\langle u, y \rangle = \langle [y, u]_{\mathrm{m}}, y - \langle u, y \rangle u \rangle.$$

On the other hand, it is obvious that

$$2\widetilde{c}(\alpha^2 - \beta^2)|_o = 2\widetilde{c}(o)(\langle y, y \rangle - \langle u, y \rangle^2).$$

Plugging the above two equations into (7.51), we get that at the origin $o$,

$$\langle [y, u]_{\mathrm{m}}, y - \langle u, y \rangle u \rangle = 2\widetilde{c}(o)\left(\langle y, y \rangle - \langle u, y \rangle^2\right).$$

Setting $y = u$ and taking into account the fact that $\langle u, u \rangle < 1$, we get

$$\widetilde{c}(o) = 0.$$

Thus

$$\left\langle [y,u]_{\mathrm{m}}, y - \langle u,y \rangle u \right\rangle = 0.$$

Replacing $y$ by $y + u$ in the above equation yields

$$\left\langle [y,u]_{\mathrm{m}}, u + y - \langle u, u+y \rangle u \right\rangle = 0.$$

From the above two equations and the fact that $\langle u,u \rangle < 1$ we deduce that

$$\langle [y,u]_{\mathrm{m}}, u \rangle = 0, \quad \langle [y,u]_{\mathrm{m}}, y \rangle = 0.$$

This proves the lemma.                                 $\square$

Note that in the above lemma we also have

$$C_{ni}^{j} + C_{nj}^{i} = 0 = C_{ni}^{n}, \tag{7.54}$$

and $t_i(o) = s_m s^m{}_i = 0$. Moreover, we also have

$$s_i(o) = c s_{ni} = \frac{c^2}{2}(\Gamma_{ni}^{n} - \Gamma_{nn}^{i}) = \frac{1}{2}\langle [u,u_i]_{\mathrm{m}}, u \rangle = 0.$$

Now we can give the proof of Theorem 7.6. First, we need to perform some complicated computations. Set

$$\Upsilon(ljk) := \frac{\partial}{\partial x^k}(\Gamma_{nj}^{s} a_{sl})\Big|_{o}.$$

Plugging (7.46) and (7.47) into $\Upsilon(ljk)$ yields

$$\Upsilon(ljk) = \frac{\partial \Gamma_{nj}^{l}}{\partial x^k} + \Gamma_{nj}^{s}(\Gamma_{ks}^{l} + \Gamma_{kl}^{t}\delta_{ts}) \tag{7.55}$$

$$= f(k,l)C_{kl}^{s}\Gamma_{nj}^{t}\delta_{ts} + f(k,j)C_{kj}^{a}\langle \nabla_{\hat{u}_n}\hat{u}_a, \hat{u}_l \rangle + C_{kn}^{s}\langle \nabla_{\hat{u}_s}\hat{u}_j, \hat{u}_l \rangle + \hat{u}_k\langle \nabla_{\hat{u}_n}\hat{u}_j, \hat{u}_l \rangle. \tag{7.56}$$

By (7.46), (7.48), (7.49), (7.50), (7.54), and (7.55), we have

$$\Upsilon(000) = f(0,0)C_{00}^{s}\Gamma_{n0}^{t}\delta_{ts} + f(0,0)C_{00}^{s}\langle \nabla_{\hat{u}_n}\hat{u}_s, \hat{u}_0 \rangle + f(0,0)C_{00}^{\lambda}\langle \nabla_{\hat{u}_n}\hat{u}_{\lambda}, \hat{u}_0 \rangle$$

$$+ C_{0n}^{s}\langle \nabla_{\hat{u}_s}\hat{u}_0, \hat{u}_0 \rangle + \hat{u}_0\langle \nabla_{\hat{u}_n}\hat{u}_0, \hat{u}_0 \rangle$$

$$= f(0,0)C_{00}^{s}(\langle \nabla_{\hat{u}_n}\hat{u}_0, \hat{u}_s \rangle + \langle \nabla_{\hat{u}_n}\hat{u}_s, \hat{u}_0 \rangle)$$

$$+ f(0,0)C_{00}^{\lambda}C_{\lambda n}^{0} - C_{0n}^{s}C_{s0}^{0} + C_{0a}^{0}C_{n0}^{a}$$

$$= f(0,0)C_{00}^{s}(C_{n0}^{s} + C_{ns}^{0}) + f(0,0)C_{00}^{\lambda}C_{\lambda n}^{0} + C_{0\lambda}^{0}C_{n0}^{\lambda} = f(0,0)C_{00}^{\lambda}C_{\lambda n}^{0}. \tag{7.57}$$

Combining (7.46) with the fact that $C^i_{\lambda n} = 0$, we get that $\Upsilon(000) = 0$. Furthermore, we also have

$$
\begin{aligned}
\frac{2}{c}\frac{\partial s_{k0}}{\partial x^i}\Big|_o &= \frac{\partial(\Gamma^s_{n0}a_{sk} - \Gamma^s_{nk}a_{s0})}{\partial x^i} = \Upsilon(k0i) - \Upsilon(0ki) \\
&= f(i,k)C^s_{ik}\Gamma^t_{n0}\delta_{st} + f(i,0)C^a_{i0}\langle\nabla_{\hat{u}_n}\hat{u}_a,\hat{u}_k\rangle + C^s_{in}\langle\nabla_{\hat{u}_s}\hat{u}_0,\hat{u}_k\rangle \\
&\quad -(f(i,0)C^s_{i0}\Gamma^t_{nk}\delta_{st} + f(i,k)C^a_{ik}\langle\nabla_{\hat{u}_n}\hat{u}_a,\hat{u}_0\rangle + C^s_{in}\langle\nabla_{\hat{u}_s}\hat{u}_k,\hat{u}_0\rangle) \\
&\quad +\hat{u}_i\langle\nabla_{\hat{u}_n}\hat{u}_0,\hat{u}_k\rangle - \hat{u}_i\langle\nabla_{\hat{u}_n}\hat{u}_k,\hat{u}_0\rangle \\
&= f(i,0)(C^a_{i0}\langle\nabla_{\hat{u}_n}\hat{u}_a,\hat{u}_k\rangle - C^s_{i0}\Gamma^t_{nk}\delta_{st}) \\
&\quad -f(i,k)(C^a_{ik}\langle\nabla_{\hat{u}_n}\hat{u}_a,\hat{u}_0\rangle - C^s_{ik}\Gamma^t_{n0}\delta_{st}) \\
&\quad +C^s_{in}(\langle\nabla_{\hat{u}_s}\hat{u}_0,\hat{u}_k\rangle - \langle\nabla_{\hat{u}_s}\hat{u}_k,\hat{u}_0\rangle) \\
&\quad +\hat{u}_i\langle\nabla_{\hat{u}_n}\hat{u}_0,\hat{u}_k\rangle - \hat{u}_i\langle\nabla_{\hat{u}_n}\hat{u}_k,\hat{u}_0\rangle.
\end{aligned}
$$

By (7.46) and (7.48), we have

$$
\langle\nabla_{\hat{u}_s}\hat{u}_0,\hat{u}_k\rangle - \langle\nabla_{\hat{u}_s}\hat{u}_k,\hat{u}_0\rangle = C^s_{k0}.
$$

Then by (7.49), we get

$$
\begin{aligned}
C^a_{i0}\langle\nabla_{\hat{u}_n}\hat{u}_a,\hat{u}_k\rangle - C^s_{i0}\Gamma^t_{nk}\delta_{st} &= C^s_{i0}(\langle\nabla_{\hat{u}_n}\hat{u}_s,\hat{u}_k\rangle - \langle\nabla_{\hat{u}_n}\hat{u}_k,\hat{u}_s\rangle) + C^\lambda_{i0}\langle\nabla_{\hat{u}_n}\hat{u}_\lambda,\hat{u}_k\rangle \\
&= C^s_{i0}C^n_{ks} + C^a_{i0}C^n_{ka_\lambda} = C^a_{i0}C^n_{ka}.
\end{aligned}
$$

From (7.50) we easily deduce that

$$
\hat{u}_i\langle\nabla_{\hat{u}_n}\hat{u}_0,\hat{u}_k\rangle - \hat{u}_i\langle\nabla_{\hat{u}_n}\hat{u}_k,\hat{u}_0\rangle = C^n_{ia}C^a_{0k} + C^s_{0k}C^t_{in}\delta_{st}.
$$

Combining the above four equations, we see that $C^n_{\lambda i} = 0$ and that

$$
\frac{\partial s_{k0}}{\partial x^i}\Big|_o = \frac{c}{2}\left(f(i,k)C^s_{ik}C^n_{s0} + f(i,0)C^s_{i0}C^n_{ks} + C^s_{0k}C^n_{is}\right). \tag{7.58}
$$

*Proof of Theorem 7.9.* Since the "only if" part is obvious, we just prove the "if" part. Let $G/H$ be a Ricci-quadratic homogeneous Randers space. Taking the local coordinate system as (7.8), we have seen that (7.53) holds. In particular, we have $C^n_{ni} = 0$. Note that it is also true that

$$
C^n_{na} = 0.
$$

By (7.58) we have

$$
\frac{\partial s_{n0}}{\partial x^0}\Big|_o = \frac{c}{2}\left(f(0,n)C^s_{0n}C^n_{s0} + f(0,0)C^s_{00}C^n_{ns} + C^s_{0n}C^n_{0s}\right) = 0.
$$

Differentiating (7.51) and taking into account the fact that $\widetilde{c}(o) = 0 = s_0(o)$, we deduce from (7.52) that

$$2\widetilde{c}_0(\alpha^2 - \beta^2)|_o = c\frac{\partial(\Gamma_{n0}^k a_{k0})}{\partial x^0} + 2c\beta\frac{\partial s_{n0}}{\partial x^0} = c\Upsilon(000) + 2c\beta\frac{\partial s_{n0}}{\partial x^0}.$$

Then it follows from the above two equations and (7.57) that

$$\widetilde{c}_0 = 0.$$

Thus

$$A_0(o) = \frac{1}{2}\widetilde{c}_0(o) = 0.$$

On the other hand, we have

$$s_{ij}(o) = \frac{c}{2}(\Gamma_{nj}^i - \Gamma_{ni}^j) = \frac{c}{2}C_{ij}^n.$$

Thus

$$s^k{}_{0|k}(o) = \sum_k s_{k0|k} = \sum_k (\frac{\partial s_{k0}}{\partial x^k} - \Gamma_{kk}^l s_{l0} - \Gamma_{0k}^l s_{kl})$$

$$= \frac{c}{2}\sum_{k=1}^n (f(k,0) - 1)C_{k0}^a C_{ka}^n$$

$$+ \sum_{1\leq k,l\leq n}\frac{c}{2}\left(C_{l0}^n C_{kl}^k - C_{kl}^n\left(f(0,k)C_{0k}^l - \frac{1}{2}\left(C_{0k}^l + C_{0l}^k + C_{kl}^0\right)\right)\right)$$

$$= \frac{c}{2}\sum_{1\leq k,l\leq n}\left(C_{l0}^n C_{kl}^k + \frac{1}{2}C_{kl}^n C_{kl}^0\right).$$

Therefore (7.52) reduces to

$$\frac{c}{2}\sum_{1\leq k,l\leq n}\left(C_{l0}^n C_{kl}^k + \frac{1}{2}C_{kl}^n C_{kl}^0\right) = 0.$$

Setting $y = u$ in the above equation and taking into account (7.53), we get

$$C_{kl}^n = 0, \quad k,l = 1,\ldots,n,$$

that is,

$$\langle [u_k, u_l]_{\mathfrak{m}}, u\rangle = 0, \quad k,l = 1,\ldots,n.$$

By Proposition 7.9, $F$ must be a Berwald metric. This completes the proof of the theorem. $\qquad\square$

## 7.6 Homogeneous Randers Spaces of Positive Flag Curvature

In this section we will give a classification of homogeneous Randers spaces with (almost) isotropic S-curvature and positive flag curvature. Let us first explain the motivation of this study and survey some results in the Riemannian case.

It is one of the central problems in Riemannian geometry to determine how large the classes of manifolds with positive/nonnegative sectional, Ricci, or scalar curvature are. Up to now, this problem has been fairly successfully solved in scalar curvature. Also, a large number of interesting examples of Riemannian metric with positive Ricci curvature have been constructed. So far, the only known obstruction to positive Ricci curvature comes from obstructions to positive scalar curvature and from the classical Bonnet–Myers theorem. For example, it is known that a compact homogeneous manifold has an invariant Riemannian metric with positive Ricci curvature if and only its fundamental group is finite.

The most difficult part of the above problem lies in the sectional curvature case. Although very few obstructions to positive sectional curvature are known, only several classes of examples of Riemannian manifolds with positive curvature have been constructed. Among these examples, homogeneous manifolds constitute the most important part. Besides the compact rank-one symmetric Riemannian spaces, Berger [24] found two normal homogeneous Riemannian manifolds with positive curvature in 1961, with respective dimensions 7 and 13; In 1972, Wallach [159] found three homogeneous Riemannian manifolds, with dimensions 6, 12, 24, which have positive sectional curvature and are not normal; Later, Aloff–Wallach [6] constructed an infinite class of 7-dimensional homogeneous Riemannian manifolds with positive curvature. As proved by Bergery [23], the above examples are all the connected simply connected homogeneous Riemannian manifolds with positive curvature. More recently, several classes of inhomogeneous Riemannian manifolds with positive sectional curvature have been constructed, see for example Eschenberg [70], Wilking [169], Bazakian [22], Grove–Wilking–Ziller [78] and Grove–Verdiani–Ziller [79]. However, the problem is far from being completely understood. For example, the classical problem of Hopf, asking whether there is a Riemannian metric on the manifold $S^2 \times S^2$ with positive curvature, is still unsolved.

Recently we have initiated the study of this important problem for Finsler spaces. As in the Riemannian case, the first step is to find examples of homogeneous Finsler spaces with positive flag curvature and to classify them. However, the problem is rather complicated in the general case. So we first considered the Randers case. Our main result is a classification of the homogeneous Randers spaces with almost isotropic S-curvature and positive flag curvature. Using that, we have presented a large number of Randers spaces with nonconstant positive flag curvature.

We first establish a principle for classifying homogeneous Randers spaces with almost isotropic S-curvature and positive flag curvature. For this we need a result of Huang–Mo. In [91], Huang–Mo studied the property of change of flag curvature of a Finsler metric under the influence of a vector field. They found that the flag curvature will decrease under a navigation. In particular, they proved the following:

**Proposition 7.8 (Huang-Mo [91]).** *Let $(M,F)$ be a Finsler manifold and $\tilde{u}$ a homothetic vector field with dilation $\sigma$ with $F(x,\tilde{u}) < 1$. Let $\tilde{F}$ be the Finsler metric produced by navigation problem by $F$ and $\tilde{u}$. Then the flag curvature of $\tilde{F}$ (resp. $F$), denoted by $\tilde{K}(y,v)$ (resp. $K(y,v)$), satisfies*

$$\tilde{K}(y,v) = K(\tilde{y},v) - \sigma^2,$$

*where $\tilde{y} = y - \tilde{F}(x,y)\tilde{u}$.*

A special case is that in which the homothetic vector field is a Killing vector field. In this case, If $F$ has positive flag curvature, then $\tilde{F}$ also has positive flag curvature. For our purposes, we must first give a characterization of vanishing S-curvature for homogeneous Randers spaces using the navigation data. We now deduce the following theorem.

**Theorem 7.11.** *Let $G/H$ be a reductive homogeneous manifold with a reductive decomposition of the Lie algebra:*

$$\mathfrak{g} = \mathfrak{h} + \mathfrak{m}. \tag{7.59}$$

*Suppose $F$ is a homogeneous Randers space with navigation data $(h,W)$. Then $F$ has almost isotropic S-curvature if and only if $W$ is a Killing vector field with respect to $h$, if and only if the the linear map $\mathrm{ad}(w)_\mathfrak{m}$, where $w = W|_o \in \mathfrak{m}$, is skew-symmetric with respect to $h$, i.e.,*

$$\langle [w,x]_\mathfrak{m}, y \rangle_h + \langle x, [w,y]_\mathfrak{m} \rangle_h = 0, \quad \forall x, y \in \mathfrak{m},$$

*or equivalently*

$$\langle [w,x]_\mathfrak{m}, x \rangle_h = 0, \quad \forall x \in \mathfrak{m},$$

*where $\langle , \rangle_h$ denotes the inner product on $\mathfrak{m}$ induced by $h$. In this case, the S-curvature of $F$ is necessarily vanishing.*

*Proof.* Since both $h$ and $W$ are invariant under $G$, it suffices to prove the theorem at the origin $o = H$. If we write $F$ in its defining form $F = \alpha + \beta$ and let $\langle , \rangle$, $u$ be as in Corollary 7.1, then we need only prove that $\mathrm{ad}(u)_\mathfrak{m}$ is skew-symmetric with respect to $\langle , \rangle$ if and only if $\mathrm{ad}(w)_\mathfrak{m}$ is skew-symmetric with respect to $\langle , \rangle_h$. We now prove the "only if" part. The proof of the "if" part is similar and will be omitted. For simplicity we select an orthonormal basis of $\mathfrak{m}$ with respect to $h$, $v_1, v_2, \ldots, v_n$, such that $w = \sqrt{\mu} v_n$, where $\mu = h(w,w) \neq 0$. Then $\mathrm{ad}(w)_\mathfrak{m}$ is skew-symmetric with respect to $\langle , \rangle_h$ if and only if

$$\langle [w,v_i]_\mathfrak{m}, v_j \rangle_h + \langle v_i, [w,v_j]_\mathfrak{m} \rangle_h = 0, \quad 1 \leq i,j \leq n.$$

Equivalently, suppose the matrix of the linear map $\mathrm{ad}(w)_\mathfrak{m}$ with respect to the basis $v_1, v_2, \ldots, v_n$ is $B = (b_{ij})_{n\times n}$. Then $\mathrm{ad}(w)_\mathfrak{m}$ is skew-symmetric if and only if $b_{ij} = -b_{ji}$. On the other hand, by (7.4) and (7.5), we have

$$\langle v_i, v_j \rangle = \frac{\delta_{ij}}{\lambda} + \frac{\delta_{in}}{\lambda} \frac{\delta_{jn}}{\lambda}, \quad 1 \leq i,j \leq n,$$

and $u = -w/\lambda$, where $\lambda = 1 - \mu$. Then

$$\langle [u, v_i]_\mathfrak{m}, v_j \rangle = \left\langle -\frac{1}{\lambda} \left( \sum_{l=1}^n b_{il} v_l \right), v_j \right\rangle = -\frac{1}{\lambda} \left( \sum_{l=1}^n b_{il} \langle v_l, v_j \rangle \right)$$

$$= -\frac{1}{\lambda} \left( \sum_{l=1}^n b_{il} \left( \frac{\delta_{lj}}{\lambda} + \frac{\delta_{ln}}{\lambda} \frac{\delta_{jn}}{\lambda} \right) \right) = -\frac{1}{\lambda} \left( \frac{b_{ij}}{\lambda} + \frac{b_{in} \delta_{jn}}{\lambda^2} \right).$$

Similarly,

$$\langle v_i, [u, v_j]_\mathfrak{m} \rangle = -\frac{1}{\lambda} \left( \frac{b_{ji}}{\lambda} + \frac{b_{jn} \delta_{in}}{\lambda^2} \right).$$

Note that $[w, v_n] = [\sqrt{\mu} v_n, v_n] = 0$. Therefore $b_{nl} = -b_{ln} = 0$, $l = 1, 2, \ldots, n$. From this and taking into account the fact that the matrix $B$ is skew-symmetric, we deduce that

$$\langle [u, v_i]_\mathfrak{m}, v_j \rangle + \langle v_i, [u, v_j]_\mathfrak{m} \rangle = 0, \quad 1 \le i, j \le n.$$

Since $v_1, v_2, \ldots, v_n$ form a basis of $\mathfrak{m}$, we have

$$\langle [u, x]_\mathfrak{m}, y \rangle + \langle x, [u, y]_\mathfrak{m} \rangle = 0, \quad \forall x, y \in \mathfrak{m}.$$

This completes the proof of the theorem.                                □

Now we summarize the above to obtain a principle for classifying homogeneous Randers spaces with almost S-curvature and positive flag curvature as follows:

**Theorem 7.12.** *Let $G/H$ be a reductive homogeneous manifold with a reductive decomposition of Lie algebra as in (7.59). Then there exists a $G$-invariant non-Riemannian Randers metric on $G/H$, with almost isotropic S-curvature and positive flag curvature, if and only if*

1. *There exists a $G$-invariant Riemannian metric $\alpha$ on $G/H$ with positive sectional curvature.*
2. *There exists a nonzero vector $u \in \mathfrak{m}$ such that*

$$\mathrm{Ad}(h)(u) = u, \quad \forall h \in H,$$

*satisfying*

$$\langle [u, x]_\mathfrak{m}, y \rangle + \langle x, [u, y]_\mathfrak{m} \rangle = 0, \quad \forall x, y \in \mathfrak{m},$$

*where $\langle , \rangle$ is the inner product on $\mathfrak{m}$ induced by $\alpha$ when we identify $\mathfrak{m}$ with the tangent space $T_o(G/H)$.*

Now we are ready to give the classification. Our first step is to summarize the results on the classification of the connected simply connected compact homogeneous Riemannian manifolds with positive curvature. The results are listed in Table 7.2. Note that in this list, we not only present all the Lie group pairs $(G, H)$,

**Table 7.2** Compact homogeneous Riemannian manifolds with positive curvature

| Even dimensions | Isotropy representation |
|---|---|
| $S^{2n} = SO(2n+1)/SO(2n)$ | Irreducible |
| $\mathbb{C}P^m = SU(m+1)/U(m)$ | Irreducible |
| $\mathbb{H}P^k = Sp(k+1)/(Sp(k) \times Sp(1))$ | Irreducible |
| $\mathrm{Cay}P^2 = F_4/Spin(9)$ | Irreducible |
| $F^6 = SU(3)/T^2$ | $\mathfrak{m} = \mathfrak{m}_1 \oplus \mathfrak{m}_2 \oplus \mathfrak{m}_3$ |
| $F^{12} = Sp(3)/(Sp(1) \times Sp(1) \times Sp(1))$ | $\mathfrak{m} = \mathfrak{m}_1 \oplus \mathfrak{m}_2 \oplus \mathfrak{m}_3$ |
| $F^{24} = F_4/Spin(8)$ | $\mathfrak{m} = \mathfrak{m}_1 \oplus \mathfrak{m}_2 \oplus \mathfrak{m}_3$ |
| Odd dimensions | Isotropy representation |
| $S^{2n+1} = SO(2n+2)/SO(2n+1)$ | Irreducible |
| $M^7 = SO(5)/SO(3)$ | Irreducible |
| $M^{13} = SU(5)/(Sp(2) \times_{\mathbb{Z}_2} S^1)$ | $\mathfrak{m} = \mathfrak{m}_1 \oplus \mathfrak{m}_2$ |
| $N_{(1,1)} = SU(3) \times SO(3)/U^*(2)$ | $\mathfrak{m} = \mathfrak{m}_0 \oplus \mathfrak{m}_1 \oplus \mathfrak{m}_2$ |
| $N_{(k,l)} = SU(3)/S^1_{(k,l)}$, $\gcd(k,l) = 1$, $kl(k+l) \neq 0$ | $\mathfrak{m} = \mathfrak{m}_0 \oplus \mathfrak{m}_1 \oplus \mathfrak{m}_2$ |

but also give some partial information of the isotropy representation of the corresponding coset spaces. However, we will omit the explicit computation involved here. We just point out that from this table, it is easy to determine whether the coset space $G/H$ has a nonzero $G$-invariant vector field.

Let us explain in some detail the manifolds in this list. The first four rows in the even-dimensional case and the first row in the odd-dimensional case constitute all the simply connected rank-one Riemannian symmetric spaces. Since the isotropic representation is transitive on the unit sphere with respect to the standard metric, every invariant Finsler metric on such a coset space must be the standard Riemannian metric (up to a positive scalar). However, it is known that there are other compact Lie groups admitting effective and transitive actions on these spaces, and this will produce other coset spaces that are diffeomorphic to rank-one symmetric spaces. On these coset spaces, there may exist some invariant metrics other than the standard metrics, which have positive curvature. These metrics have been classified by Verdianni–Ziller. We will deal with this problem below. The spaces $M^7$ and $M^{13}$ are normal and were found to have positive curvature by Berger. The spaces $F^6$, $F^{12}$, and $F^{24}$ are not normal, and they were found to admit invariant Riemannian metrics with positive curvature by Wallach. Finally, the 7-dimensional spaces $N_{(k,l)}$, where the subgroups $S^1_{(k,l)}$ are defined by

$$S^1_{(k,l)} = \left\{ \begin{pmatrix} e^{2\pi\sqrt{-1}k\theta} & 0 & 0 \\ 0 & e^{2\pi\sqrt{-1}l\theta} & 0 \\ 0 & 0 & e^{-2\pi\sqrt{-1}(k+l)\theta} \end{pmatrix} \middle| \theta \in \mathbb{R} \right\},$$

with $kl(k+l) \neq 0$ and $\gcd(k,l) = 1$, were found to admit invariant Riemannian metrics with positive curvature by Aloff–Wallach. It was erroneously thought that all these spaces were not normal. However, Wilking proved in [169] that when $(k,l) = (1,1)$, the homogeneous Riemannian manifold is normal in some special case and can also be written as $SU(3) \times SO(3)/U^*(2)$. It was proved by Wallach and Berard Bergery that every connected simply connected nonsymmetric homogeneous Riemannian manifold with positive curvature can be expressed as one of the nonsymmetric coset spaces in Table 7.2.

Now we explain the isotropy representation of the coset spaces in Table 7.2. If the isotropy representation of $G/H$ is irreducible, then we indicate this by writing "irreducible" in the table. In this case, there is no nonzero $H$-fixed vector in the subspace $\mathfrak{m}$. Therefore every $G$-invariant Randers metric on $G/H$ must be Riemannian. If the isotropy representation is not irreducible, then the space $\mathfrak{m}$ can be written as a direct sum:

$$\mathfrak{m} = \mathfrak{m}_0 + \mathfrak{m}_1 + \cdots + \mathfrak{m}_k,$$

where $\mathfrak{m}_0$ is the subspace of all $H$-fixed vectors in $\mathfrak{m}$ and $\mathfrak{m}_1, \mathfrak{m}_2, \ldots, \mathfrak{m}_k$ are irreducible subspaces of $\mathfrak{m}$. From the above table, we see that for the coset spaces $F^6$, $F^{12}$, $F^{24}$, $M^7$, and $M^{13}$, $\mathfrak{m}_0 = 0$. For the spaces $N_{(k,l)}$, $\mathfrak{m}_0 \neq 0$.

As we have explained before, besides the groups $G$ in Table 7.2, there are some other compact Lie groups acting effectively and transitively on compact rank-one symmetric spaces. These groups have been classified by Montgomery–Samelson [121], Borel [31], and Onishchik [126]. A survey of their results can be found in Besse [28]. It was proved by Onishchik that the only connected simply connected compact Lie group acting effectively and transitively on the symmetric space $\mathbb{H}P^k = Sp(k+1)/(Sp(k) \times Sp(1))$, resp. $\mathrm{Cay}P^2 = F_4/\mathrm{Spin}(9)$, is $Sp(k+1)$, resp. $F_4$. On the complex projective space $\mathbb{C}P^m = SU(m+1)/U(m)$, with $m = 2n - 1$ odd, the group $Sp(n)$ has an effective and transitive action through the identification $\mathbb{H}^n = \mathbb{C}^{2n}$. However, since the isotropy representation of $Sp(n)/Sp(n-1)U(1)$ is irreducible, this coset space does not admit any invariant non-Riemannian Randers metric.

Therefore, it suffices to consider only the spheres here. The compact Lie groups that admit an effective and transitive action on spheres, as well as the corresponding isotropy representations, have been summarized in Table 7.1. From this table we see that there are four cases for which the isotropy representations have nonzero fixed points:

1. $S^{2n+1} = SU(n+1)/SU(n)$;
2. $S^{2n+1} = U(n+1)/U(n)$;
3. $S^{4n+3} = Sp(n+1)/Sp(n)$;
4. $S^{4n+3} = Sp(n+1)U(1)/Sp(n)U(1)$.

Note that a Riemannian metric on $S^{2n+1}$ is $U(n+1)$-invariant if and only if it is $SU(n+1)$-invariant, so case (2) reduces to case (1). Furthermore, (4) is a special case of (3). So we need consider only cases (1) and (3).

Invariant Riemannian metrics with positive curvature on the coset space (1) have been determined by Verdianni–Ziller [158] and can be described as follows. In this case we have $\mathfrak{g} = \mathfrak{su}(n+1)$, $\mathfrak{h} = \mathfrak{su}(n)$, and $\mathfrak{m}$ has the decomposition

$$\mathfrak{m} = \mathfrak{m}_0 \oplus \mathfrak{m}_1,$$

where

$$\mathfrak{m}_0 = \mathbb{R}X, \quad X = \sqrt{-1} \begin{pmatrix} -\frac{1}{n}E & 0 \\ 0 & 1 \end{pmatrix}, \tag{7.60}$$

and

$$\mathfrak{m}_1 = \left\{ \begin{pmatrix} 0 & \alpha \\ -\overline{\alpha}' & 0 \end{pmatrix} \middle| \alpha' = (x_1, \dots, x_n) \in \mathbb{C}^n \right\}.$$

Every $SU(n+1)$-invariant Riemannian metric on $SU(n+1)/SU(n)$ can be expressed as

$$h_t(X_1, X_2) = t c_1 c_2 + \mathrm{Re}(\alpha_1' \overline{\alpha_2}), \tag{7.61}$$

with $t > 0$, where

$$X_i = c_i \sqrt{-1} \begin{pmatrix} -\frac{1}{n}E & 0 \\ 0 & 1 \end{pmatrix} + \begin{pmatrix} 0 & \alpha_i \\ -\overline{\alpha}_i' & 0 \end{pmatrix}, \quad i = 1, 2.$$

This metric has positive sectional curvature if and only if $0 < t < \frac{2(n+1)}{3n}$.

Now we consider the nonzero $SU(n)$-invariant vector $X$ (unique up to a scalar) in $\mathfrak{m}$. A direct computation shows that for every $X_1 \in \mathfrak{m}$, we have

$$[X, X_1]_\mathfrak{m} = -\sqrt{-1} \left( 1 + \frac{1}{n} \right) \begin{pmatrix} 0 & \alpha_1 \\ \overline{\alpha}_1' & 0 \end{pmatrix}.$$

Hence

$$h_t([X, X_1]_\mathfrak{m}, X_1) = \left( 1 + \frac{1}{n} \right) \mathrm{Im}(\alpha_1' \overline{\alpha_1}) = 0.$$

Therefore the vector field $\widetilde{X}$ generated by $X$ is a Killing vector field. This proves the following result.

**Theorem 7.13.** *On $S^{2n+1} = SU(n+1)/SU(n)$ every $SU(n+1)$-invariant Randers metric with almost isotropic S-curvature and positive flag curvature must be a Randers metric that solves the Zermelo navigation problem of the Riemannian metric $h_t$ in (7.61), with $0 < t < \frac{2(n+1)}{3n}$, under the influence of the vector field generated by the vector $cX$, where $X$ is defined in (7.60), and $|c| < 1/\sqrt{t}$.*

Next we consider case (3). The situation here is much more complicated. Here we can take the subspace $\mathfrak{m}$ of $\mathfrak{sp}(n+1)$ to be

$$\mathfrak{m} = \mathfrak{m}_0 \oplus \mathfrak{m}_1,$$

where

$$m_0 = \mathbb{R}X_1 \oplus \mathbb{R}X_2 \oplus \mathbb{R}X_3 \tag{7.62}$$

is the subspace of $H$-fixed vectors in $\mathfrak{m}$, and $X_i$, $i = 1,2,3$, denote the elements of $\mathbb{H}^{(n+1)\times(n+1)}$ with the only nonzero element at the $(n+1,n+1)$ entry equal to $\sqrt{2}I$, $\sqrt{2}J$, and $\sqrt{2}K$ respectively; here $I, J, K$ denote the standard imaginary units in $\mathbb{H}$, and

$$m_1 = \left\{ \begin{pmatrix} 0 & \alpha \\ -(\alpha^*)' & 0 \end{pmatrix} \middle| \alpha' = (x_1,\dots,x_n) \in \mathbb{H}^n \right\}.$$

Then, up to a positive multiple, every $\mathrm{Sp}(n+1)$-invariant Riemannian metric on $S^{4n+3} = \mathrm{Sp}(n+1)/\mathrm{Sp}(n)$ can be written as

$$g_{(t_1,t_2,t_3)}(Y,Z) = t_1 y_1 z_1 + t_2 y_2 z_2 + t_3 y_3 z_3 + \mathrm{Re}((\xi^*)'\eta), \tag{7.63}$$

where $t_1, t_2, t_3$ are positive real numbers and

$$Y = y_1 X_1 + y_2 X_2 + y_3 X_3 + \begin{pmatrix} 0 & \xi \\ -(\xi^*)' & 0 \end{pmatrix},$$

$$Z = z_1 X_1 + z_2 X_2 + z_3 X_3 + \begin{pmatrix} 0 & \eta \\ -(\eta^*)' & 0 \end{pmatrix}.$$

The condition for such a metric to have positive curvature can be stated as follows. Let

$$V_i = (t_j^2 + t_k^2 - 3t_i^2 + 2t_i t_j + 2t_i t_k - 2t_j t_k)/t_i \quad \text{and} \quad H_i = 4 - 3t_i,$$

with $(i,j,k)$ a cyclic permutation of $(1,2,3)$. Then it is shown in [158] that the homogeneous metrics $g_{(t_1,t_2,t_3)}$ on $S^{4n+3}$ have positive sectional curvature if and only if

$$V_i > 0, \quad H_i > 0, \quad 3|t_j t_k - t_j - t_k + t_i| < t_j t_k + \sqrt{H_i V_i}. \tag{7.64}$$

It is also pointed out in [158] that the set of the triples $(t_1,t_2,t_3)$ satisfying the above condition forms a nonempty slice.

Now we determine the condition for an $H$-fixed vector $X$ in $m_0$ of (7.62) to generate a Killing vector field with respect to $g_{(t_1,t_2,t_3)}$. Suppose $X = x_1 X_1 + x_2 X_2 + x_3 X_3 \in m_0$, where $x_1, x_2, x_3 \in \mathbb{R}$. Set $x = \sqrt{2}(x_1 I + x_2 J + x_3 K) \in \mathbb{H}$, $y = \sqrt{2}(y_1 I + y_2 J + y_3 K) \in \mathbb{H}$. Then we have

$$[X,Y]_m = \begin{pmatrix} 0 & \xi x \\ -(x\xi^*)' & xy - yx \end{pmatrix}.$$

Therefore $g_{(t_1,t_2,t_3)}([X,Y]_m,Y) = 0$ if and only if

$$\mathrm{Re}((\xi x)^*)'\xi + 2\sqrt{2} \begin{vmatrix} y_1t_1 & y_2t_2 & y_3t_3 \\ x_1 & x_2 & x_3 \\ y_1 & y_2 & y_3 \end{vmatrix}$$

$$= 2\sqrt{2}x_2(t_1 - t_3)y_1y_3 - 2\sqrt{2}x_3(t_1 - t_2)y_2y_1 - 2\sqrt{2}x_1(t_2 - t_3)y_3y_2$$

$$= 0.$$

This is equivalent to the following system of linear equations:

$$\begin{cases} x_2(t_1 - t_3) = 0, \\ x_3(t_1 - t_2) = 0, \\ x_1(t_2 - t_3) = 0. \end{cases}$$

If $t_1,t_2,t_3$ are distinct, then $x_1 = x_2 = x_3 = 0$. In this case there is no non-Riemannian Randers metric. If precisely two of $t_1,t_2,t_3$ are equal, for example $t_1 = t_2 \neq t_3$, then $x_1 = x_2 = 0$. In this case the metric $g_{(t_1,t_2,t_3)}$ is invariant under $\mathrm{Sp}(n)\mathrm{U}(1)$. Furthermore, if $t_1 = t_2 = 1 \neq t_3$, then the metric is invariant under $\mathrm{U}(2n+2)$. Finally, if $t_1 = t_2 = t_3$, then for every $x_i$, the condition is satisfied. In this case, the Riemannian metric is invariant under $\mathrm{Sp}(n)\mathrm{Sp}(1)$. We have thus proved the following theorem.

**Theorem 7.14.** *On the sphere $S^{4n+3} = \mathrm{Sp}(n+1)/\mathrm{Sp}(n)$, every $\mathrm{Sp}(n+1)$-invariant Randers metric with almost isotropic S-curvature and positive flag curvature must be a Randers metric that solves Zermelo's navigation problem of the Riemannian metric $g_{(t_1,t_2,t_3)}$ in (7.63), with $(t_1,t_2,t_3)$ satisfying the condition (7.64), under the influence of a vector field generated by a vector in (7.62). Moreover, the following conditions must be satisfied:*

(i) *If $t_1,t_2,t_3$ are distinct, then $x_1 = x_2 = x_3 = 0$. In this case, the Randers metric must be Riemannian.*

(ii) *If $t_i = t_j \neq t_k$, then $x_i = x_j = 0$ and $|x_k| < 1/t_k$; here $(i,j,k)$ is any cyclic permutation of $(1,2,3)$. In this case, there are non-Riemannian Randers metrics.*

(iii) *If $t_1 = t_2 = t_3 = t$, then $x_1,x_2,x_3$ can be any real numbers satisfying $|x_1|^2 + |x_2|^2 + |x_3|^2 < 1/t$. In this case, there are non-Riemannian Randers metrics.*

*Remark 7.6.* Note that if $t_i = t_j = 1 \neq t_k$, with $(i,j,k)$ a cyclic permutation of $(1,2,3)$, then the Randers metrics described in (ii) of Theorem 7.14 are exactly the same as the metrics described in Theorem 7.13, i.e., the metrics are invariant under $\mathrm{SU}(2n+2)$ (and $\mathrm{U}(2n+2)$). It is easy to check that in this case, different cyclic permutations will produce isometric Riemannian metrics and isometric Randers metrics (up to a homothety). If $t_i = t_j \neq 1$ and $t_i \neq t_k$, then the Riemannian metrics and the Randers metrics are invariant under $\mathrm{Sp}(n)\mathrm{U}(1)$, but not invariant under $\mathrm{U}(2n+2)$. If $t_i = t_j = t_k$, then the underlying Riemannian metrics are invariant under

$Sp(n)Sp(1)$. However, since $Sp(n)Sp(1)$ has no nonzero invariant vector fields, the Randers metrics are not invariant under $Sp(n)Sp(1)$, unless it is Riemannian. Note also that when $t_i = t_j = t_k = 1$, the underlying Riemannian metric is the standard Riemannian metric on $S^{4n+3}$ with constant curvature and it is invariant under $SO(4(n+1))$.

Next we consider nonsymmetric spaces. According to Table 7.2, to determine the invariant Randers metrics with almost isotropic S-curvature and positive flag curvature on a nonsymmetric homogeneous space, we need consider only the spaces $N_{(k,l)}$, with $kl(k+l) \neq 0$ and $\gcd(k,l) = 1$. For $(k,l) = (1,1)$, the group $SO(3)$ has a free isometric action on $N_{(1,1)}$. Therefore $N_{(1,1)} = SU(3) \times SO(3)/U^*(2)$. But for our purposes, this expression is not necessary, since an $SU(3) \times SO(3)$-invariant Randers metric is automatically $SU(3)$-invariant. Therefore in the following we consider only the coset spaces $N_{k,l} = SU(3)/S^1_{(k,l)}$, where $kl(k+l) \neq 0$ and $\gcd(k,l) = 1$.

We first consider the case $k \neq l$. In this case the isotropy representation has a decomposition as

$$\mathfrak{m} = V_0 \oplus V_{2k+l} \oplus V_{2l+k} \oplus V_{k-l},$$

where

$$V_{2k+l} = \left\{ \begin{pmatrix} 0 & 0 & z \\ 0 & 0 & 0 \\ -\bar{z} & 0 & 0 \end{pmatrix} \Big| z \in \mathbb{C} \right\},$$

$$V_{2l+k} = \left\{ \begin{pmatrix} 0 & 0 & 0 \\ 0 & 0 & z \\ 0 & -\bar{z} & 0 \end{pmatrix} \Big| z \in \mathbb{C} \right\},$$

$$V_{k-l} = \left\{ \begin{pmatrix} 0 & z & 0 \\ -\bar{z} & 0 & 0 \\ 0 & 0 & 0 \end{pmatrix} \Big| z \in \mathbb{C} \right\},$$

and

$$V_0 = \left\{ \sqrt{-1} \begin{pmatrix} a & 0 & 0 \\ 0 & b & 0 \\ 0 & 0 & -(a+b) \end{pmatrix} \Big| (2k+l)a + (2l+k)b = 0, a,b \in \mathbb{R} \right\}.$$

Now the Lie algebra of $S^1_{(k,l)}$ is $\mathbb{R}h_{(k,l)}$, where

$$h_{(k,l)} = \left\{ \begin{pmatrix} k\sqrt{-1} & 0 & 0 \\ 0 & l\sqrt{-1} & 0 \\ 0 & 0 & -(k+l)\sqrt{-1} \end{pmatrix} \right\}.$$

Set

$$V_1 = V_0 \oplus V_{k-l}, \quad V_2 = V_{2k+l} \oplus V_{2l+k},$$

and define

$$q_0(X,Y) = -\mathrm{Re}(\mathrm{tr}(XY)), \quad X,Y \in \mathfrak{su}(3).$$

Then $q_0$ is an Ad($SU(3)$)-invariant inner product on $\mathfrak{su}(3)$; hence it defines a bi-invariant Riemannian metric on $SU(3)$. It is easy to check that

$$q_0(h_{k,l}, V_1) = 0, \quad q_0(V_1, V_2) = 0.$$

Now a direct computation shows that

$$[V_1, V_1] \subset \mathbb{R} h_{k,l} + V_1, \quad [V_1, V_2] \subset V_2, \quad [V_2, V_2] \subset \mathbb{R} h_{k,l} + V_1. \tag{7.65}$$

For $X, Y \in \mathfrak{m}$, set $X = X_1 + X_2$, $Y = Y_1 + Y_2$, with $X_i, Y_i \in V_i$, and define

$$q_t(X,Y) = (1+t)q_0(X_1, Y_1) + q_0(X_2, Y_2) = q_0(X,Y) + t q_0(X_1, Y_1). \tag{7.66}$$

It is shown in [6] that if $-1 < t < 0$, then $q_t$ defines an $SU(3)$-invariant Riemannian metric on $SU(3)/S^1_{(k,l)}$ with positive curvature.

Now we can prove the following result.

**Proposition 7.9.** *1. If $x \in V_1$, then $(\mathrm{ad}\,x)_\mathfrak{m}$ is skew-symmetric with respect to $q_t$. 2. If $y \in V_2$, and $(\mathrm{ad}\,y)_\mathfrak{m}$ is skew-symmetric with respect to $q_t$, then $y = 0$.*

*Proof.* Given $x \in V_1$ and $v_i \in V_i$, $i = 1, 2$, it is easily seen from (7.65) that

$$q_t([x, v_1]_\mathfrak{m}, v_2) = 0,$$
$$q_t([x, v_2]_\mathfrak{m}, v_1) = 0,$$
$$q_t([x, v_1]_\mathfrak{m}, v_1) = (1+t)q_0([x, v_1] + c h_{k,l}, v_1) = (1+t)q_0([x, v_1], v_1) = 0,$$
$$q_t([x, v_2]_\mathfrak{m}, v_2) = q_t([x, v_2], v_2) = q_0([x, v_2], v_2) = 0.$$

From the above equalities, (1) follows.

On the other hand, suppose $y \in V_2$ and $v_1, v_2$ are as above. If $\mathrm{ad}(y)_\mathfrak{m}$ is skew-symmetric, then we have

$$q_t([y, v_1]_\mathfrak{m}, v_1) = 0 = q_t([y, v_2]_\mathfrak{m}, v_2) = 0$$

and

$$q_t([y, v_1]_\mathfrak{m}, v_2) + q_t([y, v_2]_\mathfrak{m}, v_1) = q_0([y, v_1], v_2) + (1+t)q_0([y, v_2] + c_1 h_{k,l}, v_1)$$
$$= t q_0([y, v_2], v_1) = t q_0([v_2, v_1], y) = 0.$$

This means that $y$ is perpendicular to the subspace $[V_1, V_2]$ of $\mathfrak{su}(3)$, with respect the inner product $q_0$. However, an easy computation shows that $[V_1, V_2]$ is actually the whole space $\mathfrak{su}(3)$. Therefore $y = 0$ and the lemma is proved. $\qquad\square$

Next we consider the coset space $N_{(1,1)} = SU(3)/S^1_{(1,1)}$. This is only slightly different from the above case. In this case, we define the following subspaces:

$$V' = \left\{ \begin{pmatrix} 0 & 0 & z \\ 0 & 0 & 0 \\ -\bar{z} & 0 & 0 \end{pmatrix} \middle| z \in \mathbb{C} \right\}.$$

$$V'' = \left\{ \begin{pmatrix} 0 & 0 & 0 \\ 0 & 0 & z \\ 0 & -\bar{z} & 0 \end{pmatrix} \middle| z \in \mathbb{C} \right\}.$$

$$V''' = \left\{ \begin{pmatrix} 0 & z & 0 \\ -\bar{z} & 0 & 0 \\ 0 & 0 & 0 \end{pmatrix} \middle| z \in \mathbb{C} \right\}.$$

Moreover, set

$$V_0 = \left\{ \sqrt{-1} \begin{pmatrix} a & 0 & 0 \\ 0 & -a & 0 \\ 0 & 0 & 0 \end{pmatrix} \middle| a \in \mathbb{R} \right\}.$$

Then the tangent space $\mathfrak{m}$ can be decomposed into the direct sum of irreducible $S^1_{(1,1)}$-submodules as

$$\mathfrak{m} = V_0 \oplus V' \oplus V'' \oplus V'''.$$

Define

$$V_1 = V_0 \oplus V''', \quad V_2 = V' \oplus V'',$$

and $q_0$, $q_t$ as above. Then $q_t$ generates an $SU(3)$-invariant Riemannian metric on $N_{(1,1)}$. This metric has positive curvature if and only if $-1 < t < 0$. Moreover, Proposition 7.9 still holds in this case.

Let us determine the elements $w \in \mathfrak{m}$ that generate an invariant Killing vector field on $N_{(k,l)}$. By Proposition 7.9, we need consider only the elements in $V_1$. Since in both cases the subgroup $S^1_{(k,l)}$ is connected, it suffices to determine the elements in $V_1$ that commute with $\mathbb{R}h^1_{(k,l)}$. If $(k, l) \neq (1, 1)$, then it is easily seen that an element $w \in V_1$ satisfies

$$[h^1_{(k,l)}, w] = 0$$

if and only if $w \in V_0$. If $(k, l) = (1, 1)$, then a direct computation shows that $[h^1_{(k,l)}, V_1] = 0$.

We now summarize the above as the following theorem.

**Theorem 7.15.** *1. If $(k,l) \neq (1,1)$, then an SU(3)-invariant Randers metric F on $N_{(k,l)}$ has almost isotropic S-curvature and positive flag curvature if and only if F solves Zermelo's navigation problem of the Riemannian metric defined by $q_t$ in (7.66), with $-1 < t < 0$, under the influence of a vector field generated by an element $w = \mathrm{diag}(a\sqrt{-1}, b\sqrt{-1}, -(a+b)\sqrt{-1}) \in \mathfrak{m}$, where $a,b$ are real numbers satisfying the conditions*

$$(2k+l)a + (2l+k)b = 0 \quad and \quad a^2 + b^2 + (a+b)^2 < \frac{1}{1+t}.$$

*In this case there is a family of such metrics depending on two parameters.*

*2. On $N_{(1,1)}$ an SU(3)-invariant Randers metric F has almost isotropic S-curvature and positive flag curvature if and only if it solves Zermelo's navigation problem of the Riemannian metric defined by $q_t$ in (7.65), with $t \in (-1,0)$, under the influence of a vector field generated by an element*

$$w = \begin{pmatrix} a\sqrt{-1} & b+c\sqrt{-1} & 0 \\ -b+c\sqrt{-1} & -a\sqrt{-1} & 0 \\ 0 & 0 & 0 \end{pmatrix} \in \mathfrak{m},$$

*where $a,b,c$ are real numbers satisfying the condition $a^2 + b^2 + c^2 < \frac{1}{2(1+t)}$. In this case, there is a family of such metrics depending on four parameters.*

At this point, all non-Riemannian homogeneous Randers metrics of almost isotropic S-curvature and positive flag curvature have been determined. It remains to determine which among these metrics are isometric. We now prove the following result.

**Theorem 7.16.** *Let $(M,F_i)$, $i = 1,2$, be two connected simply connected non-Riemannian homogeneous Randers spaces with almost isotropic S-curvature and positive flag curvature solving Zermelo's navigation problem of the same Riemannian metric h, under the influence of an invariant vector field generated by an H-invariant vector $w_1$ and $w_2$, respectively. Then $(M,F_1)$ is isometric to $(M,F_2)$ if and only if $w_1$ and $w_2$ have the same length with respect to h.*

*Proof.* Non-Riemannian homogeneous Randers metrics with almost isotropic S-curvature and positive flag curvature have been described in Theorems 7.13, 7.14, and 7.15. Hence it suffices to check the assertion case by case.

*Case (1).* Metrics in Theorem 7.13. We note that the sphere $S^{2n+1}$ is identified with the coset space $SU(n+1)/SU(n)$ through the standard action of $SU(n+1)$ on $\mathbb{C}^{n+1} = \mathbb{R}^{2n+2}$, with $SU(n)$ being the isotropic subgroup at the point $p = (1,0,0,\ldots,0)$. Under this identification, the $SU(n+1)$-invariant vector fields on $S^{2n+1}$ can be described very clearly. At the point $p = (1,0,\ldots,0)$, the vector $X = (\sqrt{-1},0,\ldots,0)$ is invariant under $SU(n)$. This vector $X$ generates an $SU(n+1)$-invariant vector field, which is equal to $\sqrt{-1}\alpha$ at the point $\alpha \in S^{2n+1}$. (This can be easily seen from the fact that for every unit $\alpha \in \mathbb{C}^{n+1}$, we can find a unitary matrix $A$, with $\det(A) = 1$, such that the first column of $A$ is exactly $\alpha$.) There is another special diffeomorphism $\sigma$ of $\mathbb{C}^n$, taking the complex conjugate of each entry of a vector, which keeps the sphere as well as the point $p$ invariant. Note also that the metric $h$ in (7.63) is also invariant under $\sigma$. However, from the above description it is easily seen that the action of $\sigma$ sends the vector field $\widetilde{X}$ generated by $X$ to $-\widetilde{X}$, which is generated by $-X$. Therefore, in this case, the Randers metrics with navigation data $(h, c\widetilde{X})$ and $(h, -c\widetilde{X})$ must be isometric. This proves the assertion in this case.

*Case (2).* Metrics in (ii) of Theorem 7.14. By Remark 7.6, the cyclic permutation is irrelevant. Hence we can assume that $t_2 = t_3 \neq t_1$. In this case the underlying Riemannian metric $g_{(t_1, t_2, t_2)}$ is invariant under the action of $\mathrm{Sp}(n)\mathrm{U}(1)$, where $\mathrm{U}(1)$ acts in the standard way on the subspace $\mathbb{R}^2 = \mathbb{R}J + \mathbb{R}K$ (as rotations). Similarly to Case (1), the sphere $S^{4n+3}$ is identified with the coset space $\mathrm{Sp}(n+1)/\mathrm{Sp}(n)$ through the standard action of $\mathrm{Sp}(n+1)$ on $\mathbb{H}^{n+1} = \mathbb{C}^{2n+2} = \mathbb{R}^{4n+4}$, with $\mathrm{Sp}(n)$ as the isotropy subgroup at $p = (1, 0, \ldots, 0)$. The invariant vector field $\widetilde{X}_1$ generated by $X_1$ in (7.62) is exactly the vector field described in Case (2) above, when we identify $\mathbb{H}^{n+1}$ with $\mathbb{C}^{2(n+1)}$. In other words, $\widetilde{X}_1$ is actually an invariant vector field invariant under the action of $\mathrm{SU}(2n+2)$. (However, the metric $g_{(t_1, t_2, t_3)}$ is generically not invariant under $\mathrm{SU}(2n+2)$, unless $t_2 = t_3 = 1$.) The diffeomorphism $\sigma$ in Case (1) is also an isometry with respect to the metric $g_{(t_1, t_2, t_3)}$. Therefore, similarly to Case (1), the Randers metrics produced by $g_{(t_1, t_2, t_3)}$ and the vectors $cX_1$ and $-cX_1$, $c \in \mathbb{R}$, are isometric. This proves the assertion in Case (2).

*Case (3).* Metrics in (iii) of Theorem 7.14. As we pointed out in Remark 7.6, in this case the underlying Riemannian metric is invariant under the group $\mathrm{Sp}(n+1)\mathrm{Sp}(1)$, where $\mathrm{Sp}(n+1)$ acts in the standard way on $\mathbb{H}^{n+1}$ (left multiplication) and $\mathrm{Sp}(1)$ acts on $\mathbb{H}^{n+1}$ by right multiplication (to each entry). The isotropic subgroup at $p = (1, 0, \ldots, 0)$ is $\mathrm{Sp}(n)\mathrm{Sp}(1)$ and the isotropy action of the subgroup $\mathrm{Sp}(1)$ on the subspace $\mathfrak{m}_0$ is just the standard action of $\mathrm{Sp}(1) = \mathrm{Spin}(3)$ on $\mathbb{R}^3$ (see [77]). Since the action of $\mathrm{Sp}(1) = \mathrm{Spin}(3)$ on $\mathbb{R}^3$ is transitive on the unit sphere, and the actions of $\mathrm{Sp}(n+1)$ and $\mathrm{Sp}(1)$ are commutative, two vectors $w_1$, $w_2$ with the same length with respect to $g_{(t_1, t_2, t_3)}$ must produce isometric Randers metrics. This proves the theorem in this case.

*Case (4).* Metrics in Theorem 7.15 on $N_{(k,l)}$, with $(k, l) \neq (1, 1)$. In this case we use a diffeomorphism of $N_{(k,l)}$ induced by an automorphism of $\mathrm{SU}(3)$. Since $\mathrm{SU}(3)$ is simply connected, it suffices to describe the automorphism on the Lie algebra level. Note that $\mathfrak{su}(3)$ has an involutive automorphism $\omega$ defined by $\omega(X) = \overline{X}$. It is easily seen that $\omega$ sends each diagonal matrix of $\mathfrak{su}(3)$ to its negative and keeps the subspace $V_{k-l}$ invariant. This automorphism induces an automorphism of $\mathrm{SU}(3)$, denoted by $\widetilde{\omega}$, that keeps the subgroup $S^1_{(k,l)}$ invariant. Then $\widetilde{\omega}$ induces a diffeomorphism $\bar{\omega}$ of $N_{(k,l)}$. Note that the metric $q_0$ is in fact invariant under every automorphism of $\mathfrak{su}(3)$, and $\widetilde{\omega}$ keeps the subspaces $V_0$ and $V_{k-l}$ invariant. It follows from (7.65) that the differential of $\bar{\omega}$ at the origin is an isometry with respect to the inner product $q_t$. Now we make the following two assertions:

1. $\bar{\omega}$ maps every SU(3)-invariant vector field on $N_{(k,l)}$ to an SU(3)-invariant vector field on $N_{(k,l)}$.
2. $\bar{\omega}$ is an isometry of the Riemannian metric induced by $q_t$.

We first prove the first assertion. Note that in this case the SU(3)-invariant vector field is in one-to-one correspondence with the element in $V_0$, or in other words, every SU(3)-invariant vector field on $N_{(k,l)}$ must be of the form $\tilde{X}$ such that $\tilde{X}|_{gH} = dL_g X$, where $X \in V_0$ and $L_g$ is the diffeomorphism of $N_{(k,l)}$ defined by $g_1 H \mapsto g g_1 H$.

Since $X$ is the initial vector of the curve $\exp(tX)H$, $\tilde{X}|_{gH}$ is the initial vector of the curve $g \exp(tX)H$. Hence $d\bar{\omega}\tilde{X}|_{gH}$ is the initial vector of the curve $\bar{\omega}(g \exp(tX)H) = \tilde{\omega}(g \exp(tX))H = \tilde{\omega}(g)\tilde{\omega}(\exp(tX))H$. It is easy to see that this is equal to $dL_{\tilde{\omega}(g)}(\omega(X))$. This proves the first assertion. To prove the second assertion, suppose $Y_1, Y_2 \in T_{gH}(G/H)$ and denote the Riemannian metric on $N_{(k,l)}$ induced by $q_t$ by $Q_t$. Then we have $Q_t(Y_1, Y_2) = q_t(L_{g^{-1}}(Y_1), L_{g^{-1}}(Y_2))$. Select $y_1, y_2 \in \mathfrak{m}$ such that $dL_{g^{-1}}(Y_1)$, resp $dL_{g^{-1}}(Y_2)$, is the initial vector of the curve $\exp(ty_1)H$, resp. $\exp(ty_2)H$. Then $Y_1$, resp. $Y_2$, is the initial vector of the curve $g \exp(ty_1)H$, resp. $g \exp(ty_2)H$. Hence $d\bar{\omega}(Y_1)$, resp. $d\bar{\omega}(Y_2)$, is the initial vector of the curve $\tilde{\omega}(g)\tilde{\omega}(\exp(ty_1))H$, resp. $\tilde{\omega}(g)\tilde{\omega}(\exp(ty_2))H$. Therefore we have

$$Q_t(d\bar{\omega}(Y_1), d\bar{\omega}(Y_2)) = Q_t(dL_{\tilde{\omega}(g)}\omega(y_1), dL_{\tilde{\omega}(g)}\omega(y_2)) = q_t(\omega(y_1), \omega(y_2))$$
$$= q_t(y_1, y_2).$$

But the last term of the above equality is exactly

$$q_t(dL_{g^{-1}}(Y_1), dL_{g^{-1}}(Y_2)) = Q_t(Y_1, Y_2).$$

This proves the second assertion. Now by these two assertions one can easily see that if $w \in V_0$ with length $< 1$, then $w$ and $-w$ will produce isometric Randers spaces with the underlying Riemannian metric $h_t$. This completes the proof of the theorem for Case (4).

*Case (5).* Metrics in Theorem 7.15 on $N_{(1,1)}$. The proof of this case is similar to Case (4). First, the automorphism $\omega$ of the Lie algebra $\mathfrak{su}(3)$ defined above is equal to $-\mathrm{id}$ on the maximal commutative subalgebra

$$\mathbb{R}h^1_{(1,1)} \oplus V_0.$$

Moreover, this automorphism keeps $V_1$ invariant. Therefore $\omega$ induces an automorphism of SU(3) that keeps the subgroup $S^1_{(1,1)}$ invariant. As in Case (4), this means that the Randers metrics produced by $(q_t, w)$ and $(q_t, -w)$, with $w \in V_0$ and $q_t(w, w) < 1$, must be isometric. Now we consider two distinct vectors $w_1, w_2 \in V_1$

with $q_t(w_1, w_1) = q_t(w_2, w_2) < 1$. Note that $V_1$ is actually the Lie algebra $\mathfrak{su}(2)$ embedded in $\mathfrak{su}(3)$ through the standard embedding of the Lie group:

$$A \hookrightarrow \begin{pmatrix} A & 0 \\ 0 & 1 \end{pmatrix}, \quad A \in SU(2).$$

Also note that $SU(2)$ is a rank-one compact Lie group; hence maximal commutative subalgebras of $\mathfrak{su}(2)$ are exactly one-dimensional subspaces. By the conjugation of maximal subalgebras of compact Lie algebras, we conclude that for every two one-dimensional subspaces $U_1$, $U_2$ of $V_1$, there is an automorphism of $\mathfrak{su}(2)$ that sends $U_1$ onto $U_2$. Using the embedding above, the automorphism can be extended to an automorphism of $\mathfrak{su}(3)$ that keeps $h^1_{(1,1)}$ fixed and leaves the subspace $V_1$ invariant. If $w_1 = -w_2 \in V_1$, then we set $U_1 = \mathbb{R}(w_1)$ and $U_2 = V_0$. The above argument then shows that there is an automorphism $\sigma$ such that $\sigma(w_1) \in V_0$ and $\sigma(h^1_{(1,1)}) = h^1_{(1,1)}$. Moreover, we have $\sigma(V_1) \subset V_1$. Applying the same argument as in Case (4) to the automorphism $\omega \circ \sigma$, we easily show that the Randers metrics produced by $(q_t, w_1)$ and $(q_t, w_2)$ are isometric. If $w_1$ and $w_2$ are linearly independent, then we set $U_1 = \mathbb{R}w_1$ and $U_2 = \mathbb{R}w_2$. The above argument shows that there exists an automorphism $\sigma'$ of $\mathfrak{su}(3)$ such that $\sigma'(U_1) = U_2$, $\sigma'(h^1_{(1,1)}) = h^1_{(1,1)}$, and $\sigma'(V_1) \subset V_1$. Then a similar argument as above shows that the Randers metrics produced by $(q_t, w_1)$ and $(q_t, w_2)$ are isometric. This proves the assertion for the last case, and the theorem is completely proved. $\qquad\square$

Let us summarize the results in Theorems 7.13–7.16 as follows:

**Theorem 7.17.** *Let $(G/H, F)$ be a connected simply connected non-Riemannian homogeneous Randers space with almost isotropic S-curvature and positive flag curvature. Then $(G/H, F)$ must be isometric to one of the following:*

1. *As a coset space, $G/H = SU(n+1)/SU(n)$ and the metric $F$ is a Randers metric that solves Zermelo's navigation problem of the Riemannian metric $h_t$ in (7.61), under the influence of a vector field generated by $cX \in \mathfrak{m}_0$, where $X$ is defined in (7.60), and $0 < |c| < 1/\sqrt{t}$. Two pairs $(h_t, cX)$ and $(h_{t'}, c'X)$ correspond to isometric Randers metrics if and only if $t = t'$ and $|c| = |c'|$.*

2. *As a coset space, $G/H = Sp(n+1)/Sp(n)$ and the metric $F$ is a Randers metric that solves Zermelo's navigation problem of the Riemannian metric $g_{(t_1, t_2, t_3)}$ in (7.63), with $t_1 \neq t_2 = t_3 \neq 1$ satisfying the condition (7.64), under the influence of a vector field generated by $cX_1 \in \mathfrak{m}_0$, where $X_1$ is defined in (7.62), and $0 < |c| < 1/t_1$. Two pairs $(g_{(t_1, t_2, t_3)}, cX_1)$ and $(g_{(t_1', t_2', t_3')}, c'X_1)$ correspond to isometric Randers metrics if and only if $(t_1, t_2, t_3) = (t_1', t_2', t_3')$ and $|c| = |c'|$.*

3. *As a coset space, $G/H = Sp(n+1)/Sp(n)$ and the metric $F$ is a Randers metric that solves Zermelo's navigation problem of the Riemannian metric $g_{(t_1, t_2, t_3)}$ in (7.63), with $t_1 = t_2 = t_3 = t > 0$ satisfying the condition (7.64), under the influence of a vector field generated by $aX_1 + bX_2 + cX_3 \in \mathfrak{m}_0$ in (7.62), with*

$0 < |a^2 + b^2 + c^2| < 1/t$. *Two pairs* $(g_{(t,t,t)}, aX_1 + bX_2 + cX_3)$ *and* $(g_{(t',t',t')}, a'X_1 + b'X_2 + c'X_3)$ *correspond to isometric Randers metrics if and only if* $t = t'$ *and* $a^2 + b^2 + c^2 = (a')^2 + (b')^2 + (c')^2$.

4. $G/H = SU(3)/S^1_{(k,l)}$, *with* $(k,l) \neq (1,1)$, $\gcd(k,l) = 1$, *and* $kl(k+l) \neq 0$, *and* $F$ *is a Randers metric that solves Zermelo's navigation problem of the Riemannian metric* $q_t$ *in* (7.66), *with* $-1 < t < 0$, *under the influence of a vector field generated by an element* $w_a = \mathrm{diag}(a\sqrt{-1}, b\sqrt{-1}, -(a+b)\sqrt{-1}) \in \mathfrak{m}$, *where* $a, b$ *are real numbers satisfying the conditions* $(2k + l)a + (2l + k)b = 0$ *and* $0 < a^2 + b^2 + (a+b)^2 < 1/(1+t)$. *Two pairs* $(h_t, w_a)$ *and* $(h_{t'}, w_{a'})$ *correspond to isometric Randers metrics if and only if* $t = t'$ *and* $|a| = |a'|$.

5. $G/H = SU(3)/S^1_{(1,1)}$ *and* $F$ *is a Randers metric that solves Zermelo's navigation problem of the Riemannian metric defined by* $q_t$ *in* (7.66), *with* $t \in (-1,0)$, *under the influence of a vector field generated by an element*

$$w_{(a,b,c)} = \begin{pmatrix} a\sqrt{-1} & b + c\sqrt{-1} & 0 \\ -b + c\sqrt{-1} & -a\sqrt{-1} & 0 \\ 0 & 0 & 0 \end{pmatrix} \in \mathfrak{m},$$

*where* $a, b, c$ *are real numbers satisfying the condition* $0 < a^2 + b^2 + c^2 < \frac{1}{2(1+t)}$. *Two pairs* $(h_t, w_{(a,b,c)})$ *and* $(h_{t'}, w_{(a',b',c')})$ *correspond to isometric Randers metrics if and only if* $t = t'$ *and* $a^2 + b^2 + c^2 = (a')^2 + (b')^2 + (c')^2$.

*Moreover, Randers metrics in different cases (1) through (5) cannot be isometric.*

For the proof, just note that metrics in (ii) of Theorem 7.14, with $t_1 \neq t_2 = t_3 = 1$, are invariant under the group $SU(4n + 2)$ and are isometric to metrics in Theorem 7.13.

## 7.7  Homogeneous Randers Spaces of Negative Flag Curvature

We now consider homogeneous Randers spaces with negative flag curvature and isotropic S-curvature. As we have seen in the above section, there is a large class of non-Riemannian homogeneous Randers spaces with isotropic S-curvature and positive flag curvature. One may hope that there also exist some examples of negative flag curvature. However, in contrast to the positive case, we will obtain a rigidity theorem. In fact, the rigidity theorem holds even in the more generalized case for a negative Ricci scalar.

**Theorem 7.18.** *Let* $(M, F)$ *be a connected homogeneous Randers space. If* $F$ *has almost isotropic S-curvature and negative Ricci scalar, then* $F$ *must be Riemannian.*

Here by "negative Ricci scalar" we mean that its Ricci scalar is everywhere less than 0. We now recall some results on homogeneous Riemannian manifolds with negative sectional curvature. It is well known that the rank-one Riemannian symmetric spaces of noncompact type must be of negative curvature. By the discussion of Chap. 2, all these spaces are simply connected. Kobayashi generalized this result by showing that every homogeneous Riemannian manifold with nonpositive sectional curvature and negative Ricci curvature must be simply connected; see Sect. 4.4.

Wolf studied the bounded isometries of homogeneous Riemannian manifolds of nonpositive sectional curvature and proved that each such Riemannian metric can be realized as a left-invariant metric on a connected solvable Lie group [174]. Based on this result, Heintze thoroughly studied homogeneous Riemannian manifolds with negative sectional curvature in [81]. It turns out that besides noncompact rank-one symmetric spaces, there is a large number of such spaces. Moreover, E. Heintze also provided a possible method to classify the homogeneous Riemannian manifolds with negative sectional curvature. However, up to now there is no classification result on homogeneous Riemannian manifolds with negative Ricci curvature.

Now we turn to the proof of Theorem 7.18. It is a corollary of the following more general theorem:

**Theorem 7.19.** *Let $(M,F)$ be a connected homogeneous Randers space that solves Zermelo's navigation problem of a Riemannian metric $h$, under the influence of an external vector field $W$. If $F$ has almost isotropic $S$-curvature and the Ricci scalar of $W$ is negative somewhere, i.e., there exists $x \in M$ such that $\mathrm{Ric}(x,W) < 0$, then $F$ must be Riemannian.*

To prove this theorem, we need the following lemma.

**Lemma 7.7.** *Let $F = \alpha + \beta$ be a Minkowski Randers norm on a real vector space $V$, and $(h,W)$ its navigation data, with $W \neq 0$. Set $\lambda = 1 - h(W,W)$. Suppose $w_1 = \frac{W}{|W|}, w_2, \ldots, w_n$ is an orthonormal basis of $V$ with respect to the inner product $h$. Then $w_1, w_2, \ldots, w_n$ is an orthogonal basis of $V$ with respect to the inner product $g_{w_1}(,)$, where $g$ is the fundamental form of $F$.*

*Proof.* Since the metric matrix of $h$ with respect to $w_1, w_2, \ldots, w_n$ is the identity matrix, the metric matrix $(\widetilde{a}_{ij})$ of $\alpha$ (viewed as an inner product) with respect to the same basis satisfies

$$\widetilde{a}_{ij} = \frac{\delta_{ij}}{\lambda} + \frac{\delta_{i1}\delta_{j1}|W|^2}{\lambda^2}, \quad i,j = 1,2,\ldots,n,$$

where $\lambda = 1 - |W|^2$, and the components of $\beta$ are

$$\widetilde{b}_i = -\frac{\delta_{i1}|W|}{\lambda}, \quad i = 1,2,\ldots,n.$$

Set

$$\widetilde{l}_i = \frac{\widetilde{a}_{ij}y^j}{\alpha(y)} \quad \text{and} \quad l_i = \widetilde{l}_i + \widetilde{b}_i.$$

Then the fundamental tensor of $F$ has the expression (see [16, p. 284])

$$g_{ij}(y) = \frac{F(y)}{\alpha(y)}(\widetilde{a}_{ij}(y) - \widetilde{l}_i(y)\widetilde{l}_j(y)) + l_i(y)l_j(y).$$

Considering the value at $y = w_1$, we easily get

$$\alpha(w_1) = \frac{1}{\lambda}, \quad F(w_1) = \frac{1 - |W|}{\lambda}, \quad \widetilde{l}_i(w_1) = \frac{\delta_{i1}}{\lambda}, \quad l_i(w_1) = \frac{\delta_{i1}(1 - |W|)}{\lambda}.$$

From this we directly compute to get

$$g_{ij}(w_1) = \begin{cases} \dfrac{(1 - |W|)^2}{\lambda^2}, & \text{if } i = j = 1; \\ \dfrac{(1 - |W|)\delta_{ij}}{\lambda}, & \text{if } i \neq 1 \text{ or } j \neq 1. \end{cases}$$

From this the assertion follows.    $\square$

*Proof of Theorem 7.19.* Let $G$ be the unity component of the full group $I(M,F)$ of isometries of $(M,F)$ and let $H$ be the isotropy subgroup of $G$ at $x$. Since $(M,F)$ is homogeneous, $I(M,F)$ is transitive on $M$. Thus $G$ is also transitive on $M$ (see [83]). Therefore $M$ can be written as $M = G/H$ and both $h$ and $W$ are invariant under $G$. Let $\mathfrak{g} = \mathfrak{h} + \mathfrak{m}$ be a decomposition of the Lie algebra with $\mathrm{Ad}(h)\mathfrak{m} \subset \mathfrak{m}$ and denote by $\langle , \rangle_h$ the inner product on $\mathfrak{m}$ induced by $h$. Since $F$ has almost isotropic $S$-curvature, by Theorem 7.11, $W$ is a Killing vector field with respect to $h$ and the corresponding vector $w \in \mathfrak{m}$ satisfies

$$\langle [w, v_1]_{\mathfrak{m}}, v_2 \rangle_h + \langle v_1, [w, v_2]_{\mathfrak{m}} \rangle_h = 0, \quad \forall v_1, v_2 \in \mathfrak{m}. \tag{7.67}$$

We need only prove that $w = 0$. Suppose, to the contrary, that this is not true. Consider the Ricci scalar of $F$ at the tangent vector $w$. Select $n - 1$ ($n = \dim M$) vectors $v_1, v_2, \ldots, v_{n-1}$ such that $\frac{w}{|w|}, v_1, v_2, \ldots, v_{n-1}$ form an orthonormal basis of $\mathfrak{m}$ with respect to $h$. (Note: not with respect to the fundamental tensor of $F$.) Then by Lemma 7.7, $\frac{w}{|w|}, v_1, \ldots, v_{n-1}$ is an orthogonal basis of $\mathfrak{m}$, with respect to the inner product $g_{\frac{w}{|w|}}(,)$, where $g$ is the fundamental tensor of $F$. Therefore we have

$$\mathrm{Ric}(x, w) = \sum_{i=1}^{n-1} K(w, v_i),$$

where $K$ denotes the flag curvature of $F$. If we denote the sectional curvature of $h$ by $K_h$, then by Proposition 7.8, we have

$$K(w,v_i) = K_h(w - |w|w, v_i) = K_h((1-|w|)w, v_i) = K_h(w,v_i), \quad i = 1,2,\ldots,n-1,$$

since $|w| = \sqrt{h(w,w)} < 1$. Therefore we have

$$\text{Ric}(x,w) = \sum_{i=1}^{n-1} K_h(w,v_i) = \text{Ric}_h(w,w)/h(w,w),$$

where $\text{Ric}_h$ denotes the Ricci scalar of $h$. By assumption, we have $\text{Ric}_h(w,w) < 0$. Now by Corollary 2.1, we have (setting $v_n = \frac{w}{|w|}$)

$$\text{Ric}_h(w,w) = -\frac{1}{2}\sum_{i=1}^{n} |[w,v_i]_\mathfrak{m}|^2 - \frac{1}{2}\sum_{i=1}^{n} \langle[w,[w,v_i]]_\mathfrak{m}, v_i\rangle_h$$

$$- \sum_{i=1}^{n} \langle[w,[w,v_i]_\mathfrak{h}]_\mathfrak{m}, v_i\rangle_h + \frac{1}{4}\sum_{i,j=1}^{n} \langle[v_i,v_j]_\mathfrak{m}, w\rangle_h^2 - \langle[z,w]_\mathfrak{m}, w\rangle_h,$$

Where $z$ denotes the unique element in $\mathfrak{m}$ such that $\langle z, v\rangle_h = \text{tr}(\text{ad}\,v)$, and $\text{ad}\,v$ denotes the adjoint action of $v$ in the Lie algebra $\mathfrak{g}$. Let us compute each term of the above formula. By (7.67), we have

$$-\frac{1}{2}\sum_{i=1}^{n} \langle[w,[w,v_i]]_\mathfrak{m}, v_i\rangle_h = \frac{1}{2}\sum_{i=1}^{n} \langle[w,v_i]_\mathfrak{m}, [w,v_i]_\mathfrak{m}\rangle_h.$$

Thus the first term and the second term sum to 0. Since $w$ is invariant under $\text{Ad}(H)$, we have $[w,\mathfrak{h}] = 0$. Thus the third term is equal to 0. By (7.67), we also have

$$\langle[z,w]_\mathfrak{m}, w\rangle_h = -\langle[w,z]_\mathfrak{m}, w\rangle_h = \langle z, [w,w]_\mathfrak{m}\rangle = 0.$$

Therefore we are left with

$$\text{Ric}_h(w,w) = \frac{1}{4}\sum_{i,j=1}^{n} \langle[v_i,v_j]_\mathfrak{m}, w\rangle_h^2.$$

This is a contradiction to $\text{Ric}_h(x,w) < 0$. Therefore $w$ must be 0 and $F$ is Riemannian at $x$. Since $F$ is homogeneous, $F$ is Riemannian everywhere. This completes the proof of the theorem. □

**Corollary 7.2.** *Let* $(M,F)$ *be a connected homogeneous Randers space. If* $F$ *has almost isotropic S-curvature and negative flag curvature, then* $F$ *must be Riemannian.*

*Remark 7.7.* In [139], Shen proved the following result: Let $(M,F)$ be a complete Finsler space with nonpositive flag curvature. If $(M,F)$ has almost constant S-curvature and bounded mean Cartan torsion, then $F$ must be Riemannian where the flag curvature is negative. Since a homogeneous with almost isotropic S-curvature must be with constant S-curvature, and its mean Cartan torsion must be bounded, Corollary 7.2 also follows from this result of Shen. However, the approaches are very different. Note also that Theorems 7.18 and 7.19 cannot be deduced from the above result of Shen.

# References

1. Abate, M., Patrizio, G.: Finsler metrics – a global approach. Lecture notes in Math., vol. 1591. Springer, Berlin (1994)
2. Akabar-Zadeh, H.: Sur les espaces de Finsler à courbures sectionnelles constantes. Acad. Roy. Belg. Bull. Cl. Sci. (5) **74**, 281–322 (1988)
3. Akhiezer, D.N., Vinberg, E.B.: Weakly symmetric spaces and spherical varieties. Transformation Groups **4**, 3–24 (1999)
4. Alekseevskii, D.V.: Classification of quaternionic spaces with a transitive solvable group of motions. Izv. Akad. Nauk. SSSR Ser. Mat. **9**, 315–362 (1975) (English translation: Math. USSR-Izv **9**, 297–339 (1975))
5. Alekseevskii, D.V., Kinmel'fel'd, B.N.: Structure of homogeneous Riemannian manifolds with zero Ricci curvature. Funct. Anal. Appl. **9**, 95–102 (1975)
6. Aloff, S., Wallach, N.: An infinite family of distinct 7-manifolds admitting positively curved Riemannian structures. Bull. Am. Math. Soc. **81**, 93–97 (1975)
7. Ambrose, W., Singer, I.M.: A theorem on holonomy. Trans. Am. Math. Soc. **75**, 428–443 (1953)
8. Ambrose, W., Singer, I.M.: On homogeneous Riemannian manifolds. Duke Math. J. **25**, 647–669 (1958)
9. An, H., Deng, S.: Invariant $(\alpha, \beta)$-metrics on homogeneous manifolds. Monatsh. Math. **154**, 89–102 (2008)
10. Antonelli, P.L., Ingarden, R.S., Matsumoto, M.: The Theory of Sprays and Finsler spaces with Applications in Physics and Biology. Kluwer Academic, Dordrecht (1993)
11. Bacso, S., Matsumoto, M.: Randers spaces with the h-curvature tensor H dependent on position alone. Publ. Math. Debrecen **57**, 185–192 (2000)
12. Baez, J.C.: The octonions. Bull. Am. Math. Soc. **39**, 145–205 (2002)
13. Bao, D.: Randers space forms. Periodica Mathematica Hungarica **48**, 3–15 (2004)
14. Bao, D.: On two curvature-driven problems in Finsler geometry. Adv. Pure Math. **48**, 19–71 (2007)
15. Bao, D., Chern, S.S.: On a notable connection in Finsler geometry. Houston J. Math. **19**, 135–180 (1993)
16. Bao, D., Chern, S.S., Shen, Z.: An Introduction to Riemann-Finsler Geometry. Springer, New York (2000)
17. Bao, D., Lackey, B.: Randers surfaces whose Laplacians have completely positive symbol. Nonlinear Anal. A **38**, 27–40 (1999)
18. Bao, D., Robles, C.: On Randers spaces of constant flag curvature. Rep. Math. Phys. **51**, 9–42 (2003)

19. Bao, D., Robles, C.: Ricci and flag curvatures in Finsler geometry. In: Bao, D., Bryant, R.L., Chern, S.S., Shen, Z. (eds.) A Sampler of Riemannian-Finsler Geometry, pp. 197–260. Cambridge University Press, London (2004)

20. Bao, D., Shen, Z.: Finsler metrics with constant positive curvature on the sphere $S^3$. J. Lond. Math. Soc. **66**, 453–467 (2002)

21. Bao, D., Robles, C., Shen, Z.: Zermelo navigation on Riemannian manifolds. J. Differ. Geom. **66**, 377–435 (2004)

22. Bazaikin, Y.V.: On a certain class of 13-dimensional Riemannian manifolds with positive curvature. Siberian Math. J. **37**, 1219–1237 (1996)

23. Bérard Bergery, L.: Les variétés Riemannienes homogènes simplement connexes de dimension impair à courbure strictement positive. J. Math. Pures Appl. **55**, 47–68 (1976)

24. Berger, M.: Les variétés riemanniennes homogènes normales simplement connexes à courbure strictement positive. Ann. Scuola Norm. Sup. Pisa (3) **15**, 179–246 (1961)

25. Berndt, J., Vanhecke, L.: Geometry of weakly symmetric spaces. J. Math. Soc. Jpn. **48**, 745–760 (1996)

26. Berndt, J., Kowalski, O., Vanhecke, L.: Geodesics in weakly symmetric spaces. Ann. Glob. Anal. Geom. **15**, 153–156 (1997)

27. Berndt, J., Ricci, F., Vanhecke, L.: Weakly symmetric groups of Heisenberg type. Differ. Geom. Appl. **8**, 275–284 (1998)

28. Besse, A.: Einstein Manifolds. Springer, Berlin (2008)

29. Bochner, S., Montgomery, D.: Locally compact groups of differentiable transformations. Ann. Math. **47**, 639–653 (1946)

30. Bogoslovski, Yu.G.: Status and perspectives of theory of local anisotropic space-time. Physics of Nuclei and Particles. Moscow State University, Russia (1997) (in Russian)

31. Borel, A., Some remarks about Lie groups transitive on spheres and tori. Bull. Am. Math. Soc. **55**, 580–587 (1940)

32. Brinzei, N.: Projective relations for m-th root metric spaces. Available at the website: http://arxiv1.library.cornell.edu/abs/0711.4781v2/, preprint. Accessed 2007

33. Brion, M.: Classification des espaces homogènes sphériques. Compositio Math. **63**, 189–208 (1989)

34. Bröcker, T., tom Dieck, T.: Representations of Compact Lie Groups. Springer, New York (1995)

35. Busemann, H., Phadke, B.: Two theorems on general symmetric spaces. Pac. J. Math. **92**, 39–48 (1981)

36. Cahen, M., Parker, M.: Pseudo-Riemannian symmetric spaces. Mem. Am. Math. Soc. **24**(229) (1980)

37. Cartan, E.: Sur une classe remarquable d'espaces de Riemann. Bull. Soc. Math. France **54**, 214–264 (1926)

38. Cartan, E.: Sur la détermination d'un système orthogonal complet dans un espace de Riemann symétrique clos. Rend. Circ. Mat. Palermo **53**, 217–252 (1929)

39. Chen, X., Deng, S.: A new proof of a theorem of H. C. Wang. Balkan J. Geom. Appl. **16**(2), 25–26 (2011)

40. Chen, B.Y., Nagano, T.: Totally geodesic submanifolds of symmetric spaces I. Duke Math. J. **44**, 745–755 (1977)

41. Chen, B.Y., Nagano, T.: Totally geodesic submanifolds of symmetric spaces II. Duke Math. J. **45**, 405–425 (1978)

42. Chern, S.S.: On the Euclidean connections in a Finsler space. Proc. Nat. Acad. Sci. USA **29**, 33–37 (1943)

43. Chern, S.S.: Finsler geometry is just Riemannian geometry without the quadratic restriction. Notices Am. Math. Soc. **43**, 959–963 (1996)

44. Chern, S.S., Shen, Z.: Riemann-Finsler Geometry. World Scientific, Singapore (2004)

45. D'Atri, J.E., Ziller, W.: Naturally reductive metrics and Einstein metrics on compact Lie groups. Mem. Am. Math. Soc. **18**(215) (1979)

46. Deicke, A.: Über die Finsler Räume mit $A_i = 0$. Arch. Math. **4**, 45–51 (1953)

47. Deng, S.: Fixed points of isometries of a Finsler space. Publ. Math. Debrecen **72**, 469–474 (2008)
48. Deng, S.: The S-curvature of homogeneous Randers spaces. Differ. Geom. Appl. **27**, 75–84 (2009)
49. Deng, S.: An algebraic approach to weakly symmetric Finsler spaces. Can. J. Math. **62**, 52–73 (2010)
50. Deng, S.: Invariant Finsler metrics on polar homogeneous spaces. Pac. J. Math. **247**, 47–74 (2010)
51. Deng, S.: On the classification of weakly symmetric Finsler spaces. Israel J. Math. **181**, 29–52 (2011)
52. Deng, S.: Clifford-Wolf translations of Finsler spaces of negative flag curvature, preprint. arXiv: 1204.1048, 2012/arxiv.org
53. Deng, S.: Finsler metrics and the degree of symmetry of a closed manifold. Indiana U. Math. J. **60**, 713–727 (2011)
54. Deng, S.: On the symmetry of Riemannian manifolds. J. Reine Angew. Math., published online. doi:10.1515/CRELLE.2012.040
55. Deng, S., Hou, Z.: The group of isometries of a Finsler space. Pac. J. Math. **207**, 149–155 (2002)
56. Deng, S., Hou, Z.: Invariant Finsler metrics on homogeneous manifolds. J. Phys. A: Math. Gen. **37**, 8245–8253 (2004)
57. Deng, S., Hou, Z.: Minkowski symmetric Lie algebras and symmetric Berwald spaces. Geometriae Dedicata **113**, 95–105 (2005)
58. Deng, S., Hou, Z.: On locally and globally symmetric Berwald spaces. J. Phys. A: Math. Gen. **38**, 1691–1697 (2005)
59. Deng, S., Hou, Z.: Invariant Finsler metrics on homogeneous manifolds II: Complex structures. J. Phys. A: Math. Gen. **39**, 2599–2609 (2006)
60. Deng, S., Hou, Z.: Invariant Randers metrics on homogeneous Riemannian manifold. J. Phys. A: Math. Gen. **37**, 4353–4360 (2004); Corrigendum, J. Phys. A: Math. Gen. **39**, 5249–5250 (2006)
61. Deng, S., Hou, Z.: Homogeneous Finsler spaces of negative curvature. J. Geom. Phys. **57**, 657–664 (2007)
62. Deng, S., Hou, Z.: On symmetric Finsler spaces. Israel J. Math. **166**, 197–219 (2007)
63. Deng, S., Hou, Z.: Positive definite Minkowski Lie algebras and bi-invariant Finsler metrics on Lie groups. Geometriae Dedicata **136**, 191–201 (2008)
64. Deng, S., Hou, Z.: Homogeneous Einstein–Randers spaces of negative Ricci curvature. C. R. Math. Acad. Sci. Paris **347**, 1169–1172 (2009)
65. Deng, S., Hou, Z.: Naturally reductive homogeneous Finsler spaces. Manuscripta Math. **131**, 215–219 (2010)
66. Deng, S., Hou, Z.: Weakly symmetric Finsler spaces. Comm. Contemp. Math. **12**, 309–323 (2010)
67. Deng, S., Wang, X.: The S-curvature of homogeneous $(\alpha, \beta)$-metrics. Balkan J. Geom. Appl. **15**(2), 39–48 (2010)
68. Deng, S., Xu, M.: Clifford–Wolf translations of Finsler spaces. Forum Math. (2012), published online, doi:10.1515/forum-2012-0032
69. Deng, S., Xu, M.: Clifford–Wolf translations of homogeneous Randers spheres, preprint. arXiv: 1204.5232, 2012/arxiv.org
70. Eschenburg, J.: New examples of manifolds of positive curvature. Invent. Math. **66**, 469–480 (1982)
71. Foulon, P.: Curvature and global rigidity in Finsler manifolds. Houston J. Math. **28**, 263–292 (2002)
72. Gorbatsevich, V.V.: Invariant intrinsic Finsler metrics on homogeneous spaces and strong subalgebras of Lie algebras. Siberian Math. J. **49**, 36–47 (2008)
73. Gordon, C.S.: Naturally reductive homogeneous Riemannian manifolds. Can. J. Math. **37**, 467–487 (1985)

74. Gordon, C.S.: Homogeneous Riemannian manifolds whose geodesics are orbits. In: Topics in Geometry, in Memory of Joseph D'Atri, pp. 155–174. Birkhäuser, Basel (1996)
75. Gordan, C.S., Kerr, M.: New homogeneous Einstein metrics of negative Ricci curvature. Geometriae Dedicata **19**, 75–101 (2001)
76. Greub, W., Halperin, S., Vanstone, R.: Connections, Curvature and Cohomology, II. Academic, New York (1973)
77. Grove, K., Ziller, W.: Cohomogeneity one manifolds with positive Ricci curvature. Invent. Math. **149**, 619–664 (2002)
78. Grove, K., Wilking, B., Ziller, W.: Positively curved cohomogeneity one manifolds and 3-Sasaki geometry. J. Differ. Geom. **78**, 33–111 (2008)
79. Grove, K., Verdiani, L., Ziller, W.: An exotic $T_1 S^4$ with positive curvature, Geometric and Functional Analysis, **21**, 499–524 (2011)
80. Heber, J.: Noncompact homogeneous Einstein manifolds. Invent. Math. **133**, 279–252 (1998)
81. Heintze, E.: On homogeneous manifolds of negative curvature. Math. Ann. **211**, 23–34 (1974)
82. Helgason, S.: Totally geodesic spheres in compact symmetric spaces. Math. Ann. **165**, 309–317 (1966)
83. Helgason, S.: Differential Geometry, Lie groups and Symmetric Spaces, 2nd edn. Academic, New York (1978)
84. Hsiang, W.Y.: The natural metric on $SO(n + 2)/SO(n)$ is the most symmetric. Bull. Am. Math. Soc. **73**, 55–58 (1967)
85. Hsiang, W.C., Hsiang, W.Y.: The degree of symmetry of homotopy spheres. Ann. Math. **89**, 52–67 (1969)
86. Hu, Z., Deng, S.: Invariant fourth-root metrics on the Grassmannian manifolds. J. Geom. Phys. **61**, 18–25 (2011)
87. Hu, Z., Deng, S.: Homogeneous Randers spaces of isotropic S-curvature and positive flag curvature. Math. Z. **270**, 989–1009 (2012)
88. Hu, Z., Deng, S.: On flag curvature of homogeneous Randers spaces. Can. J. Math. (2012), published online, doi:10.4153/CJM-2012-004-6
89. Hu, Z., Deng, S.: Ricci-quadratic homogeneous Randers spaces, preprint. arxiv: 1207.1786/arxiv.org
90. Hu, Z., Deng, S.: Three-dimensional homogeneous Finsler spaces. Math. Nachr. **285**, 1243–1254 (2012). doi:10.1002/mana.201100100
91. Huang, L., Mo, X.: On curvature decreasing property of a class of navigation problems. Publ. Math. Debrecen **71**, 991–996 (2007)
92. Ichijyō, Y.: Finsler spaces modeled on a Minkowski space. J. Math. Kyoto Univ. **16**, 639–652 (1976)
93. Ingarden, R.S.: On physical applications of Finsler geometry. Contemp. Math. **196**, 213–223 (1996)
94. Jensen, G.R.: Einstein metrics on principal fibre bundles. J. Differ. Geom. **8**, 599–614 (1973)
95. Kaplan, A.: Fundamental solution for a class of hypo-elleptic PDE generated by composition of quadratic forms. Trans. Am. Math. Soc. **258**, 147–153 (1980)
96. Kath, I., Olbrich, M.: Metric Lie algebras with maximal isotropic centre. Math. Z. **246**, 23–53 (2004)
97. Kath, I., Olbrich, M.: On the structure of pseudo-Riemannian symmetric spaces. Transformation Groups **14**, 847–885 (2009)
98. Kim, C.W.: Locally symmetric positively curved Finsler spaces. Arch. Math. (Basel) **88**, 378–384 (2007)
99. Kobayashi, S.: Fixed points of isometries. Nagoya Math. J. **13**, 63–68 (1958)
100. Kobayashi, S.: Homogeneous Riemannian manifolds of negative curvature. Tohoku Math. J. **14**, 413–415 (1962)
101. Kobayashi, S.: Transformation Groups in Differential Geometry. Springer, Berlin (1972)
102. Kobayashi, S., Nomizu, K.: Foundations of Differential Geometry, vol. 1, 1963, vol. 2, 1969. Interscience Publishers, New York (1969)

103. Koszul, J.L.: Exposés sue les espaces homogènes symétriques. Pub. Soc. Math. Sao Paulo **77**, (1959)
104. Kowalski, O.: Spaces with volume-preserving symmetries and related classes of Riemannian manifolds. Rend. Sem. Mat. Univ. Politec. Torino, Facicolo Speciale, Settembre, pp. 131–158 (1983)
105. Kowalski, O., Vanhecke, L.: Riemannian manifolds with homogeneous geodesics. Boll. Un. Mat. Ital. B **7**(5), 189–246 (1991)
106. Krämer, M.: Sphärische untergruppen in Kompakten zusammenhängenden Liegruppen. Compositio Math. **38**, 129–153 (1979)
107. Ku, H., Mann, L., Sicks, J., Su, J.: The degree of symmetry of a product manifold. Trans. Am. Math. Soc. **146**, 133–149 (1969)
108. Ku, H., Mann, L., Sicks, J., Su, J.: The degree of symmetry of a homotopy real projective space. Trans. Am. Math. Soc. **161**, 51–61 (1971)
109. Kuiper, N.H.: Groups of motions of order $\frac{1}{2}n(n-1)+1$ in Riemannian $n$-spaces. Indag. Math. **48**, 313–318 (1955)
110. Lauret, J.: Homogeneous nilmanifolds attached to representations of compact Lie groups. Manuscripta Math. **99**, 287–309 (1999)
111. Lauret, J.: Einstein solvmanifolds are standard. Ann. Math. **172**, 1859–1877 (2010)
112. Lawson, H., Yau, S.T.: Scalar curvature, non-abelian group actions, and the degree of symmetry of exotic spheres. Comment. Math. Helv. **49**, 232–244 (1974)
113. Li, B., Shen, Z.: On projectively flat fourth root metrics. Can. Math. Bull. **55**, 138–145 (2012)
114. Mather, J.: Differentiable invariant. Topology **16**, 145–155 (1977)
115. Matsumoto, M.: Foundations of Finsler Geometry and Special Finsler Spaces. Kaiseisha Press, Japan (1986)
116. Matveev, V.S., Troyanov, M.: The Binet–Legender ellipsoid in Finsler geometry. Math. DG arxiv: 1104–1647.v1, preprint. Accessed 2012
117. Mazur, S.: Quelques propriétés characteristiques des espaces Euclidiens. C. R. Acad. Sci. Paris Ser. I **207**, 761–764 (1938)
118. Mestag, T.: Relative equilibria of invariant lagrangian systems on a Lie group. In: Differential Geometry Methods in Mechanics and Field Theory, pp. 115–129. Academic, New York (2007)
119. Mikityuk, I.V.: On the integrability of invariant Hamiltonian systems with homogeneous configuration spaces. Math. Sb. **57**, 527–546 (1987)
120. Milnor, J.: Curvature of left invariant Riemannian metrics on Lie groups. Adv. Math. **21**, 293–329 (1976)
121. Montgomery, D., Samelson, H.: Transformation groups of spheres. Ann. Math. **44**, 454–470 (1943)
122. Montgomery, D., Yang, C.T.: The existence of a slice. Ann. Math. **65**, 108–116 (1957)
123. Myers, S.B., Steenrod, N.: The group of isometries of a Riemannian manifold. Ann. Math. **40**, 400–416 (1939)
124. Nguyêñ, H.: Characterizing weakly symmetric spaces as Gelfand pairs. J. Lie Theor. **9**, 285–291 (1999)
125. Obata, M.: On $n$-dimensional homogeneous spaces of Lie groups of dimension greater than $\frac{1}{2}n(n-1)$. J. Math. Soc. Jpn. **7**, 371–388 (1955)
126. Onishchik, A.L.: Transitive compact transformation groups. Math. Sb. **60**, 447–485 (1963)
127. Palais, R.S.: A global formulation of the Lie theory of transformation groups. Mem. Am. Math. Soc. **22m** (1957)
128. Palais, R.S.: On the differentiability of isometries. Proc. Am. Math. Soc. **8**, 805–7807 (1957)
129. Pavlov, D.G. (ed.): Space-Time Structure. Collected papers, TETRU (2006)
130. Popov, V.L., Vinberg, E.B.: Invariant theory. In: Algebraic Geometry IV. Springer, Berlin (1994)
131. Randers, G.: On an asymmetrical metric in the four-space of general relativity. Phys. Rev. **59**, 195–199 (1941)

132. del Riego, L.: Tenseurs de Weyl d'un spray de directions, Thesis, Université Scientifique et Medicale de Grenoble (1973)
133. Robles, C.: Einstein metrics of Randers type, doctoral dissertation, University of British Colombia (2003)
134. Schultz, R.: Semifree circle actions and the degree of symmetry of homotopy spheres. Am. J. Math. **93**, 829–839 (1971)
135. Selberg, A.: Harmonic analysis and discontinuous groups in weakly symmetric Riemannian spaces with applications to Dirichlet series. J. Indian Math. Soc. B. **20**, 47–87 (1956)
136. Shankar, K.: Isometry groups of homogeneous spaces with positive sectional curvature. Differ. Geom. Appl. **14**, 57–78 (2001)
137. Shen, Z.: Finsler manifolds of constant positive curvature. Contemp. Math. **196**, 83–93 (1996)
138. Shen, Z.: Volume comparison and its applications in Riemann–Finsler geometry. Adv. Math. **128**, 306–328 (1997)
139. Shen, Z.: Differential Geometry of Spray and Finsler Spaces. Kluwer Academic, Dordrecht (2001)
140. Shen, Z.: On projectively related Einstein metrics in Riemann–Finsler geometry. Math. Ann. **320**, 625–647 (2001)
141. Shen, Z.: On R-quadratic Finsler spaces. Publ. Math. Debrecen **58**, 263–274 (2001)
142. Shen, Z.: Two-dimensional Finsler metrics with constant flag curvature. Manuscripta Math. **109**, 349–366 (2002)
143. Shen, Z.: Finsler metrics with $K = 0$ and $S = 0$. Can. J. Math. **55**, 112–132 (2003)
144. Shen, Z.: Landsberg curvature, S-curvature and Riemann curvature. In: Bao, D., Bryant, R., Chern, S.S., Shen, Z. (eds.) A Sampler of Finsler Geometry, pp. 303–355, MSRI Publications, no. 50. Cambridge University Press, London (2004)
145. Shen, Z.: Finsler manifolds with non-positive flag curvature and constant S-curvature. Math. Z. **249**, 625–639 (2005)
146. Shen, Z.: Some open problems in Finsler geometry. Available on Z. Shen's home page at www.math.iupui.edu. Preprint (2009)
147. Shen, Z., Xing, H.: On Randers metrics of isotropic S-curvature. Acta Math. Sinica **24**, 789–796 (2008)
148. Shimada, H.: On Finsler spaces with the metric $L = (a_{i_1 i_2 \cdots i_m} y^{i_1} y^{i_2} \cdots y^{i_m})^{\frac{1}{m}}$. Tensor N.S. **33**(3), 365–372 (1979)
149. Simons, J.: On the transitivity of holonomy systems. Ann. Math. **76**, 143–171 (1962)
150. Spiro, A.: Chern's orthonormal frame bundle of a Finsler space. Houston J. Math. **25**, 641–659 (1999)
151. Szabó, Z.I.: Generalized spaces with many isometries. Geometriae Dedicata **11**, 369–383 (1981)
152. Szabó, Z.I.: Positive definite Berwald spaces. Tensor N. S. **38**, 25–39 (1981)
153. Szabó, Z.I.: Spectral geometry for operator families on Riemannian manifolds. Sympos. Pure Math. **54**, 615–665 (1993)
154. Tits, J.: Sur certaine d'espaces homogènes de groupes de Lie. Acad. Roy. Belg. Cl. Sci. Mém. Coll. **29**(3) (1955)
155. Tits, J.: Espaces homogènes complexes. Comment. Math. Helv. **37**, 111–120 (1963)
156. Van Danzig, D., Van Der Waerden, B.L.: Über metrische homogene Räume. Abh. Math. Sem. Univ. Hamburg **6**, 367–376 (1928)
157. Varadarajan, V.S.: Lie Groups, Lie Algebras and Their Representations. Springer, New York (1984)
158. Verdianni, L., Ziller, W.: Positively curved homogeneous metrics on spheres. Math. Z. **261**, 473–488 (2009)
159. Wallach, N.: Compact homogeneous Riemannian manifolds with strictly positive curvature. Ann. Math. **96**, 277–295 (1972)
160. Walter, T.H.: Einstein metrics on solvable groups. Math Z. **206**, 457–471 (1991)

161. Wang, H., Deng, S.: Some homogeneous Einstein–Randers spaces. Nonlinear Anal. **72**, 4407–4414 (2010)
162. Wang, H., Deng, S.: Left invariant Einstein-Randers metrics on compact Lie groups. Can. Math. Bull. (2011), published online, doi:10.4153/CMB-2011-145-6
163. Wang, H.C.: Finsler spaces with completely integrable equations of Killing. J. Lond. Math. Soc. **22**, 5–9 (1947)
164. Wang, H.C.: Two point homogeneous spaces. Ann. Math. **55**, 177–191 (1952)
165. Wang, M., Ziller, W.: On normal homogeneous Einstein manifolds. Ann. Sci. Ec. Norm. Sup. **18**, 563–633 (1985)
166. Wang, M., Ziller, W.: Existence and non-existence of homogeneous Einstein metrics. Invent. Math. **84**, 177–194 (1986)
167. Wang, M., Ziller, W.: On isotropy irreducible Riemannian manifolds. Acta Math. **116**, 223–261 (1991)
168. Wang, H., Huang, L., Deng, S.: Homogeneous Einstein-Randers metrics on spheres. Nonlinear Anal. **74**, 6295–6301 (2011)
169. Wilking, B.: The normal homogeneous space $SU(3) \times SO(3)/U^*(2)$ has positive sectional curvature. Proc. Am. Math. Soc. **127**, 1191–1194 (1999)
170. Wilson, E.N.: Isometry groups on homogeneous nilmanifolds. Geometriae Dedicata **12**, 337–346 (1982)
171. Wolf, J.A.: Sur la classification des variétés riemanniennes homogènes à courbure constante. C. R. Acad. Sci. Paris Ser. I **250**, 3443–3445 (1960)
172. Wolf, J.A.: Locally symmetric homogeneous spaces. Comment. Math. Helv. **37**, 65–101 (1962–1963)
173. Wolf, J.A.: Curvature in nilpotent Lie groups. Proc. Am. Math. Soc. **15**, 271–274 (1964)
174. Wolf, J.A.: Homogeneity and bounded isometries in manifolds of negative curvature. Illinois J. Math. **8**, 14–18 (1964)
175. Wolf, J.A.: The geometry and structure of isotropic irreducible homogeneous spaces. Acta Math. **20**, 59–148 (1968); Correction, Acta Math. **152**, 141–142 (1984)
176. Wolf, J.A.: Harmonic Analysis on Commutative Spaces. Amer. Math. Soc. vol. 142 (2007)
177. Wolf, J.A.: Spaces of Constant Curvature, 6th edn. Amer. Math. Soc. Chelsea. American Mathematical Society, Providence (2011)
178. Xu, B.: The degree of symmetry of certain compact smooth manifolds. J. Math. Soc. Jpn. **55**, 727–737 (2003)
179. Yakimova, O.: Weakly symmetric spaces of semisimple Lie groups. Moscow Univ. Math. Bull. **57**(2), 37–40 (2002)
180. Yakimova, O.: Weakly symmetric Riemannian manifolds with a reductive isometry group. Sb. Math. **195**(3–4), 599–614 (2004)
181. Yano, K.: On $n$-dimensional Riemannian manifolds admitting a group of motions of order $\frac{1}{2}n(n-1)+1$. Trans. Am. Math. Soc. **74**, 260–279 (1953)
182. Ziller, W.: Homogeneous Einstein metrics on spheres. Math. Ann. **259**, 351–358 (1982)
183. Ziller, W.: Weakly symmetric spaces. In: Topics in Geometry, in memory of Joseph D'Atri. Progress in Nonlinear Diff. Eqs. 20, pp. 355–368. Birkhäuser, Basel (1996)

# Index